*"The **MicroStation** Re,*_____ _____ __ _ ___ __
*HAVE BOOK for anyone using MicroStation. If you can
only afford to invest in one book, this (along with the
optional disk) is it. Highly Recommended!"*

John Franklin
Instructor
Republic Research Training Centers, Inc.

*"This is super! An excellent quick reference for MicroSta-
tion. Even the hard core user will find this useful."*

Mac Otis
MicroStation Training and Consultant
Intergraph Corporation

CAD Books from OnWord Press

MicroStation 4.X

The MicroStation 4.X Delta Book
INSIDE MicroStation
INSIDE MicroStation Companion Workbook
INSIDE MicroStation Companion Workbook
Instructor's Guide
MicroStation Reference Guide
MicroStation Productivity Book
101 MDL Commands
Bill Steinbock's Pocket MDL Programmer's Guide
MDL Guides
MicroStation For AutoCAD Users

MicroStation 3.X

INSIDE MicroStation
MicroStation Reference Guide
MicroStation Productivity Book

Other CAD Titles

The CAD Rating Guide
The One Minute CAD Manager

Books By Pen & Brush Publishers
Distributed by OnWord Press

The Complete Guide to MicroStation 3D
Programming With MDL
Programming With User Commands

More MDL Titles Available from OnWord Press

OnWord Press Products are Available Directly From

1. Your Local MicroStation Dealer or Intergraph Education Center
2. Your Local Bookseller
3. In Australia, New Zealand, and Southeast Asia from:
   Pen & Brush Publishers
   2nd Floor, 94 Flinders Street
   Melbourne Victoria 3000
   Australia
   Phone 61 (0)3 818 6226
   Fax   61(0)3 818 3704

4. .... Or Directly From OnWord Press
   (see ordering information on the last page.)

101 MDL Commands

Bill Steinbock
First Edition, Release 4.X, Supports all 4.X and 3.X Platforms including DOS, Unix, Mac, and VMS

This is the book you need to get started with MDL!

With MicroStation 4.0 comes the MicroStation Development Language or MDL. MDL is a powerful programming language built right in to MicroStation.

MDL can be used to add productivity to MicroStation or to develop complete applications using MicroStation tools. Virtually all of the 3rd party applications vendors are already using MDL for their development.

Now you can too, with 101 MDL Commands.

The first part of this book is a 100 page introduction to MDL including a guide to how source code is created, compiled, linked, and run. This section includes full discussion of Resource Files, Source Codes, Include Files, Make Files, dependencies, conditionals, interference rules, command line options and more.

Learn to control MicroStation's new GUI with dialog boxes, state functions, element displays and file control.

The second part of the book is 101 actual working MDL commands ready-to-go. Here you will find about 45 applications with over 101 MDL tools. Some of these MDL commands replace user commands costing over $100 a piece in the 3.0 market!

Here's a sampling of the MDL applications in the book and on the book:

MATCH - existing element parameters
Creation - create all the new element types from MDL
Multi-line - convert existing lines and linestrings to multi-line elements
CALC - Dialog box calculator
DATSTMP - Places and updates filename and in-drawing date stamp
PREVIEW - previews a design file within a dialog box
Text - complete text control - underline, rotate, resize, upper/lower, locate text
   string, import ASCII columns, extract text
Fence - complete fence manipulations including patterning, group control,
   circular fence and more

3D surfaces - complete projection and surface of revolution control
Cell routines - place along, place view dependent cell, scale cell, extract to cell library
Dialog boxes - make your own using these templates!
Search Criteria - delete, fence, copy, etc based on extensive search criteria.

Use these MDL commands to get you started with the power of MicroStation MDL. You can put these tools to work immediately, or use the listings to learn about MDL and develop your own applications.

⊛ Optional 101 MDL Commands Disk: **US$101.00** (Australia A$155.00) Includes all of the MDL commands from the book in executable form, ready to be loaded and used.

Price: **US$49.95** (Australia A$85.00)
Pages: 680
Illustrations: 75+
ISBN: 0-934605-61-0

⊛ Optional 101 MDL Commands Source Disk 1: **US$59.95** Includes:

Element Creation Functions
Program 1: ELLIPSE
Program 2: EXTRUDE
Program 3: GRID
Program 4: LEADER
Program 5: PLBLKC
Program 6: POLY
Program 7: REVOLVE

⊛ Optional 101 MDL Commands Source Disk 2: **US$59.95** Includes:

Element Manipulation
Program 8: COPYA
Program 9: CVT2VD
Program 10: MOVSEG
Program 11: SETZ
Program 12: UNPATLIN
Fence Commands
Program 17: CIRFEN
Program 18: FENGG
Program 19: FENSEA

⊛ Optional 101 MDL Commands Source Disk 3: **US$59.95** Includes:

Text Functions
Program 20: CHNGFT
Program 21: IATXT
Program 22: LOCKTXT
Program 23: TEXTMOD
Program 24: TEXTTX
Program 25: ULTXT
Program 26: UPPER

⊛ Optional 101 MDL Commands Source Disk 4: **US$59.95** Includes:

Cell Routines
Program 13: CELLALNG
Program 14: CELLMOD
Program 15: CELLPLOT
Program 16: CELLX
Active Parameters
Program 29: MATCH
Program 30: SHOW

⊛   Optional 101 MDL Commands Source Disk 5: **US$59.95** Includes:

Utilities
Program 35: CALC
Program 36: CLOCK
Program 37: DATSTMP
Program 38: PREVW
Returning Values to the Design
Program 27: AREATXT
Program 28: NE

⊛   Optional 101 MDL Commands Source Disk 6: **US$59.95** Includes:

Placing Elements With External Data Files
Program 31: TEXTCOL
Program 32: XYPLOT
Program 33: XYZCELL
Program 34: XYZLINE

⊛   **ALL SIX 101 MDL Commands Source Disks:** *ONLY* **US$299.95**

Bill Steinbock's
Pocket MDL Programmers Guide

Bill Steinbock
First Edition, Release 4.X, Supports all 4.X and 3.X Platforms including DOS,
Unix, Mac, and VMS

Intergraph/Bentley's MDL documentation is over 1000 pages!

Bill's Steinbock's Pocket MDL Programmers Guide gives you all the MDL tools
you need to know for most applications in a brief, easy-to-read format.

All the MDL tools, all the parameters, all the definitions, all the ranges -- in a
short and sweet pocket guide.

If you're serious about MDL, put the power of MicroStation MDL in your hands
with this complete quick guide.

Includes all the MDL commands, tables, indexes, and a quick guide to completing
MDL source for MDL compilation.

Price: **US$24.95**   (Australia A$39.95)
Pages: 256
Illustrations: 75+
ISBN: 0-934605-32-7

MDL-Guides

CAD Perfect
First Edition, Release 4.X, Supports all MDL 4.X Platforms Runs under DOS only.

1000 Pages of Intergraph MDL Documentation On-Line at Your Fingertips!

The Intergraph MDL Documentation is voluminous, to say the least. MDL-GUIDES puts the MDL and MicroCSL documentation in a hypertext TSR for reference access while you are programming or debugging in MDL.

This terminate-and-stay-ready program was sanctioned by Intergraph as the practical way to find out all the MDL information you need in an easy format.

The program is environment friendly, works with high DOS memory space to leave room for your other applications, and is quick.

The package includes the hypertext software, the complete set of MDL and MicroCSL Intergraph documentation in hypertext format, and a proper set of installation instructions.

MDL-GUIDES
For DOS Formats only (Other formats available on request from CAD Perfect)
Includes Disk and User's Manual
Disk Includes all MDL Documentation formatted for use on-line
Price: **US$295.00**
ISBN 0-934605-71-8

Programming With MDL

Mach Dinh-Vu
For Intergraph MicroStation
Supports all 4.X Platforms including DOS, Unix, Mac, and VMS. Will work with all 4.X versions of MicroStation.
(Published by Pen & Brush Publishers, Distributed by OnWord Press)

Programming With MDL is an indispensable tool for MDL command newcomers and programmers alike. This book serves as both a tutorial guide and handbook to the ins-and-outs of MDL.

Step-by-step explanations and examples help you create MDL programs to speed the design and drafting process. Learn how to attach "intelligence" to the drawing with or without database links. Take control of your menu, dialog, and command environment and customize it for your own application.

MicroStation Development Language is already built into your copy of MicroStation or IGDS. Put it to work for you today with Programming With MDL.

⊛ Programming With MDL Disk: **$40.00** (Australia A$55.00) Includes all of the MDL Command examples in the book in a ready-to-use form. Use them as they are, or modify them with your own editor to get a jump-start on MDL programming.

Price: **US$49.95** (Australia A$85.00)
Pages: 320
Illustrations, Tables, Examples: 120+
ISBN: 0-934605-59-9 (Australia ISBN: 0-646-01679-2)

Order MDL Tools From OnWord Press Now!

Ordering Information:

OnWord Press Products are Available From

1. Your Local MicroStation Dealer
2. Your Local Bookseller
3. In Australia, New Zealand, and Southeast Asia from:
   Pen & Brush Publishers
   2nd Floor, 94 Flinders Street
   Melbourne Victoria 3000
   Australia
   Phone 61(0)3 818 6226
   Fax   61(0)3 818 3704

4. .... Or Directly From OnWord Press:

To Order From OnWord Press:

Three Ways To Order from OnWord Press

1. Order by FAX 505/587-1015
2. Order by PHONE: 1-800-CAD NEWS™ Outside the U.S. and Canada
   call 505/587-1010.
3. Order by MAIL: OnWord Press/CAD NEWS Bookstore, P.O. Box 500,
   Chamisal NM 87521-0500 USA.

Shipping and Handling Charges apply to all orders: 48 States: $4.50 for the first item, $2.25 each additional item. Canada, Hawaii, Alaska, Puerto Rico: $8.00 for the first item, $4.00 for each additional item. International: $46.00 for the first item, $15.00 each additional item. **Diskettes are counted as additional items.** New Mexico delivery address, please add 5.625% state sales tax.

Rush orders or special handling can be arranged, please phone or write for details. Government and Educational Institution POs accepted. Corporate accounts available.

MDL Books and Tools

Use This Form If Ordering Directly From OnWord Press

Quantity	Title	Price	Extension
	101 MDL Commands	$49.95	
	101 MDL Commands Disk/Executables	$101.00	
	101 MDL Commands Disk/Source Disk 1	$59.95	
	101 MDL Commands Disk/Source Disk 2	$59.95	
	101 MDL Commands Disk/Source Disk 3	$59.95	
	101 MDL Commands Disk/Source Disk 4	$59.95	
	101 MDL Commands Disk/Source Disk 5	$59.95	
	101 MDL Commands Disk/Source Disk 6	$59.95	
	ALL SIX 101 MDL Commands Source Disks	$299.95	
	Bill Steinbock's MDL Pocket Programmer's Guide	$24.95	
	MDL-GUIDES	$295.00	
	Programming With MDL	$49.95	
	Programming With MDL Disk	$40.00	
	Shipping & Handling*		
	5.625% Tax - State of New Mexico Delivery Only		
	Total		

AD CODE M44

Name _____

Company _____

Street _____
(No P.O. Boxes Please)

_____

City, State _____

Country, Postal Code_____

Phone _____

Fax _____

If Ordering Disks, Please Note

Disk Type _____

Payment Method

___ Cash ___ Check ___ Amex

___ MasterCARD ___VISA

_____
Card Number

Expiration Date _____

_____
Signature

FAX TO: 505/587-1015

or MAIL TO:

OnWord Press
Box 500
Chamisal NM 87521 USA

MICROSTATION™
REFERENCE GUIDE

Everything You Want to Know About MicroStation
— Fast!

By John Leavy

SECOND EDITION

MICROSTATION™ REFERENCE GUIDE

Everything You Want to Know About MicroStation — Fast!

By John Leavy

Published by:

OnWord Press P.O. Box 500 Chamisal, NM 87521 USA

Copyright © 1990, 1991 Computer Graphic Solutions, Inc.

Second Edition, 1991

10 9 8 7 6 5 4 3 2

Printed in the United States of America

Library of Congress Cataloging-in-Publication Data

        Leavy, John,
        MicroStation Reference Guide

        Includes index.

        1. MicroStation (computer program) I. Title
        91-60280
        ISBN 0-934605-55-6

Trademarks

Warning and Disclaimer

This book is designed to provide data about all the MicroStation commands. Every effort has been made to make this book complete and as accurate as possible; however, no warranty or fitness is implied.

The information is provided on an "as-is" basis. The author and OnWord Press shall have neither liability nor responsibility to any person or entity with respect to any loss or damages in connection with or rising from the information contained in this book.

About the Author

John Leavy has been closely associated with Intergraph's products for more than twelve years. He managed several CAD systems as a user during the late 1970's. For the first half of the 1980's he was employed by Intergraph Corporation as a Senior Customer Engineer, and later as a technical manager in the Midwest Region. On the road his responsibilities included teaching new customers, trouble-shooting problems, and performance issues. While in the office, he helped train new employees, supported sales demos, answered configuration questions, and managed the System Implementation program for the Midwest.

In 1986 John left Intergraph Corporation to start Computer Graphic Solutions, Inc., a consulting firm based north of Chicago, Illinois. Today, John divides his time among consulting on Intergraph product solutions, teaching, free-lance writing assignments and speaking at conferences.

Thanks for the Help

John would like to thank Bentley Systems, Inc. for contributing MicroStation. Applause to Katie Bartlett, Peter Huftalen, Grif Roberts and Sandy Shaughnessy of Bentley, for ferreting out answers for the 2nd Edition. Thanks to Intergraph Corporation for MicroStation 32, and the assistance by the techs in Huntsville.

A special appreciation goes out to Daniel Raker and David Talbott for their support, ideas, encouragement, and exhausting behind-the-scenes labor. Without them this project would not be a reality. Thanks also goes out to Michelle Noel for her effort in pulling the pieces of this book together, and for the great job she did on layout and design.

Applause also goes out to the many MicroStation users who shared their years of know-how, CAD adventures, tips, short cuts, joys, pain and suffering.

The author would also like to recognize his peers at Intergraph's offices in Arlington Heights who lent their time and talents. Thanks for making me feel at home while I wandered around the offices doing research, and gathering facts.

Thanks to my children Doug, Dan, and Sei for giving up their dad these past months, and for walking softly past the office door. Without their rooting, it would have been preposterous to sit and stare at the dumb computer screen for one more hour.

A very personal thank you goes out to my front line editor, valued critic, co-writer, best friend, and wife, Kay. Honey, you listened, read through the worst, and helped turn this book into the best. Thanks.

Cover Art

Image by Jerry D. Flynn and Robert P. Humeniuk, Design Visualization Group, McDonnell Douglas Space Systems Company, Kennedy Space Center, Florida. The image was created using MicroStation and rendered using MicroStation 4.0 and Pixar's Photo Realistic RenderMan.

A Sony GDM-1950 Monitor and Number Nine Computer 9GX level 3 video board was used for display of images. Originals were shot from screen using an Olumpus OM4 35mm camera at F8 with Kodak Kodacolor 200 slid film.

The cover art artists would especially like to thank Tony Clarey at Number Nine for the 9GX video board, and also Raymond Bentley and Brett Yeagley at Bentley Systems, Inc. for support during beta testing.

OnWord Press.....

OnWord Press is dedicated to the fine art of practical software user's documentation.

In addition to the authors who developed the material for this book, other members of the OnWord Press team make the book end up in your hands.

Thanks to: David Talbott, Michelle Noel, John Messeder, Dan Raker, Jean Nichols, and Sheila Miller.

Rejoice in everything you have put your hand to,
because the Lord your God has blessed you.
Duet. 12:7

Table of Contents

Introduction

Welcome to MicroStation Reference Guide. This Guide makes retrieving information — easy. Whether you're a rookie or a seasoned CAD veteran, this guide is designed with you in mind.

If you're the old pro you want data in the form of details, options, command syntax; and you want the data pronto. This Guide's alphabetic listing of commands makes finding information a breeze.

If, on the other hand, you're new to CAD, you probably want more fundamental information. You may need to know what a certain command does, how to use it and where the command is found on a menu or palette. The Tips, Notes, Errors and remedies give you the benefit of experience. This shortens the learning time.

Either one, rookie or veteran, will find the MicroStation Reference Guide a handy edition sitting next to your workstation.

Why this book is a cut above

Here are some of the reasons why the MicroStation Reference Guide is invaluable to anyone using MicroStation:

➤ All the commands, definitions, settings and short-cuts for the different MicroStation platforms are found in one place.

➤ Index A lists the commands and settings by association.

➤ This book shows all the ways to select each command, including key-ins, palette selections and selections in sidebar and paper menus.

➤ Command window prompts for each entry makes learning easier.

➤ Error messages and remedies point out the pitfalls.

➤ Valuable tips offer the fruits of years of experience.

What's New in the Second Edition

The most obvious addition to the second edition is palettes. Each command that can be activated by selecting an icon in a palette contains a picture of the palette with the appropriate icon selected.

Another change for the better is MicroStation's MDL interface. This program extension creates a seamless interface to invoke commands from outside MicroStation's basic group of commands. For example, the Cutter Tools application was designed to help you place cross, tee and corner joints on the new multi-line elements. B-spline curves and surfaces can be generated with another new MDL tool application called Splines.

This time around the Side-bar Menu Hierarchy is missing from up front. Thanks to the Graphic User Interface (GUI) developed by Bentley, MicroStation now runs in a OSF Motif compatible windowing environment. Side-bar menus may soon play a secondary role when choosing a menu-set for entering commands.

The old Appendix A, which listed default seed file settings of a generic drawing file has fallen by the wayside. MicroStation 4.x is delivered with 14 application specific seed files including architectural, mapping and mechanical seed files. Each discipline can choose its own default seed files.

Beyond these changes there have been a number of typographical improvements in the command listing area of the book. These are intended to make the command entries more readable. Finally there are a couple of new appendices, and the old ones have been revised.

How to read MicroStation's display screen

Sometimes the biggest part of conquering a new software package is understanding what the software is trying to tell you in the first place. If you're unfamiliar with MicroStation's screen format, stop here and read through the next few paragraphs.

The following diagram shows a MicroStation screen:

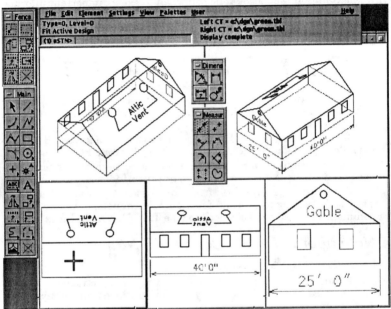

MicroStation's Display Screen

What you see is View 1, the Main palette on the left side of the screen, the command window at the top of the screen and additional palettes.

View 1 is one of eight possible views. You can stack these eight views on top of each other (cascade them) or you can have all eight of them on the screen at once (tiled). Each view acts as a separate camera pointed towards the action. One camera, or view, can be zoomed in on a particular detail, while another is used to survey the overall drawing.

The size of views can be adjusted to accommodate palettes or sidebar menus.

One of the most important areas on the screen is the command window, which in MicroStation's default setup appears at the top of the screen (although you can move it around). This is where you and MicroStation talk to each other. The command window is divided into six information fields. These are:

Status field - in the upper left corner, displays information about the active drawing settings.

Command field - below the status field, lists tool names of the active command.

USTN> - in the bottom left corner, is where commands are entered, settings are changed and where MicroStation's questions are answered.

Message field - in the upper right corner, keeps track of the active settings for Level, Area, Weight, Line Style, Class, and Color.

Prompt field - prompts for inputs required by the active command.

Error field - is used by MicroStation to raise a red flag if something goes awry.

Keep an eye on all three screen areas for effective communication with MicroStation. This book tells you when to expect important information in these fields.

How to Find Commands in This Book

Keep this guide within arm's reach. When a question comes up, you can turn to any quick index in this book to find what you are looking for.

Here is the way the book is organized:

Alphabetically by command: All of the commands as well as common nicknames and definitions can be found alphabetically in the main body of the book.

Short-cuts: Appendix B has a complete listing of keyboard accelerators (called speed keys), and Appendix C lists alternate key-ins.

Alphabetical index: All of the command names, common nicknames and definitions are listed alphabetically in Index B.

Commands grouped by function: Index A gives a listing of all commands pertaining to a certain function. For example, if you are looking for commands related to TEXT, look for the TEXT listing in Index A.

A Note about "Legal" Command Names and Typographic Conventions

As MicroStation has grown in popularity and functionality its command language has taken on many quirks in the naming conventions applied to commands.

In this book you will find "legal" commands as well as common nicknames for commands. Legal commands are defined as names that, when keyed at the prompt line, will execute.

Each command listing begins with the name of the command in white text on a black box. Only entries made in all caps are legal command names. Nicknames, written in upper and lower case, will not activate the command.

Nicknames are slang or colloquial names for commands. For example, "Grid Lock" is really a nickname for the legal command "LOCK GRID."

Definitions like "Active" are ideas you should understand, but they are not legal MicroStation commands. User commands are small programs that run within the MicroStation program. To invoke user commands, you typically key "UC=abcd," where "abcd" is the name of the user command.

How to Read Through a Command Listing

With the exception of a few concepts and definitions (written in upper and lower case), all uppercase entries are command key-ins. The syntax of the entry is the formal key-in syntax. You can activate these commands by simply keying in the command name.

Each entry in this book consists of a description, information about activating the command, command window prompts (which explain inputs required to complete the command) and special notes concerning the command. The following sample entry illustrates the essential parts of a command listing.

See Also: FENCE CHANGE tools

FENCE CHANGE LOCK   FEN CH L

Drawing Tool: Lock Fence Contents
    Secures, or locks up, a group of elements. Once locked, the elements cannot be manipulated or changed in any way. This item requires one input point to start processing the fence contents.

Item Selection
    On sidebar menu pick: n/a
    For window, point to: n/a
    On paper menu see: n/a
Command Window Prompts:
  1. Select the item

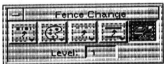

     Accept/Reject Fence Contents
  2. **Pick a point in any view**
     Processing fence contents
    To exit item: Exits by itself
  See Also: CHANGE LOCK, CHANGE UNLOCK, FENCE CHANGE UNLOCK

FENCE CHANGE STYLE   FEN CH ST

Drawing Tool: Change Fence Contents to Active Style
    Alters the line style of elements. First place a fence around the target elements. Then select FENCE CHANGE STYLE and place one input point to start the fence.

Each command entry consists of the following parts:

1. **Primary entry** - appears in reverse lettering. All caps means it is a legal command key-in.

2. **Short cut** - appears to the right of the primary entry. This can be either an alternative key-in or a minimal version of the command key-in.

3. **Type** - indicates whether the command is a view control, drawing tool, setting or other type of entry, and provides the tool name or description.

4. **Description** - a general description of what the command does with special attention paid to information required by the command.

5. **Item Selection** - describes the various ways to activate a command, including:
      Sidebar menu selection,
      Selecting an icon in a palette,
      Paper menu selection.

6. **Command window prompts** - lists messages that appear in the prompt field of the command window. These messages prompt for inputs to complete the command.

7. **Additional information** - other information including:
      Note - other pertinent information about the requirements and function of the command.
      Tip - Special hints about using a command.
      Error - error messages you might see, and ways to fix them.
      See Also - related commands.

Some commands may also contain an entry indicating the command only works on the PC, Mac or MicroStation 32 platforms.

Quick Reference to Important Commands and Definitions

Use this list to help become acquainted with some of Microstation's important commands and key Definitions.

Topic of Interest:	Turn to:
Attaching menus	ATTACH MENU
Creating graphics	PLACE tools
Changing how elements are created	ACTIVE settings
Changing how elements look	ACTIVE settings
Short key-ins	Appendix B
Handling groups of elements	FENCE, GRAPHIC GROUP, WSET
List of commands by function	Index A
MicroStation Directory Highlights	Appendix A
Moving around on the screen	VIEW and WINDOW
Oops...not that element	UNDO and REDO
Patterning	Patterns, CROSSHATCH, HATCH
Placing the same feature over and over	CREATE CELL
Multi-line cutter tools	Cutter Tools
3D rendering	RENDER commands
Take more control of the drawing	Locks, various SET items
Alternate key-ins	Appendix C

MicroStation Reference Disk

On-Line Help for MicroStation

If you would like to have the information contained in this book available to you on-line while working in MicroStation, order the MicroStation Reference Guide Disk, a companion to this book. The MicroStation Reference Guide Disk is available from your MicroStation Dealer or OnWord Press (See below).

How To Use the Reference Guide Disk

The on-line help is a compiled resource file that can be used instead of the standard help resource file included with MicroStation. Once installed, the user may display help for any MicroStation command through MicroStation's Help menu.

Installing the On-line Help

For detailed installation instructions on all three MicroStation platforms, refer to the README.HLP file on the MicroStation Reference Guide disk.

Ordering the MicroStation Reference Guide Disk

To order, enclose a check, money order or MasterCARD, VISA, or American Express card information. The price of the optional disk is U.S., $14.95 plus shipping. Shipping and handling is $4.50 in the continental U.S., $8.00 in AK, HI, PR or Canada, and $12.00 for the rest of the world.

OnWord Press
P.O. Box 500
Chamisal, NM 87521 USA

!

Key-in: Exclamation point

Key in the exclamation point (!) and strike Enter to suspend the design session and transfer control to the computer's operating system. You can then execute operating system commands and run external applications.

To display a quick directory listing, key in the exclamation character immediately before an appropriate listing command. For instance, in DOS key in !DIR \CGSI*.DGN to list the design files in the CGSI directory. In UNIX, key in !ls /usr/cgsi/*.dgn to list the same files.

📁 *Note:* Once you're done at the operating system level, type EXIT to resume the design session.

See Also: %, DOS, UNIX

%

Key-in: Percent sign

Use the percent (%) character to suspend the design session and execute one operating system command or application. A window will open, the command will execute, and the design session will automatically resume.

For example, to delete a file key in the percent character before the appropriate delete command. DOS users key in %DEL \CGSI\LEAVY.DGN. In UNIX, key in %rm /usr/cgsi/leavy.dgn.

See Also: !, DOS, UNIX

AA=

Alternate Key-in

Alternate key-in for ACTIVE ANGLE. Use it to set the angle of the elements when creating geometry in a drawing file.

See Also: ACTIVE ANGLE

AC=

Alternate Key-in

Alternate key-in for ACTIVE CELL. Use it to set the name of the cell when placing these figures in a drawing file. Also, it reactivates the previous PLACE CELL drawing tool.

See Also: ACTIVE CELL

Active

Definition

Active Settings affect the process of creating elements, lines, circles, arcs and text in a drawing file. Active Settings perform two jobs. They show the current settings, and they change the values of the Active Settings.

For example, to see what the ACTIVE COLOR value is, key in ACTIVE COLOR. To change the ACTIVE COLOR to blue, key in ACTIVE COLOR BLUE.

Each Active Setting listed in this section includes the setting name, a short description, the key-in necessary to show or change the current setting and a few examples.

📁 *Note:* Many of the Active Settings have an "alternate key-in". ACTIVE ANGLE's alternate key-in is AA=. Each alternate key-in is alphabetically arranged in the main listing and in Appendix C.

☞ *Tip:* You can change or display any Active Setting without exiting the drawing tool in which you are working.

See Also: ACTIVE settings

ACTIVE ANGLE   AA=

Setting: Active Angle

Determines the angle of the elements when placing them in the drawing file. You can change the angle format display from the Data Readout portion of the

Design Options settings box. For example, when using conventional, MicroStation measures the active angle counterclockwise from the display screen's horizontal axis. Two other angle settings are azimuth or bearing.

Item Selection
On sidebar menu pick: Params Active Angle
For window, point to: Settings Active Angle
On paper menu see: n/a
To show, key in: ACTIVE ANGLE
   Angle=0.0000°
To set, key in: ACTIVE ANGLE nnn
   Where nnn is a number between -360 and +360 degrees.

Example(s):
AA=145.10
AA=-45   for conventional
AA=35° 2′ 23″   for azimuth
AA=N 25° 8′ W   for bearing
AA=$ to display the current angle
See Also: ACTIVE ANGLE settings, LOCK ANGLE

ACTIVE ANGLE PT2   ACT AN PT2

Setting: Active Angle by Two Points
   Changes the ACTIVE ANGLE setting based on two input points. The two inputs points can be placed on an element or anywhere in a view.
Item Selection
On sidebar menu pick: n/a
For window, point to: n/a
On paper menu see: MEASURE
Command Window Prompts:
   1. Select the item
      Enter angle vertex
   2. Pick a point
      Enter endpoint of angle leg
   3. Pick a second point
To exit item: **Make next selection**
See Also: ACTIVE ANGLE settings

ACTIVE ANGLE PT3   ACT AN PT3

Setting: Active Angle by Three Points
   Changes the ACTIVE ANGLE setting based on three input points. The three points can be placed on an element or in any view.
Item Selection
On sidebar menu pick: n/a
For window, point to: n/a
On paper menu see: MEASURE
Command Window Prompts:
   1. Select the item
      Enter endpoint of angle leg
   2. Pick first angle leg
      Enter angle vertex
   3. Pick angle vertex
      Enter endpoint of angle leg
   4. Pick second angle leg
To exit item: **Make next selection**
See Also: ACTIVE ANGLE settings

ACTIVE AREA HOLE   ACT AR

Setting: Active Area Hole
　　Places hole elements for shading or patterning. The active area setting status appears in the Command Window: LVL=50,HOLE,WT=0,LC=SOL,PRI,CO=4. The phrase HOLE shows that HOLE is the active choice.

Item Selection
　On sidebar menu pick: Params | Active | Area | Hole
　For window, point to: Element   Attributes   Area:
　On paper menu see: UTILITIES
　To show, key in: ACTIVE AREA
　　Active Area = HOLE
　To set, key in: ACTIVE AREA HO
See Also: ACTIVE AREA SOLID

ACTIVE AREA SOLID   ACT AR

Setting: Active Area Solid
　　The default ACTIVE AREA setting. The ACTIVE AREA setting status appears in the Command Window: LVL=50,SOLID,WT=0,LC=SOL,PRI,CO=4. The phrase SOLID shows that SOLID is the active choice.

Item Selection
　On sidebar menu pick: Params | Active | Area | Solid
　For window, point to: Element   Attributes   Area:
　On paper menu see: UTILITIES
　To show, key in: ACTIVE AREA
　　Active Area = SOLID
　To set, key in: ACTIVE AREA SO
See Also: ACTIVE AREA HOLE

ACTIVE AXIS   ACT AX

Setting: Active Axis
　　When placing input points, this setting determines the angle of the point relative to the ACTIVE AXORIGIN. The LOCK AXIS setting must be active.

Item Selection
　On sidebar menu pick: Params | Active | Axis | Key
　For window, point to: Settings   Locks   Full   Increment:
　On paper menu see: n/a
　To show, key in: ACTIVE AXIS
　　Axis Lock = 90.00
　To set, key in: ACTIVE AXIS nnn
　　Where nnn is a number between -360 and +360 degrees.
See Also: ACTIVE AXORIGIN, LOCK AXIS

ACTIVE AXORIGIN   ACT AXO

Setting: Active AXOrigin
　　Determines the angle for the base X and Y-axis when the LOCK AXIS setting is active.

Item Selection
　On sidebar menu pick: Params | Active | Axis | Orig
　For window, point to: Settings   Locks   Full   Start Angle:
　On paper menu see: n/a
　To show, key in: ACTIVE AXORIGIN
　　Axis Lock Origin = 0.00
　To set, key in: ACTIVE AXORIGIN nn
　　Where nn is a number between 0 and 90 degrees.

Example(s):
　ACT AXO=45   sets the AXORIGIN to 45 degrees
See Also: ACTIVE AXIS, LOCK AXIS

ACTIVE BACKGROUND   AC B

Setting: Active Background
Displays images behind any drawing file. For example, display the proposed car design streaming down some stretch of highway, or inset the tool model at a construction site.

Item Selection
On sidebar menu pick: n/a
For window, point to: n/a
On paper menu see: n/a
AC B filename

✍ *Note:* Use the SET BACKGROUND setting to turn the background display on or off.
See Also: SET BACKGROUND

ACTIVE BSPLINE CLOSED   ACT BS C

Setting: Active B-spline Closed
Sets B-spline mode for closed curves.

Item Selection
On sidebar menu pick: | Params | | BSpl |
For window, point to: Element   B-splines   Curves   Type:
On paper menu see: n/a
To show, open: B-splines settings box
To set, key in: ACTIVE BSPLINE CLOSED

✳ *Error:* Invalid character - **If this message appears, type MDL LOAD SPLINES before reselecting this item.**
See Also: ACTIVE BSPLINE settings, CONSTRUCT BSPLINE tools, PLACE BSPLINE tools

ACTIVE BSPLINE OPEN   ACT BS OP

Setting: Active B-spline Open
Sets B-spline mode for open curves.

Item Selection
On sidebar menu pick: | Params | | BSpl |
For window, point to: Element   B-splines   Curves   Type:
On paper menu see: n/a
To show, open: B-splines settings box
To set, key in: ACTIVE BSPLINE OPEN

✳ *Error:* Invalid character - **If this message appears, type MDL LOAD SPLINES before reselecting this item.**
See Also: ACTIVE BSPLINE settings, CONSTRUCT BSPLINE tools, PLACE BSPLINE tools

ACTIVE BSPLINE ORDER   ACT BS OR

Setting: Active B-spline Order
Determines the order of any future B-spline curve created. The order number must fall between 2 and 15. This setting can be viewed or changed through the B-splines settings box or by a key-in.

Item Selection
On sidebar menu pick: | Params | | BSpl |
For window, point to: Element   B-splines   Curves   Order:
On paper menu see: n/a
To show, open: B-spline settings box
To set, key in: ACTIVE BSPLINE ORDER nn
Where nn is a whole number from 2 to 15 inclusive.

✳ *Error:* Invalid character - **If this message appears, type MDL LOAD SPLINES before reselecting this item.**

See Also: ACTIVE BSPLINE settings, CONSTRUCT BSPLINE tools, PLACE BSPLINE tools

ACTIVE BSPLINE POLE   ACT BS POLE

Setting: Active B-spline Pole
   Determines the number of poles of any future B-spline curve. The number of poles must fall between order number and 101.

Item Selection
   On sidebar menu pick: `Params`   `BSpl`
   For window, point to: Element   B-splines   Curves   Poles:
   On paper menu see: n/a
   To show, open: **B-spline settings box**
   To set, key in: **ACTIVE BSPLINE POLE nnn**
   Where nnn is a whole number between the order number and 101.

*   *Error:* Invalid character - **If this message appears, type MDL LOAD SPLINES before reselecting this item.**
See Also: ACTIVE BSPLINE settings, CONSTRUCT BSPLINE tools, PLACE BSPLINE tools

ACTIVE BSPLINE POLYGON   ACT BS POLY

Setting: Active B-spline Polygon
   Sets B-spline mode for polygon display. Turn this setting off when the polygon display is not needed.

Item Selection
   On sidebar menu pick: `Params`   `BSpl`
   For window, point to: Element   B-splines   Display Control Polygon
   On paper menu see: n/a
   To show, open: **B-splines setting box**
   To set, key in: **ACTIVE BSPLINE POLYGON on/off**
   Where on/off is either on or off.

*   *Error:* Invalid character - **If this message appears, type MDL LOAD SPLINES before reselecting this item.**
See Also: ACTIVE BSPLINE settings, CONSTRUCT BSPLINE tools, PLACE BSPLINE tools

ACTIVE BSPLINE TOLERANCE   ACT BS T

Setting: Active B-spline Tolerance
   Pertains to placing B-splines using the least square option. The value is expressed in working units (MU:SU). When the distance between input points and the B-spline curve exceeds the tolerance value, tolerance errors are reported.

Item Selection
   On sidebar menu pick: `Params`   `BSpl`
   For window, point to: Element   B-splines   Tolerance:
   On paper menu see: n/a
   To show, open: **B-splines settings box**
   To set, key in: **ACTIVE BSPLINE TOLERANCE ww:ww**
   Where ww:ww is valid working unit.

*   *Error:* Invalid character - **If this message appears, type MDL LOAD SPLINES before reselecting this item.**
See Also: ACTIVE BSPLINE settings, CONSTRUCT BSPLINE tools, PLACE BSPLINE tools

ACTIVE BSPLINE UCLOSED   ACT BS UC

Setting: Active B-spline U Closed
   Sets B-spline mode so that the u-direction is closed.

Item Selection
On sidebar menu pick: | Params | | BSpl |
For window, point to: Element B-splines Surfaces U Type:
On paper menu see: n/a
To show, open: **B-splines settings box**
To set, key in: **ACTIVE BSPLINE UCLOSED**

✳ *Error:* Invalid character - **If this message appears, type MDL LOAD SPLINES before reselecting this item.**
See Also: ACTIVE BSPLINE settings, CONSTRUCT BSPLINE tools, PLACE BSPLINE tools

ACTIVE BSPLINE UOPEN   ACT BS UOP

Setting: Active B-spline U Open
Sets B-spline mode so that the u-direction is open.
Item Selection
On sidebar menu pick: | Params | | BSpl |
For window, point to: Element B-splines Surfaces U Type:
On paper menu see: n/a
To show, open: **B-splines settings box**
To set, key in: **ACTIVE BSPLINE UOPEN**

✳ *Error:* Invalid character - **If this message appears, type MDL LOAD SPLINES before reselecting this item.**
See Also: ACTIVE BSPLINE settings, CONSTRUCT BSPLINE tools, PLACE BSPLINE tools

ACTIVE BSPLINE UORDER   ACT BS UOR

Setting: Active B-spline U Order
Determines the U-order of any future B-spline surface created. The order number must fall between 2 and 15.
Item Selection
On sidebar menu pick: | Params | | BSpl |
For window, point to: Element B-splines Surfaces U Order:
On paper menu see: n/a
To show, open: **B-splines settings box**
To set, key in: **ACTIVE BSPLINE UORDER nn**
Where nn is a whole number from 2 to 15 inclusive.

✳ *Error:* Invalid character - **If this message appears, type MDL LOAD SPLINES before reselecting this item.**
See Also: ACTIVE BSPLINE settings, CONSTRUCT BSPLINE tools, PLACE BSPLINE tools

ACTIVE BSPLINE UPOLE   ACT BS UP

Setting: Active B-spline U Pole
Determines the number of poles in the u-direction when creating B-spline surfaces. The number of poles must fall between the order number and 101.
Item Selection
On sidebar menu pick: | Params | | BSpl |
For window, point to: Element B-splines Surfaces U Poles:
On paper menu see: n/a
To show, open: **B-splines settings box**
To set, key in: **ACTIVE BSPLINE UPOLE nnn**
Where nnn is a whole number between the order number and 101.

✳ *Error:* Invalid character - **If this message appears, type MDL LOAD SPLINES before reselecting this item.**
See Also: ACTIVE BSPLINE settings, CONSTRUCT BSPLINE tools, PLACE BSPLINE tools

---

ACTIVE BSPLINE URULES   ACT BS UR

Setting: Active B-spline U Rules
Determines the number of u-rule lines of any future B-spline surface created.
The number of rule lines must fall between two and 256.

Item Selection
On sidebar menu pick: [Params] [BSpl]
For window, point to: Element   B-splines   Surfaces   U Rules:
On paper menu see: n/a
To show, open: B-splines settings box
To set, key in: ACTIVE BSPLINE URULES
Where nn is a whole number between the order number and 256.

✻ *Error:* Invalid character - **If this message appears, type MDL LOAD SPLINES before reselecting this item.**
See Also: ACTIVE BSPLINE settings, CONSTRUCT BSPLINE tools, PLACE BSPLINE tools

---

ACTIVE BSPLINE VCLOSED   ACT BS VC

Setting: Active B-spline V Closed
Sets the B-spline so that the v-direction is closed.

Item Selection
On sidebar menu pick: [Params] [BSpl]
For window, point to: Element   B-splines   Surfaces   V Type:
On paper menu see: n/a
To show, open: B-splines settings box
To set, key in: ACTIVE BSPLINE VCLOSED

✻ *Error:* Invalid character - **If this message appears, type MDL LOAD SPLINES before reselecting this item.**
See Also: ACTIVE BSPLINE settings, CONSTRUCT BSPLINE tools, PLACE BSPLINE tools

---

ACTIVE BSPLINE VOPEN   ACT BS VOP

Setting: Active B-spline V Open
Sets the B-spline mode so the v-direction is open.

Item Selection
On sidebar menu pick: [Params] [BSpl]
For window, point to: Element   B-splines   Surfaces   V Type:
On paper menu see: n/a
To show, open: B-splines settings box
To set, key in: ACTIVE BSPLINE VOPEN

✻ *Error:* Invalid character - **If this message appears, type MDL LOAD SPLINES before reselecting this item.**
See Also: ACTIVE BSPLINE settings, CONSTRUCT BSPLINE tools, PLACE BSPLINE tools

---

ACTIVE BSPLINE VORDER   ACT BS VOR

Setting: Active B-spline V Order
Determines the v-order of any future B-spline surface created. The order
number must fall between 2 and 15.

Item Selection
On sidebar menu pick: [Params] [BSpl]
For window, point to: Element   B-splines   Surfaces   V Order:
On paper menu see: n/a
To show, open: B-splines settings box
To set, key in: ACTIVE BSPLINE VORDER nn
Where nn is a whole number from 2 to 15 inclusive.

✳ *Error:* Invalid character - **If this message appears, type MDL LOAD SPLINES before reselecting this item.**
See Also: ACTIVE BSPLINE settings, CONSTRUCT BSPLINE tools, PLACE BSPLINE tools

ACTIVE BSPLINE VPOLE   ACT BS VP

Setting: Active B-spline V Pole
Determines the number of poles in the v-direction when creating B-spline surfaces. The number of poles must fall between the order number and 101.

Item Selection
On sidebar menu pick: Params   BSpl
For window, point to: Element   B-splines   Surfaces   V Poles:
On paper menu see: n/a
To show, open: **B-splines settings box**
To set, key in: **ACTIVE BSPLINE VPOLE nnn**
Where nnn is a whole number between the order number and 101.

✳ *Error:* Invalid character - **If this message appears, type MDL LOAD SPLINES before reselecting this item.**
See Also: ACTIVE BSPLINE settings, CONSTRUCT BSPLINE tools, PLACE BSPLINE tools

ACTIVE BSPLINE VRULES   ACT BS VR

Setting: Active B-spline V Rules
Determines the number of v-rule lines of any future B-spline surface created. The number of rule lines must fall between two and 256.

Item Selection
On sidebar menu pick: Params   BSpl
For window, point to: Element   B-splines   Surfaces   V Rules:
On paper menu see: n/a
To show, open: **B-splines settings box**
To set, key in: **ACTIVE BSPLINE VRULES**
Where nn is a whole number between the order number and 256.

✳ *Error:* Invalid character - **If this message appears, type MDL LOAD SPLINES before reselecting this item.**
See Also: ACTIVE BSPLINE settings, CONSTRUCT BSPLINE tools, PLACE BSPLINE tools

ACTIVE CAPMODE   ACT CA

3D Setting: Active Capmode
Places a cap on the top and bottom of cylinders or truncated cones. Once capped, and shaded, a cylinder will appear as a solid rod instead of a hollow pipe.
Item Selection
On sidebar menu pick: Place   Cyldr   Mode
For window, point to: Palettes   3D   Type:
On paper menu see: 3D
To show, key in: **ACTIVE CAPMODE**
´ Capped Surface Placement: OFF
To set, key in: **ACTIVE CAPMODE on/off**
Where on/off is either ON or OFF.
See Also: PLACE CONE tools, PLACE CYLINDER tools

ACTIVE CELL   AC=

Setting: Active Cell
Sets the ACTIVE CELL name and reactivates the last PLACE CELL tool.
Item Selection
On sidebar menu pick: Cells   Active

For window, point to: Settings   Cells   Placement
On paper menu see: n/a
To show, key in: ACTIVE CELL
    Active Cell = SAMPLE
To set, key in: ACTIVE CELL uuu
    Where uuu is any six character user defined cell name.

Example(s):
AC=DOOR12
AC=$ to display the current cell name
See Also: PLACE CELL tools

ACTIVE CLASS CONSTRUCTION   ACT CL

Setting: Active Class Construction
    Places construction elements in the drawing file. Construction elements
    assist in building other geometry in the design file. The active class status is
    always   shown   in   the   Command   Window:
    LVL=50,SOLID,WT=0,LC=SOL,CON,CO=4. The phrase CON shows that the
    active class is CONSTRUCTION.

Item Selection
On sidebar menu pick: [Params]   [Active]   [Class]   [Cons]
For window, point to: Element   Attributes   Class:
On paper menu see: n/a
To show, key in: ACTIVE CLASS
    Active Class: CONSTRUCTION
To set, key in: ACTIVE CLASS CO

✍ *Note:* To turn off the display of construction elements use the SET CONSTRUCT
setting.
See Also: ACTIVE CLASS PRIMARY

ACTIVE CLASS PRIMARY   ACT CL

Setting: Active Class Primary
    Places primary elements in the drawing file. Primary is the default element
    class. The active class status is always shown in the Command Window:
    LVL=50,SOLID,WT=0,LC=SOL,PRI,CO=4. The phrase PRI shows that the active
    class is PRIMARY.

Item Selection
On sidebar menu pick: [Params]   [Active]   [Class]   [Prim]
For window, point to: Element   Attributes   Class:
On paper menu see: n/a
To show, key in: ACTIVE CLASS
    Active Class: PRIMARY
To set, key in: ACTIVE CLASS PR
See Also: ACTIVE CLASS CONSTRUCTION

ACTIVE COLOR   CO=

Setting: Active Color
    Determines the color of the elements placed in a drawing file. The ACTIVE
    COLOR number is always shown in the Command Window:
    LVL=50,SOLID,WT=0,LC=SOL,PRI,CO=4. The phrase CO shows that the active
    color is 4 (yellow).

Item Selection
On sidebar menu pick: [Params]   [Active]   [Color]   [Key]
For window, point to: Element   Attributes   Color:
On paper menu see: n/a
To set, key in: ACTIVE COLOR nnn
    Where nnn is a number between 0 and 255 or one of the color names.

Color Name	Value
White	0
Blue	1

Green	2
Red	3
Yellow	4
Violet	5
Orange	6

Example(s):

CO=1

CO=blue

CO=$ displays the current color number

  Key-ins for MicroStation PC only:

CO=+n   to increment some n number up from the AC-
                                TIVE COLOR setting. Example: ACTIVE
                                COLOR is 6. CO=+3 would make the ACTIVE
                                COLOR 9.

CO=-n   to decrement some n number down from the
                                ACTIVE COLOR setting. Example: ACTIVE
                                COLOR is 6. CO=-3 would make the ACTIVE
                                COLOR 3.

☞ *Tip:* Change these settings "on the fly." For instance, a line that you are entering
   with the PLACE LINE tool is blue, but should be green. Change the ACTIVE
   COLOR setting by keying in CO=GREEN. MicroStation changes the line from blue
   to green. Now you can finish drawing the line.

See Also: CHANGE COLOR, ATTACH COLORTABLE, DIMENSION COLOR

ACTIVE DATABASE   DB=

DOS Only
Setting: Active Database
        Attaches a database file to a drawing file.

Item Selection
  On sidebar menu pick: | dBase | | Setup | | Act DB |
  For window, point to: n/a
  On paper menu see: n/a
  To show, key in: **ACTIVE DATABASE**
     Control: C:\DIRNAME\FILENAME.DBF
  To set, key in: **ACTIVE DATABASE filename**
     Where filename is an existing database file.

Example(s):

DB=\CGSI\LEAVY.DBF   for DOS users

DB=/usr/cgsi/leavy.dbf   for UNIX users

DB=$ to display the current file

See Also: SET DATABASE

ACTIVE DATYPE   DA=

Setting: Active DAType
        Allows you to associate database information and text nodes.

Item Selection
  On sidebar menu pick: n/a
  For window, point to: Palettes Database DA Type:
  On paper menu see: n/a

Example(s):

DA=120   sets the displayable attribute number to 120

See Also: LOAD DA, PLACE NODE tools

ACTIVE ENTITY   AC E

Command: Active Entity
        Defines a database entity without specifying an existing row. Instead, the
        SQL INSERT statement carries the information about the row. Key in ACTIVE
        ENTITY <SQL_INSERT_STATEMENT>.

Item Selection
   On sidebar menu pick: n/a
   For window, point to: n/a
   On paper menu see: n/a
See Also: DEFINE AE, SHOW AE

ACTIVE FILL   AC FI

Setting: Active Fill
   Allows you to toggle the placement of filled shapes in 2D or 3D drawing files.
Item Selection
   On sidebar menu pick: | Params | | Active | | Fill |
   For window, point to: Element Attributes
   On paper menu see: LOCKS
   To show, key in:
     Filled Placement : OFF **ACTIVE FILL**
   To set, key in: **ACTIVE FILL on/off**
     Where on/off is either ON or OFF.
See Also: CHANGE FILL, SET FILL

ACTIVE FONT   FT=

Setting: Active Font
   Sets the lettering style, or font, when placing text in a drawing file.
Item Selection
   On sidebar menu pick: | Text | | Font | | Key: |
   For window, point to: Element Text Font
   On paper menu see: n/a
   To show, key in: **ACTIVE FONT**
     Font=3
   To set, key in: **ACTIVE FONT nnn**
     Where nnn is a whole number between 1 and 127.

Example(s):
   FT=6
   FT=$ to display the current text font
See Also: DIMENSION FONT ACTIVE, MODIFY TEXT, PLACE TEXT tools

ACTIVE GRIDMODE ISOMETRIC   ACT GRIDM

Setting: Active Gridmode Isometric
   Sets the display grid orientation to isometric.
Item Selection
   On sidebar menu pick: n/a
   For window, point to: Settings Grid Configuration:
   On paper menu see: DISPLAY
   To show, key in: **ACTIVE GRIDMODE**
     Grid Configuration: ISOMETRIC
   To set, key in: **ACTIVE GRIDMODE IS**
See Also: ACTIVE GRIDMODE settings

ACTIVE GRIDMODE OFFSET   ACT GRIDM

Setting: Active Gridmode Offset
   Allows you to offset alternate rows of grid points.
Item Selection
   On sidebar menu pick: n/a
   For window, point to: Settings Grid Configuration:
   On paper menu see: n/a
   To show, key in: **ACTIVE GRIDMODE**
     Grid Configuration: OFFSET
   To set, key in: **ACTIVE GRIDMODE OF**

See Also: other ACTIVE GRIDMODE settings

ACTIVE GRIDMODE ORTHOGONAL   ACT GRIDM

Setting: Active Gridmode Orthogonal
    MicroStation's default display grid orientation.
Item Selection
  On sidebar menu pick: n/a
  For window, point to: Settings Grid Configuration:
  On paper menu see: n/a
  To show, key in: `ACTIVE GRIDMODE`
    `Grid Configuration: ORTHOGONAL`
  To set, key in: `ACTIVE GRIDMODE OR`
See Also: ACTIVE GRIDMODE settings

ACTIVE GRIDRATIO   ACT GRIDRA

Setting: Active Gridratio
    Allows you to set up a display grid with different horizontal and vertical spacing between the grid points. Instead of a default vertical-to-horizontal display grid ratio of one-to-one, establish a new ratio by using the ACTIVE GRIDRATIO setting.
Item Selection
  On sidebar menu pick: n/a
  For window, point to: Settings Grid Aspect Ratio (Y/X):
  On paper menu see: n/a
  To show, key in: `ACTIVE GRIDRATIO`
    `Grid Ratio: 1.00`
  To set, key in: `ACTIVE GRIDRATIO www`
    Where www equals some working unit.
Example(s):
  `ACTIVE GRIDRATIO 2`   sets a vertical-to-horizontal ratio of 2 to 1
See Also: SET GRID, SET MAXGRID

ACTIVE GRIDREF   GR=

Setting: Active Gridref
    Controls the number of positional units between each reference cross on the display grid.
Item Selection
  On sidebar menu pick: n/a
  For window, point to: Settings Grid Master / Grid:
  On paper menu see: n/a
  To show, key in: `ACTIVE GRIDREF`
    `Grid=0:1,12`
  To set, key in: `ACTIVE GRIDREF www`
    Where www equals some working unit.
Example(s):
  `GR=12`
  `GR=$`   to display the current spacing
See Also: ACTIVE GRIDUNIT

ACTIVE GRIDUNIT   GU=

Setting: Active Gridunit
    Defines the number of working units between each grid unit on the display grid.
Item Selection
  On sidebar menu pick: n/a
  For window, point to: Settings Grid Reference Grid:
  On paper menu see: n/a
  To show, key in: `ACTIVE GRIDUNIT`

```
Grid=0:1,12
```
To set, key in: `ACTIVE GRIDUNIT www`
Where www equals some working unit.

Example(s):
```
GU=5
GU=$                          to display the current spacing
```
See Also: ACTIVE GRIDREF

ACTIVE INDEX   OX=

Setting: Active Index
Employed to attach a user command index to a drawing file.
Item Selection
On sidebar menu pick: n/a
For window, point to: n/a
On paper menu see: n/a
To show, key in: `ACTIVE INDEX`
```
Index file = filename
```
To set, key in: `ACTIVE INDEX filename`
Where filename is the name of a user command index.

Example(s):
```
OX=\CGSI\LEAVYUCM.NDX      for DOS users
OX=/usr/cgsi/leavyucm.ndx for UNIX users
OX=$                       to display the current index file
```
☞ *Tip:* One use of the ACTIVE INDEX setting is to map the area on the paper menu called USER COMMANDS
See Also: USERCOMMAND

ACTIVE KEYPNT   KY=

Setting: Active Keypoint
Employed to divide elements into segments for snapping purposes.
Item Selection
On sidebar menu pick: | Params | | Set | | KeyPt: |
For window, point to: Settings Locks Divisor:
On paper menu see: n/a
To show, key in: `ACTIVE KEYPNT`
```
Snap Divisor = 1
```
To set, key in: `ACTIVE KEYPNT nnn`
Where nnn is a whole number between 1 and 255.

Example(s):
```
KY=6                       sets the divisor to 6
KY=$                       to display the current setting
```
See Also: LOCK SNAP settings

ACTIVE LEVEL   LV=

Setting: Active Level
MicroStation creates the elements on this level. The active level is always shown in the Command Window: LVL=50,SOLID,WT=0,LC=SOL,PRI,CO=4. The phrase LVL=50 shows that the active level is 50.
Item Selection
On sidebar menu pick: | Levels | | Key |
For window, point to: Element Attributes Level:
On paper menu see: n/a
To set, key in: `ACTIVE LEVEL nn`
Where nn is a whole number between 1 and 63.

Example(s):
```
LV=2
LV=$                       to display the current level
```

Key-ins for MicroStation PC only:

LV=+1	increments level by one number
LV=-1	decreases level by one number
LV=+n	to increment some n number up from the active level setting. Example: ACTIVE LEVEL is 6. LV=+3 would make the ACTIVE LEVEL 9.
LV=-n	to decrement some n number down from the active level setting. Example: ACTIVE LEVEL is 6. LV=-3 would make the ACTIVE LEVEL 3.

See Also: CHANGE LEVEL, DIMENSION LEVEL settings, LOCK LEVEL, SET LEVELS

ACTIVE LINE LENGTH   LL=

Setting: Active Line Length
Controls the number of characters in a line of text of a text node.

Item Selection
On sidebar menu pick: `Text` `Size` `LL=:`
For window, point to: Element   Text   Line Length
On paper menu see: n/a
To show, key in: ACTIVE LINE LENGTH
    Line Len. = 255
To set, key in: ACTIVE LINE LENGTH nnn
Where nnn is a whole number between 1 and 255.

Example(s):
LL=10
LL=$ to display the current line length
See Also: ACTIVE LINE SPACE, PLACE NODE tools

ACTIVE LINE SPACE   LS=

Setting: Active Line Space
Sets the spacing between lines of text. Normally this spacing is half the active text size.

Item Selection
On sidebar menu pick: `Text` `Size` `LS=:`
For window, point to: Element   Text   Line Spacing
On paper menu see: n/a
To show, key in: ACTIVE LINE SPACE
    Line Spacing = 0:3
To set, key in: ACTIVE LINE SPACE www
Where www equals some working unit.

Example(s):
LS=.10
LS=$ to display the current spacing
See Also: ACTIVE LINE LENGTH, PLACE TEXT tools

ACTIVE LINEWIDTH   ACT LINEW

Setting: Active Linewidth
Sets the exact thickness of an element. This thickness is in units of resolution (UOR). When an element is given a linewidth, it appears tubular on the display screen. Turning on the SET LINEWIDTH setting causes the element to look shaded or filled.

ACTIVE LINEWIDTH differs from ACTIVE WEIGHT. ACTIVE WEIGHT sets a stroking value to an element for plotting.

Item Selection
On sidebar menu pick: n/a
For window, point to: n/a

On paper menu see: n/a
To show, key in: ACTIVE LINEWIDTH
  Line Width = 0:0
To set, key in: ACTIVE LINEWIDTH nnn
  Where nnn equals some working unit.

Example(s):
  ACT LINEW .5
See Also: ACTIVE WEIGHT, SET LINEWIDTH

ACTIVE LINKAGE   ACT LINK

Setting: Active Linkage
  Displays the current linkage status.
Item Selection
  On sidebar menu pick: dBase   Settng
  For window, point to: Settings   Database   Linkage Mode:

  On paper menu see: DATABASE
  To show, open: Database
           settings box
  See Also: ACTIVE LINKAGE settings

ACTIVE LINKAGE DUPLICATE   ACT LINK DU

Setting: Active Linkage Duplicate
  Controls the relationship between entities in a database and graphic ele-
  ments. The ACTIVE LINKAGE DUPLICATE option gives you the ability to have
  multiple linkages pointing to the same record.
Item Selection
  On sidebar menu pick: dBase   Settng
  For window, point to: Settings   Database   Linkage Mode:
  On paper menu see: n/a
  To show, key in: ACTIVE LINKAGE
    Linkage mode: DUPLICATE
  To set, key in: ACTIVE LINKAGE DU
  See Also: ACTIVE LINKAGE settings

ACTIVE LINKAGE INFORMATION   ACT LINK IN

Setting: Active Linkage Information
  Controls the relationship between entities in a database and graphic ele-
  ments. The ACTIVE LINKAGE INFORMATION option, is similar to ACTIVE
  LINKAGE DUPLICATE, except it turns on the property bit.
Item Selection
  On sidebar menu pick: dBase   Settng
  For window, point to: Settings   Database   Linkage Mode:
  On paper menu see: n/a
  To show, key in: ACTIVE LINKAGE
    Linkage mode: INFORMATION
  To set, key in: ACTIVE LINKAGE IN
  See Also: ACTIVE LINKAGE settings

ACTIVE LINKAGE NEW

ACT LINK NE

Setting: Active Linkage New
Controls the relationship between entities in a database and graphic elements. The ACTIVE LINKAGE NEW option stands for a fresh or new record.

Item Selection
On sidebar menu pick: dBase   Settng
For window, point to: Settings Database Linkage Mode:
On paper menu see: n/a
To show, key in: ACTIVE LINKAGE
  Linkage mode: NEW
To set, key in: ACTIVE LINKAGE NE
See Also: ACTIVE LINKAGE settings

ACTIVE LINKAGE NONE

ACT LINK

Setting: Active Linkage None
Controls the relationship between entities in a database and graphic elements. The ACTIVE LINKAGE NONE option declares no linkage established.

Item Selection
On sidebar menu pick: dBase   Settng
For window, point to: Settings Database Linkage Mode:
On paper menu see: n/a
To show, key in: ACTIVE LINKAGE
  Linkage mode: NONE
To set, key in: ACTIVE LINKAGE NO
See Also: ACTIVE LINKAGE settings

ACTIVE NODE

NN=

Setting: Active Node
When placing text nodes in a drawing file, MicroStation assigns a consecutive number to each node. The ACTIVE NODE setting determines the number assigned to a text node. It can be reset to zero or any other number at any time.

Item Selection
On sidebar menu pick: n/a
For window, point to: n/a
On paper menu see: n/a
To show, key in: ACTIVE NODE
  Node Num = 2
To set, key in: ACTIVE NODE nnn
Where nnn is a whole number between 1 and 32767.

Example(s):
NN=3   sets the node number to three
NN=$ to display the current node number
See Also: PLACE NODE tools

ACTIVE ORIGIN

GO=

Setting: Active Origin
Locates your global origin on the design plane. All input points placed in a drawing file coincide to the global origin.

Item Selection
On sidebar menu pick: n/a
For window, point to: n/a
On paper menu see: n/a
To show, key in: ACTIVE ORIGIN
  GO=0:0.001,0:0.001
To set, key in: ACTIVE ORIGIN www
Where www equals some working unit to determine locations on the X, Y and Z-axis.

Example(s):
```
GO=1000,1000,1000
GO=$                        to display the current origin
```
✍ *Note:* After you key in a new origin, the system will ask you to enter the monument point which coincides with the value entered. Click a RESET to that question and the ACTIVE ORIGIN becomes the lower left corner of the design plane.

See Also: WINDOW ORIGIN

ACTIVE PATTERN ANGLE   PA=

Setting: Active Pattern Angle
Controls orientation of hatching, cross hatching or patterns relative to the horizontal axis.

Item Selection
On sidebar menu pick: `Pattrn`
For window, point to: Settings   Patterning   Pattern Angle
On paper menu see: n/a
To show, key in: ACTIVE PATTERN ANGLE
   Pattern Angle = 0.00,0.00
To set, key in: ACTIVE PATTERN ANGLE nnn
   Where nnn is a number between -360 and +360.

Example(s):
```
PA=45.10
PA=45,-45                   would render a crosshatch pattern
PA=$                        to display the current pattern angle
```
See Also: ACTIVE PATTERN settings, PATTERN AREA and PATTERN LINEAR tools

ACTIVE PATTERN CELL   AP=

Setting: Active Pattern Cell
Controls the name of the cell employed in the patterning process.

Item Selection
On sidebar menu pick: n/a
For window, point to: Settings   Patterning   Name
On paper menu see: n/a
To show, key in: ACTIVE PATTERN CELL
   Active Pattern = uuu
To set, key in: ACTIVE PATTERN CELL uuu
   Where uuu is the six character user defined cell name.

Example(s):
```
AP=TREELN
AP=$                        to display the current pattern cell
```
See Also: ACTIVE PATTERN settings, PATTERN AREA and PATTERN LINEAR tools

ACTIVE PATTERN DELTA   PD=

Setting: Active Pattern Delta
Controls the distance between patterns.

Item Selection
On sidebar menu pick: `Pattrn`
For window, point to: Settings   Patterning   Spacing
On paper menu see: n/a
To show, key in: ACTIVE PATTERN DELTA
   Active Pattern Delta = 0:0, 0:0
To set, key in: ACTIVE PATTERN DELTA www
   Where www equals some working unit.

Example(s):
```
PD=.1
```

PD=$ to display the current pattern delta
See Also: ACTIVE PATTERN settings, PATTERN AREA and PATTERN LINEAR tools

ACTIVE PATTERN MATCH   ACT PA M

Setting: Match Pattern Attributes
    Changes the ACTIVE PATTERN CELL setting by identifying an existing pattern element in the drawing file. This item expects one input point to highlight the target pattern and a second point to accept that pattern.
Item Selection
    On sidebar menu pick: n/a
    For window, point to: Settings Patterning Match
    On paper menu see: PATTERNING
Command Window Prompts:
  1. Select the tool
    Identify element
  2. Click on a pattern
    Accept/Reject (select next input)
  3. Click a send time
    PATTERN:FENCE,PA=0.0000,PS=1.0000
    ⇨ A pattern display line appears.
    To exit tool: Make next selection
See Also: PATTERN AREA and PATTERN LINEAR tools

ACTIVE PATTERN SCALE   PS=

Setting: Active Pattern Scale
    Sets the scale factor when placing a pattern cell in a drawing file.
Item Selection
    On sidebar menu pick: n/a
    For window, point to: Settings Patterning Scale
    On paper menu see: n/a
    To show, key in: ACTIVE PATTERN SCALE
    Pattern Scale = 1.0000
    To set, key in: ACTIVE PATTERN SCALE nnn
    Where nnn is the scale factor.
Example(s):
  PS=.75
  PS=$ to display the current pattern scale factor
See Also: ACTIVE PATTERN settings, PATTERN AREA and PATTERN LINEAR tools

ACTIVE POINT   ACT PO

Setting: Active Point
    Defines the point placed by the PLACE POINT and CONSTRUCT POINT tools. An ACTIVE POINT is a line without length, a single text character or a cell.
Item Selection
    On sidebar menu pick: n/a
    For window, point to: Settings Cells Point
    On paper menu see: n/a
    To show, key in: ACTIVE POINT
    Active Point = Element
    To set, key in: ACTIVE POINT uuu
    Where uuu is a six character user defined cell name.
Example(s):
  PT=SPOT   to make the active point a cell called SPOT
  PT=x   to make the active point the letter x
  PT=$ to display the current active point
See Also: CONSTRUCT POINT tools, PLACE POINT

ACTIVE RCELL   AR=

Drawing Tool: Active Rcell
> Used to swap or replace cells in an existing drawing. This setting calls the PLACE CELL RELATIVE tool.

Item Selection
> **On sidebar menu pick:** n/a
> **For window, point to:** Settings   Cells
> **On paper menu see:** n/a
> **To show, key in:** ACTIVE RCELL
>   Active Cell = SAMPLE
> **To set, key in:** ACTIVE RCELL uuu
>   Where uuu is a six character user defined cell name.

Example(s):
> AR=SPOT2
> AR=$ to display the current cell name

See Also: REPLACE CELL, DELETE CELL, RENAME CELL

ACTIVE REPORT   RS=

Setting: Active Report
> Controls the name of the database report table created during a fence operation.

Item Selection
> **On sidebar menu pick:** n/a
> **For window, point to:** n/a
> **On paper menu see:** n/a

Example(s):
> RS=states:states.lst   defines a report called "states.lst" for a table
>   named "states".

See Also: FENCE REPORT

ACTIVE REVIEW   RA=

Setting: Active Review
> Controls the name of an SQL select statement employed during a review process.

Item Selection
> **On sidebar menu pick:** n/a
> **For window, point to:** n/a
> **On paper menu see:** n/a

Example(s):
> RA=select * from parts   this review statement selects all parts

See Also: REVIEW

ACTIVE SCALE   AS=

Setting: Active Scale
> Sets the scale factor when placing elements or cells in a drawing file.

Item Selection
> **On sidebar menu pick:** | Params | | Active | | Scale | | Key |
> **For window, point to:** Settings   Active Scale   1.0
> **On paper menu see:** n/a
> **To show, key in:** ACTIVE SCALE
>   XS=1.0000,YS=1.0000,ZS=1.0000
> **To set, key in:** ACTIVE SCALE nnn
>   Where nnn is the scale factor.

Example(s):
> AS=.5
> AS=/2   half the current scale factor
> AS=*2   twice the current scale factor

AS=$ to display the current scale factor
See Also: ACTIVE SCALE DISTANCE, ACTIVE XSCALE, ACTIVE YSCALE,
ACTIVE ZSCALE, LOCK SCALE, REFERENCE SCALE

ACTIVE SCALE DISTANCE   ACT SC D

Setting: Active Scale by Three Points
         Graphically sets the active scale relative to an existing element in the
         drawing file. This item expects three input points. The first point defines the origin;
         the second defines the endpoint of a line segment representing the current scale;
         and the third represents the size of this line segment after scaling.

Item Selection
  On sidebar menu pick: `Params` `Active` `Scale` `Dist`
  For window, point to: n/a
  On paper menu see: MEASURE
  To show, key in: ACTIVE SCALE DISTANCE
    XS=1.0000,YS=1.0000,ZS=1.0000
  To set, key in: ACTIVE SCALE DISTANCE
Command Window Prompts:
  1. Select the item
     Enter scale origin point
  2. Pick a point
     Enter current distance
  3. Pick a second point
     Enter new distance
  4. Pick a third point
  To exit item: Exits by itself
See Also: ACTIVE SCALE, ACTIVE XSCALE, ACTIVE YSCALE, ACTIVE ZSCALE

ACTIVE STREAM ANGLE   ACT ST AN

Setting: Active Stream Angle
         Sets the stream angle tolerance in degrees.
Item Selection
  On sidebar menu pick: `Utils` `Digitz`
  For window, point to: Settings Digitizing Angle:
  On paper menu see: n/a
  To show, key in: ACTIVE STREAM ANGLE
    Angle Tol = 5.00
  To set, key in: ACTIVE STREAM ANGLE nnn
    Where nnn is a number between -360 and +360.

Example(s):
  ACTIVE STREAM ANGLE 15   sets the stream angle to 15 degrees
See Also: ACTIVE STREAM settings

ACTIVE STREAM AREA   ACT ST AR

Setting: Active Stream Area
         Sets the stream area tolerance in master units.
Item Selection
  On sidebar menu pick: `Utils` `Digitz`
  For window, point to: Settings Digitizing Area:
  On paper menu see: n/a
  To show, key in: ACTIVE STREAM AREA
    Area Tol = 0.00e+000
  To set, key in: ACTIVE STREAM AREA nnn
    Where nnn is a floating point number.

Example(s):
  ACTIVE STREAM AREA 75.00   sets the tolerance to 75.00
See Also: ACTIVE STREAM settings

ACTIVE STREAM DELTA   SD=

Setting: Active Stream Delta
    Sets the distance between sampled input points in stream mode.

Item Selection
  On sidebar menu pick: `Place` `LnStr` `Sd=:`
  For window, point to: Settings Digitizing Delta:
  On paper menu see: n/a
  To show, key in: ACTIVE STREAM DELTA
    Stream Delta = 0:0 5/16
  To set, key in: ACTIVE STREAM DELTA www
    Where www equals some working unit.

Example(s):
  SD=.01   sets the sampled distance to .01
  SD=$ to display the current delta distance
See Also: ACTIVE STREAM settings

ACTIVE STREAM TOLERANCE   ST=

Setting: Active Stream Tolerance
    Sets the largest distance, or tolerance, saved between input points while in stream mode.

Item Selection
  On sidebar menu pick: `Place` `LnStr` `St=:`
  For window, point to: Settings Digitizing Tolerance:
  On paper menu see: n/a
  For shortcut use: ACTIVE STREAM TOLERANCE
    Stream Tol = 0.0 5/16
  To set, key in: ACTIVE STREAM TOLERANCE www
    Where www is some working unit.

Example(s):
  ST=.02   sets the tolerance to .02
  ST=3:5:500
  ST=$ to display the current tolerance
See Also: ACTIVE STREAM settings

ACTIVE STYLE   LC=

Setting: Active Style
    Determines the current style, or line code, when creating elements. ACTIVE STYLE is always shown in the Command Window: LVL=50,SOLID,WT=0,LC=SOL,PRI,CO=4. The phrase LC=SOL shows that the line style is SOLID.

Item Selection
  On sidebar menu pick: `Params` `Active` `Style` `Key`
  For window, point to: Element Attributes Style:
  On paper menu see: n/a
  To set, key in: ACTIVE STYLE nnn
    Where nnn is a whole number between 0 and 7.

Line code	Value
Solid line (SOL)	0
Dotted line (DOT)	1
Medium-dashed line (MEDD)	2
Long-dashed line (LNGD)	3
Dot-dashed line (DOTD)	4
Short-dashed line (SHD)	5
Dashed-dotted line (DADD)	6
Long-dashed line (LDSD)	7

Example(s):
  LC=3   sets the line code to long dashed
  LC=$ to display the current line style

Use these key-ins on MicroStation PC only:

LC=+n
to increment some n number up from the active style setting. Example: ACTIVE STYLE is 4. LC=+2 would make the ACTIVE STYLE 6.

LC=-n
to decrement some n number down from the active style setting. Example: ACTIVE STYLE is 6. LC=-3 would make the ACTIVE STYLE 3.

See Also: CHANGE STYLE

ACTIVE TAB   TB=

Setting: Active Tab
Determines the amount of space given to tab characters when merging text with the INCLUDE tool.
Item Selection
On sidebar menu pick: n/a
For window, point to: n/a
On paper menu see: n/a
To show, key in: ACTIVE TAB
    Tab Interval = 8
To set, key in: ACTIVE TAB nn
Where nn is a whole number between 1 and 16.

Example(s):
TB=7   sets the tab spacing to seven horizontal spaces
TB=$ to display the current tab setting

Note: FILEDESIGN does not save the ACTIVE TAB setting.
See Also: INCLUDE

ACTIVE TAG   TI=

Setting: Active Tag
Increments numerics, such as lot numbers or circuit numbers. This setting works in conjunction with the INCREMENT TEXT tool.
Item Selection
On sidebar menu pick: `Text`   `ED Fld` `Copy`   `Ti=:`
For window, point to: Palettes Main Text Tag Increment:
On paper menu see: n/a
To show, key in: ACTIVE TAG
    Tag Increment = 1
To set, key in: ACTIVE TAG nnn
Where nnn is a whole number between 1 and 1023.

Example(s):
TI=2   sets the increment factor to 2
TI=$ to display the current tag increment

Note: FILEDESIGN does not save the ACTIVE TAG setting.
See Also: INCREMENT tools

ACTIVE TERMINATOR   LT=

Setting: Active Terminator
Defines the name of the cell used by the PLACE TERMINATOR tool. Changing this setting automatically calls the PLACE TERMINATOR tool.
Item Selection
On sidebar menu pick: `Place`   `Term`   `LT=:`
For window, point to: Settings Cells Terminator
On paper menu see: n/a
To show, key in: ACTIVE TERMINATOR
    Active terminator = cellname
To set, key in: ACTIVE TERMINATOR uuu

Where uuu is a six character user defined cell name.

Example(s):
LT=ARROW   sets the terminator name to ARROW
LT=$ to display the current cell name
See Also: PLACE TERMINATOR

ACTIVE TEXT   ACT TEX

Setting: Match Text Attributes
      Sets the text attributes by identifying an existing text string in the drawing file.
Item Selection
  On sidebar menu pick: | Text | | Match |
  For window, point to: Element   Text   Match
  On paper menu see: TEXT
Command Window Prompts:
  1. Select the item
     Identify text element
  2. Pick a text string
     ⇨ The prompt does not change.
  3. Give an accept point
     ⇨ TH, TW, FT, LS change.
  To exit item: **Make next selection**
See Also: PLACE TEXT tools

ACTIVE TEXTSTYLE   ACT TEXTS choice

Mac Only

Setting: Active Textstyle
      Sets the textstyle of your Macintosh raster fonts. This font style differs from the MicroStation 32 and PC which use vector fonts.

Item Selection
  On sidebar menu pick: n/a
  For window, point to: Element   Text   Text Style:
  On paper menu see: n/a
  To show, key in: **ACTIVE TEXTSTYLE**
     Active textstyle: BOLD
  To set, key in: **ACTIVE TEXTSTYLE choice on/off**
      Where choice is bold, condense, extend, italic, normal, outline, shadow, underline. And where on/off is either ON or OFF.
See Also: SET RASTERTEXT

ACTIVE TNJ   ACT TNJ

Setting: Active TNJ
      Sets the justification of text nodes.

Item Selection
  On sidebar menu pick: | Text | | Just |
  For window, point to: Element   Text   Multi-line Text
  On paper menu see: JUSTIFICATION
  To show, key in: **ACTIVE TNJ**
     Node Just=CC
  To set, key in: **ACTIVE TNJ opt**
      Where opt is one of the following settings:

Justification	Setting
Left Margin Top	LMT
Left Margin Center	LMC
Left Margin Bottom	LMB
Left Top	LT
Left Center	LC
Left Bottom	LB
Center Top	CT

Center	CC
Center Bottom	CB
Right Top	RT
Right Center	RC
Right Bottom	RB
Right Margin Top	RMT
Right Margin Center	RMC
Right Margin Bottom	RMB

Example(s):
ACTIVE TNJ CC   sets the justification to center
See Also: PLACE TEXT tools

ACTIVE TSCALE   TS=

Setting: Active TScale
Sets scale factor used when placing an active terminator.
Item Selection
On sidebar menu pick: `Place`   `Term`   `TS=:`
For window, point to: Palettes Main Lines Scale:
On paper menu see: n/a
To show, key in: ACTIVE TSCALE
  TS=1.0000
To set, key in: ACTIVE TSCALE nnn
  Where nnn is the scale factor.

Example(s):
TS=.25
TS=$ to display the current terminator scale factor
See Also: PLACE TERMINATOR

ACTIVE TXHEIGHT   TH=

Setting: Active TXHeight
Controls the height of text placed in a drawing.
Item Selection
On sidebar menu pick: `Text`   `Size`   `TH=:`
For window, point to: Element Text Height
On paper menu see: MODIFY TEXT
To show, key in: ACTIVE TXHEIGHT
  TH=0:9,TW=0:6
To set, key in: ACTIVE TXHEIGHT www
  Where www equals some working unit.

Example(s):
TH=.3   sets the height to .3
TH=/2   half the current text height
TH=*2   twice the current text height
TH=$ to display the current text height
See Also: ACTIVE TXHEIGHT PT2, ACTIVE TXSIZE, ACTIVE TXWIDTH tools

ACTIVE TXHEIGHT PT2   ACT TXH P

Setting: Set Active Text Height by Two Points
Graphically sets ACTIVE TEXT HEIGHT based on two input points.
Item Selection
On sidebar menu pick: `Text`   `Size`   `2ptH`
For window, point to: n/a
On paper menu see: TEXT
Command Window Prompts:
1. Select the item
   Enter first point
2. Pick a point
   Enter end point

3. Pick a second point
   To exit item: Click a Reset
   See Also: ACTIVE TXHEIGHT, ACTIVE TXSIZE, ACTIVE TXWIDTH tools

ACTIVE TXJ   ACT TXJ

Setting: Active TXJ
        Sets the justification of text placed in a drawing file.

Item Selection
    On sidebar menu pick: | Text | | Just |
    For window, point to: Element Text   Justification
    On paper menu see: JUSTIFICATION
    To show, key in: ACTIVE TXJ
        Text Just = CC
    To set, key in: ACTIVE TXJ opt
        Where opt is one of the following settings:

Justification	Setting
Center Bottom	CB
Center	CC
Center Top	CT
Left Bottom	LB
Left Center	LC
Left Top	LT
Right Bottom	RB
Right Center	RC
Right Top	RT

Example(s):
    ACTIVE TXJ CC   sets the justification to center
See Also: PLACE TEXT tools

ACTIVE TXSIZE   TX=

Setting: Active TXSize
        Sets both the height and width dimensions of text to the same value.

Item Selection
    On sidebar menu pick: | Text | | Size | | TX=: |
    For window, point to: Element Text   Height & Width
    On paper menu see: n/a
    To show, key in: ACTIVE TXSIZE
        TH=0:9,TW=0:6
    To set, key in: ACTIVE TXSIZE www
        Where www equals some working unit.

Example(s):
    TX=.5   sets the height and width to .5
    TX=/2   half the current text height and width
    TX=*2   twice the current text height and width
    TX=$ to display the current text size
See Also: ACTIVE TXHEIGHT tools, ACTIVE TXWIDTH tools

ACTIVE TXWIDTH   TW=

Setting: Active TXWidth
        Controls the width of text placed in a drawing.

Item Selection
    On sidebar menu pick: | Text | | Size | | TW=: |
    For window, point to: Element Text   Width
    On paper menu see: MODIFY TEXT
    To show, key in: ACTIVE TXWIDTH
        TH=0:9,TW=0:6
    To set, key in: ACTIVE TXWIDTH www
        Where www equals some working unit.

Example(s):

TW=.3	sets the width to .3
TW=/2	half the current text width
TW=*2	twice the current text width
TW=$	to display the current text width

See Also: ACTIVE TXHEIGHT tools, ACTIVE TXSIZE, ACTIVE TXWIDTH PT2

ACTIVE TXWIDTH PT2   ACT TXW P

Setting: Set Active Text Width by Two Points
Graphically sets the ACTIVE TEXT WIDTH based on two input points.

Item Selection

On sidebar menu pick: | Text | | Size | | 2ptW |
For window, point to: n/a
On paper menu see: TEXT
Command Window Prompts:
1. Select the item
   Enter first point
2. Pick a point
   Enter end point
3. Pick a second point
To exit item: Click a Reset
See Also: ACTIVE TXHEIGHT tools, ACTIVE TXSIZE, ACTIVE TXWIDTH

ACTIVE UNITROUND   UR=

Setting: Active Unitround
Sets a value for unit roundoff. When the LOCK UNIT setting is on, each point locates itself to the nearest multiple of this roundoff.

Item Selection

On sidebar menu pick: n/a
For window, point to: Settings   Locks   Full   Distance:
On paper menu see: n/a
To show, key in: ACTIVE UNITROUND
   UR=0:1
To set, key in: ACTIVE UNITROUND www
Where www equals some working unit.

Example(s):

UR=.5	sets the round-off to .5
UR=$	to display the current roundoff

See Also: LOCK UNIT

ACTIVE WEIGHT   WT=

Setting: Active Weight
Determines the current weight or thickness when creating elements. Always shown in the Command Window: LVL=50,SOLID,WT=0,LC=SOL,PRI,CO=4. The phrase WT=0 shows that the active weight is zero.

Item Selection

On sidebar menu pick: | Params | | Active | | Weight | | Key |
For window, point to: Element   Attributes   Weight:
On paper menu see: n/a
To set, key in: ACTIVE WEIGHT nn
Where nn is a whole number between 0 and 15.

Line Weight	Value
Thinnest line	0
Thickest line	15

Example(s):

WT=3	sets the weight to 3
WT=$	to display the current line weight

Use these key-ins on MicroStation PC only:

WT=+n   to increment some n number up from the active weight setting. Example: ACTIVE WEIGHT is 4. WT=+2 would make the ACTIVE WEIGHT 6.

WT=-n   to decrement some n number down from the active weight setting. Example: ACTIVE WEIGHT is 6. WT=-3 would make the ACTIVE WEIGHT 3.

See Also: CHANGE WEIGHT, DIMENSION WEIGHT settings

ACTIVE XSCALE   XS=

Setting: Active XScale
Controls the scale factor along the X-axis for non-symmetrical scaling of elements and cells.

Item Selection
On sidebar menu pick: Params   Active   Scale   XS=
For window, point to: Settings   Active Scale   X Scale:
On paper menu see: n/a
To show, key in: ACTIVE XSCALE
   XS=1.0000,YS=1.0000,ZS=1.0000
To set, key in: ACTIVE XSCALE nnn
   Where nnn is a scale factor.

Example(s):
   XS=.4   sets the X scale factor to .4
   XS=$ to display the current X scale factor
See Also: ACTIVE YSCALE, ACTIVE ZSCALE

ACTIVE YSCALE   YS=

Setting: Active YScale
Controls the scale factor along the Y-axis for non-symmetrical scaling of elements and cells.

Item Selection
On sidebar menu pick: Params   Active   Scale   YS=
For window, point to: Settings   Active Scale   Y Scale:
On paper menu see: n/a
To show, key in: ACTIVE YSCALE
   XS=1.0000,YS=1.0000,ZS=1.0000
To set, key in: ACTIVE YSCALE nnn
   Where nnn is a scale factor.

Example(s):
   YS=.4   sets the Y scale factor to .4
   YS=$ to display the current Y scale factor
See Also: ACTIVE XSCALE, ACTIVE ZSCALE

ACTIVE ZDEPTH ABSOLUTE   AZ=

Setting: Set Active Depth
Defines the Z-depth of a target view in a three dimensional file based on a keyed-in value.

Item Selection
On sidebar menu pick: n/a
For window, point to: n/a
On paper menu see: n/a
Command Window Prompts:
   1. Select the item
      Select view
   2. Pick a view
To exit item: Click A Reset

See Also: ACTIVE ZDEPTH RELATIVE, DEPTH ACTIVE settings

ACTIVE ZDEPTH RELATIVE   DZ=

Setting: Set Active Depth (Relative)
Adjusts the Z-depth of a target view in a three dimensional file relative to a keyed-in value.

Item Selection
On sidebar menu pick: n/a
For window, point to: n/a
On paper menu see: n/a
Command Window Prompts:
1. Select the item
   Select view
2. Pick a view
To exit item: Click A Reset
See Also: ACTIVE ZDEPTH ABSOLUTE, DEPTH ACTIVE settings

ACTIVE ZSCALE   ZS=

Setting: Active ZScale
Controls the scale factor along the Z-axis for non-symmetrical scaling of elements and cells.

Item Selection
On sidebar menu pick: | Params | | Active | | Scale | | ZS= |
For window, point to: Settings   Active Scale   Z Scale:
On paper menu see: n/a
To show, key in: ACTIVE ZSCALE
   XS=1.0000,YS=1.0000,ZS=1.0000
To set, key in: ACTIVE ZSCALE nnn
   Where nnn is a scale factor.

Example(s):
ZS=.4   sets the Z scale factor to .4
ZS=$ to display the current Z scale factor
See Also: ACTIVE XSCALE, ACTIVE YSCALE

AD=

Alternate Key-in
Alternate key-in for POINT ACSDELTA. Use it to place an input point at coordinates relative to the previous tentative or data point placed in the active Auxiliary Coordinate System (ACS).
See Also: POINT ACSDELTA

AE=

Alternate Key-in
Alternate key-in for ACTIVE ENTITY. It defines a database entity.
See Also: ACTIVE ENTITY

ALIGN   AL

View Control: Align View
A 2D or 3D view control which aligns the viewing area of one view with the viewing area of another. It requires at least two input points. The first point determines the source view; the second and subsequent points align the destination views to the source view.

Item Selection
On sidebar menu pick: | Window | | View | | Align |
For window, point to: n/a
On paper menu see: n/a

Command Window Prompts:
1. Select the item
   `Select source view`
2. Pick the source view with an input point.
   `Select destination view(s)`
3. Pick a second view to align.
To exit item: `Click the RESET button twice.`

✍ *Note:* If a rotated view is aligned, the alignment takes place but the view's rotation does not change.
See Also: WINDOW view controls

AM=

Alternate Key-in
   Alternate key-in for ATTACH MENU. It attaches a menu to the workstation or PC.
See Also: ATTACH MENU

ANALYZE   Ctrl+I

Command: Analyze Element
   Displays an Element Information setting box after identifying an element. This item requires one input point to identify the element you want information about. You may also adjust the setting and have these changes applied to the element. For example, you could change the color setting, which changes the color of the selected element.
Item Selection
   On sidebar menu pick: | Utils | | Analyz |
   For window, point to: Element Info
   On paper menu see: UTILITIES
Command Window Prompts:
1. Select the item
   `Identify element`
2. Pick an element
   ⇨ The element highlights, no prompt appears
3. Give an accept point
To exit item `Make next selection`

✱ *Error:* `ANALYZE is already loaded.` - **If this message appears, ANALYZE is still running. Just click on the element in question.**
See Also: LOCELE

AP=

Alternate Key-in
   Alternate key-in for ACTIVE PATTERN CELL. Use it to set the name of the cell used in the patterning process.
See Also: ACTIVE PATTERN CELL

AR=

Alternate Key-in
   Alternate key-in for ACTIVE RCELL. Use it to set the name of a relative cell and start up the PLACE CELL RELATIVE tool.
See Also: ACTIVE RCELL

ARRAY POLAR   AR P

Drawing Tool: Polar Array
   Places a series of the same size elements in a circular array. For instance, if you are drawing a steel plate with eight bolt holes 45 degrees apart, you place the first hole and the ARRAY POLAR tool places the other seven. The tool asks you for the number of array elements to generate, the sweep the array will occupy, and whether the array elements should be rotated. The tool then expects one input

point identifying the element to array, and a second accept point to start the array process. Cells and graphic groups can also be arrayed with this tool.

Item Selection

On sidebar menu pick: `Constr` `Array` `Polar`

For window, point to: Palettes Main Copy Element

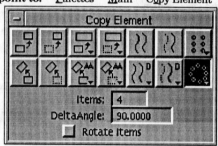

On paper menu see: COPY ELEMENT

Command Window Prompts:

1. Select the item
   `Enter number of array items`
2. Enter some whole number
   `Enter angle to fill`
3. Enter some whole number
   `(1) ROTATE (y) :`
4. Key in Y to rotate array elements
   `Identify element`
5. Pick the element to array
   `Accept/Reject`
6. Select the element again

To exit item: **Exits by itself**

See Also: ARRAY RECTANGULAR, FENCE ARRAY POLAR

ARRAY RECTANGULAR   AR

Drawing Tool: Rectangular Array

Places a series of the same size elements in a rectangular array. Two examples of rectangular arrays are building columns on a plan view drawing or tiles in an acoustic ceiling plan.

This tool asks you how many rows and columns are in the array and the spacing between those rows and columns. It then expects one input point to identify the element to array and a second accept point to start up the array process. Cells and graphic groups can also be arrayed with this tool.

Item Selection

On sidebar menu pick: `Constr` `Array` `Rctn`

For window, point to: Palettes Main Copy Element

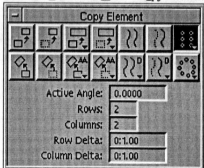

On paper menu see: COPY ELEMENT

Command Window Prompts:
1. Select the item
   ```
   Enter number of array rows
   ```
2. Enter some whole number
   ⇨ The rows go horizontally.
   ```
   Enter number of array columns
   ```
3. Enter some whole number
   ⇨ The columns go vertically.
   ```
   Enter distance between horizontal rows
   ```
4. Enter some whole number
   ```
   Enter distance between vertical columns
   ```
5. Enter some whole number
   ```
   Identify element
   ```
6. Pick the element to array
   ```
   Accept/Reject
   ```
7. Select the element again
To exit item: Exits by itself
See Also: ARRAY POLAR, FENCE ARRAY RECTANGULAR

AS=

Alternate Key-in
Alternate key-in for ACTIVE SCALE. Use it to set the scale factor when placing elements.
See Also: ACTIVE SCALE

AT=

Alternate Key-in
Alternate key-in for TUTORIAL. It displays a screen tutorial menu.
See Also: TUTORIAL

ATTACH ACS   AT AC

Setting: Select ACS
Attaches and loads a previously stored Auxiliary Coordinate System.

Item Selection
On sidebar menu pick: | Params | | ACS | | Set |
For window, point to: Settings   Auxiliary Coordinates   Attach
On paper menu see: ACS
To set, key in: ATTACH ACS uuu
Where uuu is a previously stored acs name.
See Also: DELETE ACS, SAVE ACS, SHOW ACS

ATTACH AE   AT A

Setting: Attach Active Entity
Attaches active database entities to graphic elements.

Item Selection
On sidebar menu pick: | dBase | | Tools |
For window, point to: Palettes   Database   Settings

On paper menu see: DATABASE
Command Window Prompts:
1. Select the item
   ```
   Identify element
   ```

2. Pick the element to attach.
   `Accept/Reject`
3. Pick the element again.
   ⇨ The setting waits for more elements to link.

To exit item: `Make next selection`

See Also: DETACH

ATTACH COLORTABLE
CT=

Setting: Attach Colortable

Attaches a colortable to a display screen. A single colortable can be attached to both display screens or just the right or left screen.

Item Selection

On sidebar menu pick: `Params` `Color` `Key:`

For window, point to: Settings Color Palette... File Open...

On paper menu see: n/a

To show, key in: CT=$

To set, key in: CT=filename

Where filename is a valid colortable file.

Example(s):

CT=C:\CGSI\LEAVY.TBL   for DOS users

CT=/usr/cgsi/leavy.tbl   for UNIX users

✍ *Note:* To detach a colortable, click on Settings, Color Palette., Default.

✍ *Note:* Click on Settings, Color Palette. to make your own colortables.

✍ *Note:* Click on Settings, Color Palette. to display the attached colortables.

☞ *Tip:* If you name all of your colortables with the .TBL extension, you do not have to enter it as part of the name when attaching the table.

See Also: ACTIVE COLOR, CHANGE COLOR

ATTACH COLORTABLE CREATE
AT C C

Command: Attach Colortable (Create)

Use this command to create your own color table. Key a filename in the dialog box. For example, SHADES.TBL. Performs same action as the Save As.. option in the Color Palette dialog box.

Item Selection

For window, point to: Settings Color Palette... File Save As...

On sidebar menu pick: n/a

On paper menu see: n/a

See Also: ATTACH COLORTABLE commands

ATTACH COLORTABLE WRITE
AT C W

Command: Attach Colortable (Write)

Use this command to write your color table modifications to a color table file. Key a filename in the dialog box. For example, SHADES.TBL. Performs same action as the Save As.. option in the Color Palette dialog box.

Item Selection

For window, point to: Settings Color Palette... File Save As...

On sidebar menu pick: n/a

On paper menu see: n/a

See Also: ATTACH COLORTABLE commands

ATTACH DA
AT DA

Setting: Attach Displayable Attributes

Attaches existing text node information to database records.

Item Selection

On sidebar menu pick: n/a

For window, point to: Palettes   Database

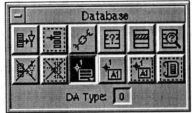

On paper menu see: DATABASE
Command Window Prompts:
1. Select the item
   Identify element
2. Pick a text node
To exit item: Make next selection
See Also: ACTIVE DA, PLACE NODE

ATTACH LIBRARY   RC=

Setting: Attach Cell Library
   Attaches a cell library to a drawing file. Standard symbols and or features
   live in cell libraries. Only one cell library can be attached to a single drawing file
   at any one time.

Item Selection
   On sidebar menu pick: n/a
   For window, point to: File   Cell Library   Attach   Name
   On paper menu see: n/a
   To show, key in: RC=$
   To set, key in: RC=filename
      Where filename is a valid cell library file.

Example(s):
   RC=C:\DGN\LEAVY.CEL   for DOS users
   RC=/usr/cgsi/leavy.cel   for UNIX users

✍ *Note:* A cell library can be attached to as many drawing files as necessary.

✍ *Note:* Attaching a second cell library detaches the first library.

☞ *Tip:* If you name your cell libraries with the .CEL extension you do not have to enter
   it as part of the library name when attaching the library.
   See Also: PLACE CELL tools, CREATE LIBRARY

ATTACH MENU   AM=

Setting: Attach Menu
   Attaches screen menus or paper menus to the PC or workstation. Menus are
   NOT attached to individual drawing files like cell libraries or databases.
      Different types of menus are attached to the workstation by adding a variable
   after the menu name. For instance, to attach the standard MicroStation paper
   command menu (called MENU), key in ATTACH MENU MENU,CM (CM stands
   for command menu).

Item Selection
   ATTACHING A COMMAND MENU:
1. AM=MENU,CM
   Identify menu origin
2. Pick lower left hand corner of menu.
   Identify upper right of <HELP> block
3. Pick your second point.
   Menu successfully attached
   DETACHING A COMMAND MENU:
1. AM=,CM
      ⇨   Same key-in, just leave out the menu name.

34 ATTACH MENU

🖉 *Note:* MicroStation's command menus reside in the C:\USTA-TION\DATA\MSMENU.CEL cell library for DOS users, and /usr/ip32/mstation/tutlib.cel for UNIX devotees. This cell library does not have to be attached to your drawing file to use ATTACH MENU.

✳ *Error:* `Cell is wrong type of menu` - If this message appears, you are attempting to attach the wrong type of menu. For example, trying to attach a sidebar menu with the CM variable instead of using the SB variable.

✳ *Error:* `CM menu MYMENU not found` - If this message appears, the menu cell cannot be found in the cell library.

ATTACHING A MATRIX MENU:
1. AM=HORIZ1,M1
```
Identify menu origin
    ⇨ Pick lower left hand corner of menu.
Identify upper right of <TX=*.J> block
    ⇨ Pick your second point.
Menu successfully attached
```
DETACHING A MATRIX MENU:
1. AM=,M1
    ⇨ Same key-in, just leave out the menu name.

🖉 *Note:* Up to three matrix menus can be attached at any one time. Use M2 and M3 to attach a second and third matrix menu to the station.

🖉 *Note:* MicroStation's matrix menus reside in the C:\USTA-TION\DATA\MSMENU.CEL cell library for DOS users, and /usr/ip32/mstation/tutlib.cel in UNIX. This cell library does not have to be attached to your drawing file to use ATTACH MENU.

✳ *Error:* `Error - menus overlap` - You may be trying to attach a second menu over one already active.

ATTACHING A SIDEBAR MENU:
1. AM=USTN,SB1
    ⇨ The system can hesitate for a second while it loads the menu into the computer's memory. The SB1 variable told MicroStation to attach the sidebar menu to VIEW 1. Sidebar menus can be attached to any view or views simultaneously.
DETACHING A SIDEBAR MENU:
1. AM=,SB1
    ⇨ Same key-in, just leave out the menu name.

🖉 *Note:* MicroStation's sidebar menu (USTN.SBM) resides in C:\USTATION\DATA in DOS or /usr/ip32/mstation/menus in UNIX.

Menu Prompts
ATTACHING A CONTROL STRIP MENU:
1. AM=USTNPTRN,CS
    ⇨ For MicroStation 32 only. The control strip menu is attached to the primary display screen. It cannot be view dependent. A control strip menu has a second name - a pull-down menu.
DETACHING A CONTROL STRIP MENU:
1. AM=,CS
    ⇨ Same key-in, just leave out the menu name.

🖉 *Note:* MicroStation's USTNPTRN control strip menu resides in a file called C:\USTATION\DATA in DOS and /usr/ip32/mstation/menus in UNIX.

ATTACHING A FUNCTION KEY MENU:
Commands can be attached to unassigned function keys (F1, F2, etc.) on the computer's keyboard.
1. AM=FKCGSI,FK
```
Menu successfully attached
```
DETACHING A FUNCTION KEY MENU:
1. AM=,FK
    ⇨ Same key-in, just leave out the menu name.

🖉 *Note:* MicroStation's function key menu (FUNCKEY.MNU) resides in \USTA-TION\DATA for DOS and /usr/ip32/mstation/menus for UNIX.
See Also: SAVE FUNCTION_KEY and SET FUNCTION to assign your own commands to the function keys.

ATTACHING A CURSOR BUTTON MENU:
Commands can be attached to unassigned cursor buttons other than the Command, Data, Tentative and Reset buttons of a tablet cursor.
1. AM=CBMENU,CB
   `Menu successfully attached`
   DETACHING A CURSOR BUTTON MENU:
1. AM=,CB
   ⇨ Same key-in, just leave out the menu name.

📝 *Note:* MicroStation's cursor button menu (CBMENU) resides in the C:\USTA-TION\DATA\MSMENU.CEL cell library for DOS users, and /usr/ip32/msta-tion/tutlib.cel in UNIX. This cell library does not have to be attached to your drawing file to use ATTACH MENU.
See Also: MC, TUTORIAL

ATTACH REFERENCE   AT R

Setting: Attach Reference File
ATTACH REFERENCE followed by a drawing name (LEAVY.DGN) opens a dialog box. In this box you can label the reference file with a common or logical name. You can also pick the attachment mode and scale factor of the reference file being attached.
Item Selection
On sidebar menu pick: n/a
For window, point to: Palettes   Reference Files
On paper menu see: n/a
Command Window Prompts:
1. Select the item
   `The Reference File dialog box appears.`
To exit item: Click OK or Cancel
See Also: REFERENCE ATTACH, REFERENCE DETACH

AX=

Alternate Key-in
Alternate key-in for POINT ACSABSOLUTE. It places absolute coordinate points for an Auxiliary Coordinate System based on key-ins.
See Also: POINT ACSABSOLUTE

AZ=

Alternate Key-in
AZ= an alternate key-in for ACTIVE ZDEPTH ABSOLUTE, sets the active depth to an absolute keyed-in value.
See Also: ACTIVE ZDEPTH ABSOLUTE

BACKUP   BA pathname

Command: Backup
Creates a duplicate copy of design file data. Regularly backing up drawing files will guard against losing data when there are power losses or system malfunctions.
The default name of the backup file is FILENAME.BAK. For instance, the file LEAVY.DGN's backup file would be named LEAVY.BAK. MicroStation creates the backup file in the default directory unless you specify a different path. You can also give the backup file a different name.
Item Selection
On sidebar menu pick: Utils   Backup
For window, point to: n/a
On paper menu see: UTILITIES
Command Window Prompts:
1. Select the item
   ⇨ No questions appear.
   ⇨ The backup file ends up in the default directory.

```
Backed up to C:\CGSI\DOUG.DGN
```
To exit item: Exits by itself

☞ *Tip:* To change the name of the backup file, key in: BACKUP NEWNAME. MicroStation will add the .BAK file extension.

☞ *Tip:* To backup the drawing file to floppy, key in: BACKUP A:

✳ *Error:* Path A:\CGSI does not exist - **If this message appears, it means BACKUP cannot find the directory name on the floppy. Try BACKUP A:**

✳ *Error:* Not ready error reading drive A - Abort, Ready, Ignore? - **If this message appears, it means the floppy drive is not ready.**

See Also: COMPRESS DESIGN, COMPRESS LIBRARY, EXIT commands, QUIT commands

Bad Things

Expression
>An expression which describes what takes place when training is not taken, books remain in their plastic, instructions not followed, time allowances not made for new users to get up to speed and suggestions by the MicroStation support staff not kept in mind. If these issues continue to exist, "Bad Things" are bound to happen.

See Also: Bentley Boys, Bingo, Brain Fault, Good Things

BEEP

Command: Beep
>Causes the computer to make a "beep" sound.

☞ *Tip:* Use BEEP in user commands to signal the end of a significant event.It is convenient to use the beep sound to gain your attention.While the computer is crunching numbers, you can work on something else. Then when the "beep" sounds, you can turn your attention back to the screen.

Bentley Boys

History
>In September of 1921 the first Bentley motorcar rolled off the assembly line, and quickly became the backbone of British motoring. John Duff drove the first production Bentley at Brooklands where he took class records at 88mph. The 1924 Indy win by racing members John Duff and Frank Clement set the pattern for Bentleys' competition motoring. Those daring men, like Duff and Clement, who raced the Bentley machines at Brooklands, Indy and Le Mans, came to be known as the "Bentley Boys".

✐ *Note:* Not to be confused with brothers who live in Pennsylvania and write computer software.

*See Also:*Bad Things, Bingo, Brain Fault, Good Things

Bingo

Expression
>An expression used when the light has dawned, the solution's found, success is at hand, the problem's solved, the software upgrade's successful, and everything is going your way.

See Also: Bad Things, Bentley Boys, Brain Fault, Good Things

Brain Fault

Expression
>This condition might happen at anytime. You're grabbing the cursor ready to draw or placing your fingers on the keyboard ready to type – the pause comes. You don't know why. Maybe you find yourself staring at the display screen and can't remember which element you were working with – the pause lengthens. Perhaps your buddy at the next station turns to you and asks: "How do change active colors again?" You turn to answer. Your lips move, but no sound come out.

You can't remember the answer. This condition is known as a brain fault. Short gaps in reality shared by those under pressure. Probably a result of too many software upgrades, long work hours, few days off, tight deadlines or steep learning curves.

See Also: Bad Things, Bentley Boys, Bingo, Good Things

CC=

Alternate Key-in
Alternate key-in for CREATE CELL. Use to file a newly created figure in a cell library.
See Also: CREATE CELL

CD=

Alternate Key-in
Alternate key-in for DELETE CELL. Use to remove cells from the attached cell library.
See Also: DELETE CELL

CHAMFER   CHAM

Drawing Tool: Chamfer
Shaves off, or chamfers, the corner of intersecting lines. The tool asks for two key-in values. The first value controls the distance between the intersecting line segments. The second key-in determines the length of the chamfer line. Use two input points to identify the lines to chamfer.

Item Selection
On sidebar menu pick: Modify   Chamfr
For window, point to: Palettes   Main   Fillets

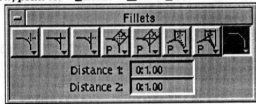

On paper menu see: FILLET
Command Window Prompts
1. Select the item
   Enter First chamfer distance
2. Key in first distance
   Enter Second chamfer distance
3. Key in second distance
   Select first chamfer segment
4. Pick an element
   Select second chamfer segment
5. Pick other element
   ⇨   The elements will highlight.
6. Give an accept point
To exit item: Make next selection
See Also: EXTEND ELEMENT tools, EXTEND LINE tools

Change

Definition
The change drawing tools modify an existing element's attributes such as color, thickness, level, line style, etc. First alter the appropriate active setting and then pick the change drawing tool to perform the operation on the one you want. Each of the change tools require two input points. The first point highlights the element for change. The next point accepts the element and makes the change.

See Also: CHANGE tools

CHANGE AREA HOLE   CHAN A H

Drawing Tool: Change Element to Active Area (Solid/Hole)
Toggles the status of a closed shape to hole. This tool expects two input points.
The first point identifies the element for change and the second point changes the
element's area status.

Item Selection
On sidebar menu pick: Modify   Change   Area
For window, point to: Palettes   Main   Change Element

On paper menu see: CHANGE
Command Window Prompts:
1. Select the item
   Identify element
2. Pick an element
   Accept/Reject (select next input)
3. Give an accept point
To exit item: Make next selection
See Also: CHANGE AREA SOLID, ACTIVE AREA tools

CHANGE AREA SOLID   CHAN A S

Drawing Tool: Change Element to Active Area (Solid/Hole)
Toggles the status of a closed shape to solid. This tool expects two input points.
The first point identifies the element for change and the second point changes the
element's area status.

Item Selection
On sidebar menu pick: Modify   Change   Area
For window, point to: Palettes   Main   Change Element

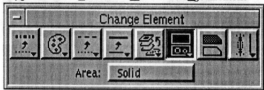

On paper menu see: CHANGE
Command Window Prompts:
1. Select the item
   Identify element
2. Pick an element
   Accept/Reject (select next input)
3. Give an accept point
To exit item: Make next selection
See Also: CHANGE AREA HOLE, ACTIVE AREA tools

CHANGE BSPLINE CURVE

Obsolete
Toggles the curve display of a particular B-spline element between on and
off. This tool requires two input points. The first point highlights the target
element. The second point changes the display of the element.

CHANGE BSPLINE ORDER   CHAN B O

Drawing Tool: Change B-spline to Active Order
      Modifies the order of a particular B-spline curve. The order number of the
curve must be greater than the number of existing poles. This MDL application
tool expects a key-in to determine the new B-spline order number and two input
points. The first point identifies the target element and the subsequent point
accepts the element for change.

Item Selection
On sidebar menu pick: Place | BSpl | Curves
For window, point to: Palettes Main B-splines Order:
On paper menu see: B-SPLINES 2D 3D
Command Window Prompts:
 1. Select the item
    Enter B-spline order
 2. Enter new order number
    Identify element
 3. Pick a B-spline curve
    Accept / Reject (select next input)
 4. Give an accept point
 To exit item: Make next selection

* *Error:* Invalid character **- If this message appears after your key in, enter
  MDL LOAD SPLINES once each design session.**

* *Error:* Error - # poles order - If this message appears, you need to lower
  the order number below the number of poles in the curve.
  See Also: CHANGE BSPLINE tools, PLACE BSPLINE tools

CHANGE BSPLINE POLYGON   CHAN B P

View Control   Change B-spline Polygon Display On/Off
      Toggles the polygon display of a particular B-spline element between on and
off. This MDL application tool requires two input points. The first point highlights
the element for change and the second point changes the element's display status.

Item Selection
On sidebar menu pick: Place | BSpl | Curves
For window, point to: Palettes Main B-splines

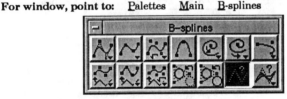

On paper menu see: B-SPLINES 2D 3D
Command Window Prompts:
 1. Select the item
    Identify element
 2. Pick a B-spline curve
    Accept / Reject (select next input)
 3. Give an accept point
 To exit item: Make next selection

* *Error:* Invalid character **- If this message appears after your key in, enter
  MDL LOAD SPLINES once each design session.**
  See Also: CHANGE BSPLINE tools, PLACE BSPLINE tools

CHANGE BSPLINE UORDER   CHAN B UO

3D Drawing Tool: Change B-spline Surface to Active U-Order
      Modifies a B-spline surface element in the u-direction to the active u-order.
This MDL application tool requires two input points. The first point identifies the
target element and the subsequent point accepts the element for change.

Item Selection
On sidebar menu pick: Place   BSpl   ChgSur
For window, point to: Palettes   3DB-splines   hange Surface

On paper menu see: CHANGES SURFACES
Command Window Prompts:
1. Select the item
    Identify element
2. Pick an element
    Accept / Reject (select next input)
3. Give an accept point
    To exit item: **Make next selection**

* *Error:* Invalid character - **If this message appears after your key in, enter MDL LOAD SPLINES once each design session.**

* *Error:* Invalid character - If this message appears, load the splines MDL application by keying in MDL LOAD SPLINES
See Also: CHANGE BSPLINE tools, PLACE BSPLINE tools

CHANGE BSPLINE URULES   CHAN B UR

3D Drawing Tool: Change B-spline Surface to Active U-Rules
    Modifies the number of rule lines of a B-spline surface element in the u-direction to the active number of u-rules. This item requires two input points. The first point identifies the target element, and the subsequent point accepts the element for change.
Item Selection
On sidebar menu pick: Place   BSpl   ChgSur
For window, point to: Palettes   3DB-splines   Change Surface

On paper menu see: CHANGES SURFACES
Command Window Prompts:
1. Select the item
    Identify element
2. Pick an element
    Accept / Reject (select next input)
3. Give an accept point
    To exit item: **Make next selection**

* *Error:* Invalid character - **If this message appears after your key in, enter MDL LOAD SPLINES once each design session.**
See Also: CHANGE BSPLINE tools, PLACE BSPLINE tools

CHANGE BSPLINE VORDER   CHAN B VO

3D Drawing Tool: Change B-spline Surface to Active V-Order
    Modifies a B-spline surface element in the v-direction to the active v-order. This MDL application tool requires two input points. The first point identifies the target element and the subsequent point accepts the element for change.

Item Selection
On sidebar menu pick: `Place` `BSpl` `ChgSur`
For window, point to: Palettes 3DB-splines Change Surface

On paper menu see: CHANGE SURFACES
Command Window Prompts:
1. Select the item
   Identify element
2. Pick an element
   Accept / Reject (select next input)
3. Give an accept point
   To exit item: Make next selection

✱ *Error:* Invalid character - If this message appears after your key in, enter MDL LOAD SPLINES once each design session.
See Also: CHANGE BSPLINE tools, PLACE BSPLINE tools

CHANGE BSPLINE VRULES   CHAN B VR

3D Drawing Tool: Change B-spline Surface to Active V-Rules
Modifies the number of rule lines of a B-spline surface element in the v-direction to the active number of v-rules. This MDL application tool requires two input points. The first point identifies the target element and the subsequent point accepts the element for change.

Item Selection
On sidebar menu pick: `Place` `BSpl` `ChgSur`
For window, point to: Palettes 3DB-splines Change Surface

On paper menu see: CHANGES SURFACES
Command Window Prompts:
1. Select the item
   Identify element
2. Pick an element
   Accept / Reject (select next input)
3. Give an accept point
   To exit item: Make next selection

✱ *Error:* Invalid character - If this message appears after your key in, enter MDL LOAD SPLINES once each design session.
See Also: CHANGE BSPLINE tools, PLACE BSPLINE tools

CHANGE CLASS CONSTRUCTION   CHAN CL C

Drawing Tool: Change Element to Active Class
Toggles the status of elements to construction. Use construction elements as visual aids in construction drawings. This tool expects two input points. The first point highlights the target element. The second point changes the element's class status.

Item Selection
On sidebar menu pick: n/a

For window, point to: Palettes   Main   Change Element

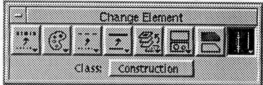

On paper menu see: CHANGE
Command Window Prompts:
1. Select the item
   Identify element
2. Pick an element
   Accept/Reject (select next input)
3. Give an accept point
To exit item: **Make next selection**
See Also: ACTIVE CLASS, CHANGE CLASS PRIMARY

CHANGE CLASS PRIMARY   CHAN CL P

Drawing Tool: Change Element to Active Class
    Toggles the status of elements to primary. Primary is the default class for elements in a drawing file. This tool expects two input points. The first point highlights the target element. The second point changes the element's class status.
Item Selection
    On sidebar menu pick: n/a
    For window, point to: Palettes   Main   Change Element

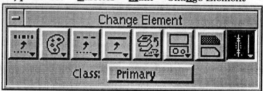

On paper menu see: CHANGE
Command Window Prompts:
1. Select the item
   Identify element
2. Pick an element
   Accept/Reject (select next input)
3. Give an accept point
To exit item: **Make next selection**
See Also: ACTIVE CLASS settings, CHANGE CLASS CONSTRUCTION

CHANGE COLOR   CHAN CO

Drawing Tool: Change Element to Active Color
    Modifies an existing element's color to that of the ACTIVE COLOR setting. This tool expects two input points. The first point identifies the target element. The second point changes the element's color.
Item Selection
    On sidebar menu pick: Modify Change Color
    For window, point to: Palettes   Main   Change Element

On paper menu see: CHANGE
Command Window Prompts:
1. Select the item
   Identify element
2. Pick an element
   Accept/Reject (select next input)
3. Give an accept point
To exit item: Make next selection
See Also: ACTIVE COLOR

CHANGE DIMENSION

CHA D

Drawing Tool: Change Dimension to Active Settings
Changes the attributes of an existing dimension element to the current dimension settings. This item requires an input point to identify the dimension data for change, and an accept point to make the alterations take place.

Item Selection
On sidebar menu pick: n/a
For window, point to: n/a
On paper menu see: n/a
Command Window Prompts:
1. Select the item
   Identify element
2. Pick dimension element
   Accept/Reject (select next input)
To exit item: Make next selection
See Also: CHANGE DIMENSION SYMBOLOGY tools

CHANGE DIMENSION SYMBOLOGY ALTERNATE

CHA D S A

Drawing Tool: Change to Alternate Dimension Symbology
Changes the symbology of an existing witness line to the alternate dimension attributes that were active when the witness line was generated. This item requires an input point to identify the target line, and an accept point for the change to take place.

Item Selection
On sidebar menu pick: n/a
For window, point to: n/a
On paper menu see: n/a
Command Window Prompts:
1. Select the item
   Identify element
2. Pick witness line
   Accept/Reject (select next input)
3. Accept witness line
To exit item: Make next selection
See Also: CHANGE DIMENSION SYMBOLOGY tools, DIMENSION WITNESS settings

CHANGE DIMENSION SYMBOLOGY STANDARD

CHA D S S

Drawing Tool: Change to Standard Dimension Symbology
Changes the symbology of an existing witness line to the dimension attributes that were active when the witness line was generated. Symbology deals with color, weight and line style.
This item requires an input point to identify the target line and an accept point for the change to take place.

Item Selection
On sidebar menu pick: n/a
For window, point to: n/a

On paper menu see: n/a
Command Window Prompts:
1. Select the item
   `Identify element`
2. Pick witness line
   `Accept/Reject (select next input)`
3. Accept witness line
To exit item: Make next selection
See Also: CHANGE DIMENSION SYMBOLOGY tools, DIMENSION WITNESS settings

CHANGE FILL   CHAN F

Drawing Tool: Change Fill
Changes the fill status of a closed shape between filled and unfilled. This tool looks for two input points. The first point identifies the target element and the second changes the element's fill status.

Item Selection
On sidebar menu pick: | Modify | | Change | | Fill |
For window, point to: Palettes   Main   Change Element

On paper menu see: CHANGE
Command Window Prompts:
1. Select the item
   `Identify element`
2. Pick an element
   `Accept/Reject (select next input)`
3. Give an accept point
To exit item: Make next selection
See Also: ACTIVE FILL, SET FILL

CHANGE LEVEL   CHAN L

Drawing Tool: Change Element to Active Level
Modifies an existing element's level to that of the ACTIVE LEVEL setting. This tool expects two input points. The first point highlights the element and the second changes the element's level assignment.

Item Selection
On sidebar menu pick: | Modify | | Change | | Level |
For window, point to: Palettes   Main   Change Element

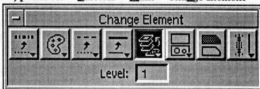

On paper menu see: CHANGE
Command Window Prompts:
1. Select the item
   `Identify element`
2. Pick an element
   `Accept/Reject (select next input)`
3. Give an accept point
To exit item: Make next selection
See Also: ACTIVE LEVEL

CHANGE LOCK

Setting: Lock Element
Secures or locks up elements in the drawing file. You cannot change or manipulate a locked element's attributes. This setting requires two input points. The first point highlights the target element to lock. The second point locks the element.

Item Selection
On sidebar menu pick: n/a
For window, point to: Edit   Lock
On paper menu see: n/a
Command Window Prompts:
1. Select the item
   Identify element
2. Click on an element
   Accept/Reject (select next input)
3. Click a second time
To exit item: Make next selection
See Also: CHANGE UNLOCK, FENCE CHANGE LOCK, FENCE CHANGE UNLOCK

CHANGE STYLE

Drawing Tool: Change Element to Active Line Style
Modifies an existing element's style or line code to that of the ACTIVE STYLE setting. This tool expects two input points. The first point identifies the element and the second changes the element's style.

Item Selection
On sidebar menu pick: Modify   Change   Style
For window, point to: Palettes   Main   Change Element

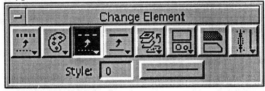

On paper menu see: CHANGE
Command Window Prompts:
1. Select the item
   Identify element
2. Pick an element
   Accept/Reject (select next input)
3. Give an accept point
To exit item: Make next selection
See Also: ACTIVE STYLE

CHANGE SYMBOLOGY

Drawing Tool: Change Element to Active Symbology
Modifies an existing element's symbology to that of the active symbology settings. (i.e. color, weight, line style) This tool looks for two input points. The first point identifies the target element; the second changes the element's symbology.

Item Selection
On sidebar menu pick: Modify   Change   Symb

For window, point to: alettes   Main   Change Element

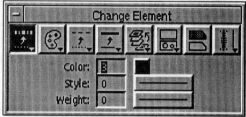

On paper menu see: CHANGE
Command Window Prompts:
1. Select the item
   `Identify element`
2. Pick an element
   `Accept/Reject (select next input)`
3. Give an accept point
To exit item: Make next selection
See Also: SET LVLSYMB

CHANGE UNLOCK   CHAN UN

Setting: Unlock Element
  Releases elements from their locked status. This setting requires two input points. The first point highlights the target element to unlock. The second point unlocks the element.

Item Selection
 On sidebar menu pick: n/a
 For window, point to: Edit   Unlock
 On paper menu see: n/a
Command Window Prompts:
1. Select the item
   `Identify element`
2. Click on an element
   `Accept/Reject (select next input)`
To exit item: Make next selection
See Also: CHANGE LOCK, FENCE CHANGE LOCK, FENCE CHANGE UNLOCK

CHANGE WEIGHT   CHAN W

Drawing Tool: Change Element to Active Line Weight
  Modifies an existing element's weight to that of the ACTIVE WEIGHT setting. This tool requires two input points. The first point identifies the target element. The second point accepts that element for change.

Item Selection
 On sidebar menu pick: Modify   Change   Weight
 For window, point to: Palettes   Main   Change Element

On paper menu see: CHANGE
Command Window Prompts:
1. Select the item
   `Identify element`
2. Pick an element
   `Accept/Reject (select next input)`
3. Give an accept point

To exit item: **Make next selection**
See Also: ACTIVE WEIGHT

CHOOSE ALL   CHO A

Drawing Tool: Choose All
　　　Normally, you select the drawing tool first, and then pick the element you want to change. For example, you might select the CHANGE COLOR tool and then click on the elements you want changed. This tool adds a twist to element selection.
　　　CHOOSE ALL gives you the ability to select all the elements, seen and unseen, in the drawing file first. Then you pick the tool for change.
　　　You know the selection process took place when small white squares, called handles, appear at the selection points on the elements.

Item Selection
On sidebar menu pick: n/a
For window, point to: Edit   Select All
On paper menu see: TOP BORDER
Command Window Prompts:
　1. Select the item
　　　⇨   The element handles appear.
To exit item   Make next selection
See Also: CHOOSE ELEMENT

CHOOSE ELEMENT   CHO

Drawing Tool: Element Selection
　　　Normally, you select the drawing tool first, and then pick the element you want to change. For example, you might select the CHANGE COLOR tool and then click on the elements you want changed. CHOOSE ELEMENT adds a twist to element selection by giving you the ability to select an element in the drawing file first. Then you pick the tool for change.
　　　You know the selection process took place when small white squares, called handles, appear at the selection points on the element.

Item Selection
On sidebar menu pick: n/a
For window, point to: Palettes   Main
On paper menu see: TOP BORDER
Command Window Prompts:
　1. Select the item
　　　⇨   The element handles appear.
To exit item   Make next selection

☞ *Tip:* Want to choose several elements easily? Select CHOOSE ELEMENT. Click a point in a view, and hold the button down. Drag the mouse around the elements you want to select. A rectangle appears. When you have all the elements surrounded, let the button up. Now pick the change operation.
See Also: CHOOSE ALL

CLIPBOARD

Mac Only
Command: Clipboard
　　　Allows information to be shared between most applications via the MicroStation clipboard.
See Also: CLIPBOARD commands.

CLIPBOARD CUT

Mac Only

Command: Cut to Clipboard
　　　Moves a section of information to the clipboard. There will be some loss of resolution when using the clipboard because the clipboard resolution is 16-bit and MicroStation is 32-bit.

See Also: CLIPBOARD commands

CLIPBOARD PASTE

Mac Only

Command: Paste From Clipboard
   Moves information from the clipboard to the drawing file. There will be some
   loss of resolution when using the clipboard because the clipboard resolution is
   16-bit and MicroStation is 32-bit.
See Also: CLIPBOARD commands

CLOSE DESIGN   Ctrl+W

Command: Close Design
   Closes the current drawing file and opens the Open Design File dialog box
   so you can then select another drawing without reloading MicroStation.
Item Selection
   On sidebar menu pick: n/a
   For window, point to: File   Close
   On paper menu see: n/a
Command Window Prompts:
   1. Select the item
      The Open Design File dialog box appears
   To exit item   Click OK or Cancel
☞ *Tip:* Click OK to return to the drawing file you just exited.
See Also: EXIT, RD=

CLOSE ELEMENT   CL E

Command: Close Element
   Closes a shape created with PLACE SHAPE or PLACE MLINE. For example,
   choose PLACE MULTI-LINE to draw a rectangular building outline. Place the
   first three sides, then key-in CLOSE ELEMENT. MicroStation will create the last
   side automatically.
Item Selection
   On sidebar menu pick: n/a
   For window, point to: n/a
   On paper menu see: n/a
See Also: CREATE SHAPE tools

Closed Cross Joint

Drawing Tool: Closed Cross Joint
   Creates a closed joint that crosses two multi-line elements. This MDL
   application tool requires three input points. The first two points identify the
   multi-line elements and the third point creates the closed joint.
Item Selection
   On sidebar menu pick: n/a
   For window, point to: Palettes   Multi-line Joints

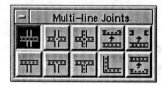

On paper menu see: n/a
Command Window Prompts:
   1. Select the item
      Identify element
   2. Pick first multi-line element

⇨   First element highlights, no new prompt

3. Pick second multi-line element
      ⇨   Second element highlights, no new prompt
4. Give an accept point
To exit item   Make next selection

🐾 *Note:* This MDL application is loaded automatically when you start the design session.
See Also: Merged Cross Joint, Open Cross Joint, PLACE MLINE

Closed Tee Joint   n/a

Drawing Tool: Closed Tee Joint
      Creates a closed tee that crosses two multi-line elements. This MDL application tool requires three input points. The first two points identify the multi-line elements and the third point creates the tee.

Item Selection
      On sidebar menu pick: n/a
      For window, point to: Palettes   Multi-line Joints

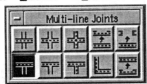

      On paper menu see: n/a
Command Window Prompts:
      1. Select the item
            Identify element
      2. Pick first multi-line element
            ⇨   First element highlights, no new prompt
      3. Pick second multi-line element
            ⇨   Second element highlights, no new prompt
      4. Give an accept point
      To exit item   Make next selection

🐾 *Note:* This MDL application is loaded automatically when you start the design session.
See Also: Merged Tee Joint, Open Tee Joint, PLACE MLINE

CM=

Alternate Key-in
      Alternate key-in for MATRIX CELL. It takes an existing cell and creates a matrix of cells.
See Also: MATRIX CELL

CO=

Alternate Key-in
      CO= is an alternate key-in for ACTIVE COLOR. Use it to control an element's color.
See Also: ACTIVE COLOR

COLORTABLE DEFAULT   COL D

Setting: Attach Default Colortable
      Attaches the default color table. This colortable can be viewed or changed through the Color Palette dialog box.
Item Selection
      On sidebar menu pick: n/a
      For window, point to: Settings   Color Palette...   Default
      On paper menu see: n/a

To show, open: `Color Palette dialog box`
To set, key in: `COLORTABLE DEFAULT`
See Also: ATTACH COLORTABLE

COLORTABLE INVERT
COL I

Mac Only
Setting: Colortable Invert
    Swaps the view background color in the color palette with the element color 0.

Item Selection
On sidebar menu pick: n/a
For window, point to: Yiew
On paper menu see: n/a
See Also: ATTACH COLORTABLE, COLORTABLE DEFAULT

COMPRESS DESIGN
COM

Command: Compress Design
    Removes deleted elements from a drawing file. Compressing the drawing file is a good habit to keep in mind each time your design session comes to an end.
Item Selection
On sidebar menu pick: `Utils` `ComprD`
For window, point to: File Compress Design
On paper menu see: TOP BORDER
Command Window Prompts:
1. Select the item
   File Compressed
To exit item: `Exits by itself`
See Also: COMPRESS LIBRARY, BACKUP, EXIT Commands, FREE

COMPRESS LIBRARY
COM C

Command: Compress Library
    Removes deleted cells from a cell library. The cell library must be attached to a drawing file before the COMPRESS LIBRARY program is run.
Item Selection
On sidebar menu pick: `Utils` `ComprC`
For window, point to: File Cell Library Compress
On paper menu see: n/a
Command Window Prompts:
1. Select the item
```
Library File : C:\DGN\SAMPLE.CEL
Index File : C:\DGN\SAMPLE.CDX
3 Cells Indexed
```
To exit item: `Exits by itself`
See Also: COMPRESS DESIGN, BACKUP, EXIT Commands, FREE

|CONNECT

Command: Connect
    Log on to an Oracle database with the |CONNECT command. After signing on, you can perform any database functions necessary.
Item Selection
On sidebar menu pick: n/a
For window, point to: n/a
On paper menu see: n/a
To log on, key in: `|CONNECT <username>/<password>`
    Where username is a valid username and password is a valid password.

Example(s):
`|CONNECT leavy/gocubsgo`

✍ *Note:* Oracle's implementation on the PC is for a single user system only. If you log onto your database in MicroStation, remember to log off, or you will not be able to access the database from DOS.
See Also: |DISCONNECT

CONSTRUCT BISECTOR ANGLE   CON B

Drawing Tool: Construct Angle Bisector
    Bisects an angle with a line. This MDL application tool requires three input points. The first point identifies one leg on the angle. The second point identifies the angle's vertex. The third point specifies the other angle leg.

Item Selection
  On sidebar menu pick: [Constr] [Bisect] [Angle]
  For window, point to: Palettes Main Lines

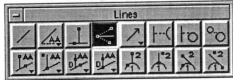

  On paper menu see: LINES
Command Window Prompts:
  1. Select the item
     Enter endpoint of angle leg
  2. Pick an angle leg
     Enter angle vertex
  3. Pick angle vertex
     Enter endpoint of angle leg
  4. Pick other angle leg
  To exit item: **Make next selection**
See Also: CONSTRUCT BISECTOR LINE

CONSTRUCT BISECTOR LINE   CON B L

Drawing Tool: Construct Line Bisector
    Bisects a line with a line. This tool requires two input points. The first point identifies the line, and the second specifies the length of the line bisector.

Item Selection
  On sidebar menu pick: [Constr] [Bisect] [Line]
  For window, point to: Palettes Main Lines

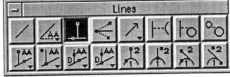

  On paper menu see: LINES
Command Window Prompts:
  1. Select the item
     Select line segment
  2. Pick a line
     Select endpoint of bisector
  3. Pick a point
  To exit item: **Make next selection**
See Also: CONSTRUCT BISECTOR ANGLE

CONSTRUCT
BSPLINE CONVERT COPY

CON BS CO CO

Drawing Tool: Convert Element to B-spline (Copy)

This MDL application tool copies any line, line string, arc, ellipse, complex chain or shape, surface of projection or surface of revolution or cone to a B-spline element.

Item Selection

On sidebar menu pick: | Place | BSpl | Curves |

For window, point to: Palettes Main B-splines

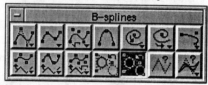

On paper menu see: B-SPLINES 2D 3D

Command Window Prompts:
1. Select the item
   Identify element
2. Pick the element
   Accept/Reject (select next input)
3. Give an accept point

To exit item: Make next selection

✱ Error: Invalid character - **If this message appears after your key in, enter MDL LOAD SPLINES or pick the tool from the B-splines palette once each design session.**

See Also: CONSTRUCT BSPLINE tools

CONSTRUCT
BSPLINE CONVERT ORIGINAL
CON BS CO

Drawing Tool: Convert Element to B-spline (Original)

This MDL application tool changes any line, line string, arc, ellipse, complex chain or shape, surface of projection or revolution, or cone to a B-spline element. The shape of the converted element remains the same.

Item Selection

On sidebar menu pick: | Place | BSpl | Curves |

For window, point to: Palettes Main B-splines

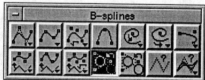

On paper menu see: B-SPLINES 2D 3D

Command Window Prompts:
1. Select the item
   Identify element
2. Pick the element
   Accept/Reject (select next input)
3. Give an accept point

To exit item: Make next selection

✱ Error: Invalid character - **If this message appears after your key in, enter MDL LOAD SPLINES or pick the tool from the B-splines palette once each design session.**

See Also: CONSTRUCT BSPLINE tools

CONSTRUCT
BSPLINE CURVE LEASTSQUARE   CON BS CU LEA

Drawing Tool: Construct B-spline Curve by Least Squares
    This MDL application tool create a B-spline curve using a line string or shape
as a template. The B-spline element uses the point along the line string or shape
to approximate the curve. The values in the B-splines setting box determine the
order. Line strings result in open B-splines, and shapes result in a closed B-spline
element.

Item Selection
    On sidebar menu pick: Place   BSpl   Curves
    For window, point to: Palettes   Main   B-splines

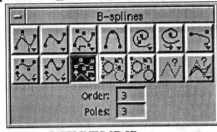

On paper menu see: B-SPLINES 2D 3D
Command Window Prompts:
   1. Select the item
      Identify element
   2. Pick the element
      Accept - Initiate construction
   3. Give an accept point
      Avg error = 0.00 Max error - 0.00
   To exit item: Make next selection

* *Error:* Invalid character **- If this message appears after your key in, enter
   MDL LOAD SPLINES or pick the tool from the B-splines palette once each
   design session.**
   See Also: CONSTRUCT BSPLINE CURVE tools

CONSTRUCT
BSPLINE CURVE POINTS   CON BS CU POI

Drawing Tool: Construct B-spline Curve by Points
    This MDL application tool creates a B-spline curve using a line string or
shape as a template. The B-spline curve passes through the points of the line string
or shape. The values in the B-splines setting box determine the order.

Item Selection
    On sidebar menu pick: Place   BSpl   Curves
    For window, point to: Palettes   Main   B-splines

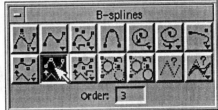

On paper menu see: B-SPLINES 2D 3D
Command Window Prompts:
   1. Select the item
      Identify element

2. Pick the element
   `Accept - Initiate construction`
3. Give an accept point
   To exit item: `Make next selection`

* *Error:* `Invalid character` **- If this message appears after your key in, enter MDL LOAD SPLINES or pick the tool from the B-splines palette once each design session.**
See Also: CONSTRUCT BSPLINE CURVE tools

CONSTRUCT
BSPLINE CURVE POLES   CON BS CU POL

Drawing Tool: Construct B-spline Curve by Poles
   This MDL application tool creates a B-spline curve using a line string or shape as a template. The values in the B-splines setting box determine the order.

Item Selection

On sidebar menu pick: | Place | BSpl | Curves |

For window, point to: Palettes Main B-splines

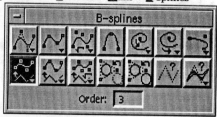

On paper menu see: B-SPLINES 2D 3D

Command Window Prompts:
1. Select the item
   `Identify element`
2. Pick the element
   `Accept - Initiate construction`
3. Give an accept point
   To exit item: `Make next selection`

* *Error:* `Invalid character` **- If this message appears after your key in, enter MDL LOAD SPLINES or pick the tool from the B-splines palette once each design session.**
See Also: CONSTRUCT BSPLINE CURVE tools

CONSTRUCT
BSPLINE SURFACE CROSS   CON BS S C

3D Drawing Tool: Construct B-spline Surface by Cross-section
   This MDL application tool transforms a B-spline surface among two to 101 lines, line string, shapes, arcs, curves, B-spline curves, chains and complex shapes.
   Use the element selection tool to identify the candidate elements in the order the surface will be transformed. Then pick this drawing tool.

Item Selection

On sidebar menu pick: | Place | BSpl | DerSur |

For window, point to: Palettes 3D B-splines Derived Surfaces

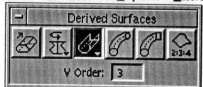

On paper menu see: DERIVED SURFACES
Command Window Prompts:
1. Select the item
   ⇨ The surface is generated.
To exit item: **Exits by itself**

✱ *Error:* Invalid character - **If this message appears after your key in, enter MDL LOAD SPLINES or pick the tool from the Derived Surfaces palette once each design session.**
See Also: CONSTRUCT BSPLINE SURFACE tools

CONSTRUCT
BSPLINE SURFACE EDGE   CON BS S E

3D Drawing Tool: Construct B-spline Surface by Edges
   This MDL application tool creates a B-spline surface out of lines, line string, shapes, arcs, curves, B-spline curves, chains and complex shapes.
   Use the element selection tool to identify the candidate elements. Then pick this drawing tool.

Item Selection
On sidebar menu pick: | Place | BSpl | DerSur |
For window, point to: Palettes   3D B-splines   Derived Surfaces

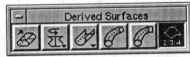

On paper menu see: n/a
Command Window Prompts:
1. Select the item
   ⇨ The surface is generated.
To exit item: **Exits by itself**

✍ *Note:* Select two elements to create a ruled surface.

✍ *Note:* Select three or four elements to create a bilinearly blended Coons patch.

✱ *Error:* Invalid character - **If this message appears after your key in, enter MDL LOAD SPLINES or pick the tool from the Derived Surfaces palette once each design session.**
See Also: CONSTRUCT BSPLINE SURFACE tools

CONSTRUCT
BSPLINE SURFACE LEASTSQUARE   CON BS S L

3D Drawing Tool: Construct B-spline Surface by Least Squares
   This MDL application tool creates a B-spline surface by approximating the vertices of a set of B-spline lines, line strings and shapes.
   Use the element selection tool to identify the candidate elements in the order the surface will follow. Then pick this drawing tool.

Item Selection
On sidebar menu pick: | Place | BSpl | SurFce |
For window, point to: Palettes   3D B-splines   Surfaces

On paper menu see: SURFACES
Command Window Prompts:
1. Select the item
   ⇨ The surface is generated.

To exit item: Exits by itself

* *Error:* Invalid character - **If this message appears after your key in, enter MDL LOAD SPLINES or pick the tool from the Surfaces palette once each design session.**

🕮 *Note:* Set the order number and the number of poles through the pop-downs when you select the tool, or in the B-splines settings box.
See Also: CONSTRUCT BSPLINE SURFACE tools

CONSTRUCT
BSPLINE SURFACE POINTS   CON BS S POI

3D Drawing Tool: Construct B-spline Surface by Points
        This MDL application tool interpolates a B-spline surface that passes through the vertices of a set of lines, line string and shapes.
        Use the element selection tool to identify the target elements in the order the surface will follow. Then pick this drawing tool.

Item Selection
On sidebar menu pick: `Place`   `BSpl`   `SurFce`
For window, point to: Palettes   3D B-splines   Surfaces

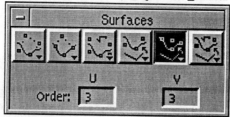

On paper menu see: SURFACES
Command Window Prompts:
  1. Select the item
        ⇨   The surface is generated.
  To exit item: Exits by itself

* *Error:* Invalid character - **If this message appears after your key in, enter MDL LOAD SPLINES or pick the tool from the Surfaces palette once each design session.**

🕮 *Note:* Set the order number and the number of poles through the pop-downs when you select the tool, or in the B-splines settings box.
See Also: CONSTRUCT BSPLINE SURFACE tools

CONSTRUCT
BSPLINE SURFACE POLES   CON BS S POL

3D Drawing Tool: Construct B-spline Surface by Poles
        This MDL application tool constructs a B-spline surface that passes through the vertices of a set of lines, line string and shapes. The vertices of the elements define the poles of the surface.
        Use the element selection tool to identify the target elements in the order the surface will follow. Then pick this drawing tool.

Item Selection
On sidebar menu pick: `Place`   `BSpl`   `SurFce`

For window, point to: Palettes   3D B-splines   Surfaces

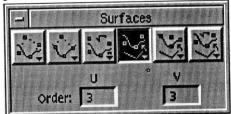

On paper menu see: SURFACES
Command Window Prompts:
1. Select the item
   ⇨   The surface is generated.
   To exit item: **Exits by itself**

✍ *Note:* Set the order number and the number of poles through the pop-downs when you select the tool, or in the B-splines settings box.

✍ *Note:* The number of poles in the U and V direction must be greater than or equal to the order number for that direction.

✳ *Error:* Invalid character - **If this message appears after your key in, enter MDL LOAD SPLINES or pick the tool from the Surfaces palette once each design session.**
See Also: CONSTRUCT BSPLINE SURFACE tools

CONSTRUCT
BSPLINE SURFACE PROJECTION   CON BS SU PR

3D Drawing Tool: Construct B-spline Surface of Projection
This MDL application tool creates a B-spline surface by projecting a planar element such as a line, line string, complex line chain or shape, arc, ellipse or B-spline curve as a template.
Item Selection
   On sidebar menu pick: [Place]   [BSpl]   [DerSur]
   For window, point to: Palettes   3D B-splines   Derived Surfaces

On paper menu see: DERIVED SURFACES
Command Window Prompts:
1. Select the item
   Identify element
2. Pick the element
   Define projection distance
3. Key-in or give a second point
   To exit item: **Make next selection**

✳ *Error:* Invalid character - **If this message appears after your key in, enter MDL LOAD SPLINES or pick the tool from the Derived Surfaces palette once each design session.**
See Also: CONSTRUCT BSPLINE SURFACE tools

CONSTRUCT
BSPLINE SURFACE REVOLUTION   CON BS SU R

3D Drawing Tool: Construct B-spline Surface of Revolution
Use this MDL application tool to create a B-spline surface by revolving a planar element such as a line, line string, complex line chain or shape, arc, ellipse

or B-spline curve based on three input points and a keyed-in sweep angle. One input point identifies the element and the other two points define the axis.

Item Selection

On sidebar menu pick: `Place` `BSpl` `DerSur`

For window, point to: Palettes   3D B-splines   Derived Surfaces

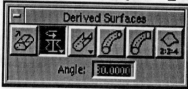

On paper menu see: DERIVED SURFACES

Command Window Prompts:
1. Select the item
   `Identify element`
2. Pick the element
   `Enter one end of axis`
3. Pick one axis point
   `Enter other end of axis`
4. Pick other endpoint

To exit item: **Make next selection**

* *Error:* `Invalid character` - **If this message appears after your key in, enter MDL LOAD SPLINES or pick the tool from the Derived Surfaces palette once each design session.**

See Also: CONSTRUCT BSPLINE SURFACE tools

CONSTRUCT
BSPLINE SURFACE SKIN
CON BS SU S

3D Drawing Tool: Construct B-spline Surface by Skin

Use this MDL application tool to create a B-spline surface by identifying a planar element such as a line, line string, complex line chain or shape, arc, ellipse or B-spline curve. The skin is formed by transforming a planar element (the initial section) along another (the trace) while creating a new element (the final section). This item requires a series of input points to define the initial section, trace element, and orientation of the skin surface.

Item Selection

On sidebar menu pick: `Place` `BSpl` `DerSur`

For window, point to: Palettes   3D B-splines   Derived Surfaces

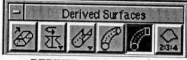

On paper menu see: DERIVED SURFACES

Command Window Prompts:
1. Select the item
   `Identify trace curve`
2. Pick the trace element
   `Select next section site or RESET to     complete`
3. Pick point or RESET
   `Identify section curve`
4. Pick section curve
   `Accept section at point to attach`
5. Pick attachment point
   `Define section orientation`
6. Pick orientation point
   `Select next section site or RESET to complete`
7. Pick point or RESET

To exit item: **Make next selection**

✱ *Error:* Invalid character **- If this message appears after your key in, enter MDL LOAD SPLINES or pick the tool from the Derived Surfaces palette once each design session.**
See Also: CONSTRUCT BSPLINE SURFACE tools

CONSTRUCT BSPLINE SURFACE TUBE   CON BS SU T

3D Drawing Tool: Construct B-spline Surface by Tube
    Use this MDL application tool to create a B-spline surface by identifying a planar element such as a line, line string, complex line chain or shape, arc, ellipse or B-spline curve. The surface is formed by interpreting the planar element (the section) along another (the trace). This item requires five input points. The first point identifies the trace curve. The next two points identify section curve and the attachment point. The last two points define the section orientation and corresponding point.

Item Selection
  On sidebar menu pick: Place   BSpl   DerSur
  For window, point to: Palettes   3D B-splines   Derived Surfaces

  On paper menu see: DERIVED SURFACES
Command Window Prompts:
  1. Select the item
     Identify trace curve
  2. Pick the trace element
     Identify section curve
  3. Pick the section curve
     Accept section at point to attach
  4. Pick attachment point
     Define section orientation
  5. Pick orientation point
     Locate corresponding point (Keyin/Data)
  6. Pick corresponding point
  To exit item: Make next selection

✱ *Error:* Invalid character **- If this message appears after your key in, enter MDL LOAD SPLINES or pick the tool from the Derived Surfaces palette once each design session.**
See Also: CONSTRUCT BSPLINE SURFACE tools

Construct Line

Definition
    There are five CONSTRUCT LINE tools. The AA1 option constructs a line at the ACTIVE ANGLE and intersects an identified line. AA2 asks for a key-in to determine the distance of the line constructed at the ACTIVE ANGLE. The AA3 option starts the ACTIVE ANGLE line at the point of identification. The AA4 selection starts the line at the identification point and asks for a key-in to determine the length of the line. Finally, the MINIMUM choice constructs a line the shortest distance between two elements.
See Also: CONSTRUCT tools

CONSTRUCT LINE AA1   CON LI AA 1

Drawing Tool: Construct Line at Active Angle to Point
    The CONSTRUCT LINE AA1 tool constructs a line at the ACTIVE ANGLE and intersects an identified line. This tool expects two input points. The first point identifies the line. The second point determines the length and location of the new line.

Item Selection
On sidebar menu pick: | Place | Line |
For window, point to: Palettes Main Lines

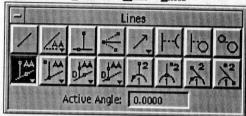

On paper menu see: LINES
Command Window Prompts:
1. Select the item
   Select line segment
2. Pick a line
   Enter endpoint
3. Pick an endpoint
To exit item: **Make next selection**

* *Error:* Err - Active angle = 0.0 - If this message appears, change the ACTIVE ANGLE setting.
See Also: CONSTRUCT LINE tools, CONSTRUCT TANGENT BETWEEN

CONSTRUCT LINE AA2   CON LI AA 2

Drawing Tool: Construct Line at Active Angle to Point (Key-in)
The CONSTRUCT LINE AA2 asks for a key-in to determine the length of the line constructed at the ACTIVE ANGLE. This tool expects a keyed in length and two input points. The first point identifies the line and the second point determines the placement of the line created at the ACTIVE ANGLE.

Item Selection
On sidebar menu pick: | Place | Line |
For window, point to: Palettes Main Lines

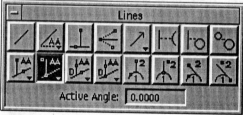

On paper menu see: LINES
Command Window Prompts:
1. Select the item
   Enter distance
2. Key in a distance in working units
   Select line segment
3. Pick a line
   Enter endpoint
4. Pick an endpoint
To exit item: **Make next selection**

* *Error:* Err - Active angle = 0.0 - If this message appears, change the ACTIVE ANGLE setting.
See Also: CONSTRUCT LINE tools, CONSTRUCT TANGENT BETWEEN

CONSTRUCT LINE AA3

Drawing Tool: Construct Line at Active Angle from Point

The CONSTRUCT LINE AA3 tool starts the ACTIVE ANGLE line at the point of identification. This item expects two input points. The first point identifies the intersection point on the line and the second point determines the length of the line created at the ACTIVE ANGLE.

Item Selection

On sidebar menu pick: Place   Line

For window, point to: Palettes   Main   Lines

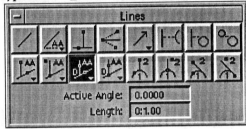

On paper menu see: LINES

Command Window Prompts:

1. Select the item
   Select line segment
2. Pick a line
   Enter endpoint
3. Pick an endpoint

To exit item   Make next selection

* **Error:** Err - Active angle = 0.0 - If this message appears, change the ACTIVE ANGLE setting.

See Also: CONSTRUCT LINE tools, CONSTRUCT TANGENT BETWEEN

CONSTRUCT LINE AA4

Drawing Tool: Construct Line at Active Angle from Point (Key-in)

The CONSTRUCT LINE AA4 tool starts a line at the identification point and asks for a key-in to determine the length of the line. This tool expects a keyed-in distance and two input points. The first point identifies the line's starting point and the second point determines the placement of the line created at the ACTIVE ANGLE.

Item Selection

On sidebar menu pick: Place   Line

For window, point to: Palettes   Main   Lines

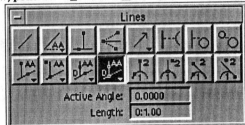

On paper menu see: LINES

Command Window Prompts:

1. Select the item
   Enter distance
2. Key in a distance in working units
   Select line segment
3. Pick a line

```
Enter endpoint
```
4. Pick an endpoint

To exit item: **Make next selection**

✱ *Error:* Err - Active angle = 0.0 - If this message appears, change the ACTIVE ANGLE setting.

See Also: CONSTRUCT LINE tools, CONSTRUCT TANGENT BETWEEN

CONSTRUCT LINE MINIMUM   CON LI M

Drawing Tool: Construct Minimum Distance Line

The CONSTRUCT LINE MINIMUM tool draws a line the shortest distance between two elements. This tool expects three input points. Two input points identify the elements involved. The third point draws the line between the two elements.

Item Selection

On sidebar menu pick: n/a

For window, point to: Palettes   Main   Lines

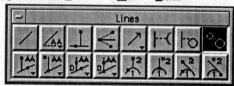

On paper menu see: LINES

Command Window Prompts:

1. Select the item
   ```
   Identify first element
   ```
2. Pick an element
   ```
   Accept, Identify 2nd element/Reject
   ```
3. Pick the second element
   ```
   Accept, Initiate min dist calculation
   ```
4. Give an accept point

To exit item: **Make next selection**

See Also: CONSTRUCT LINE tools, CONSTRUCT TANGENT BETWEEN

CONSTRUCT PERPENDICULAR FROM   CON PE F

Drawing Tool: Construct Perpendicular from Element

The CONSTRUCT PERPENDICULAR FROM tool builds a perpendicular line from an element. The element can be a line, line string, circle, arc or other circular element. This tool requires two input points. The first point identifies the perpendicular's origin on the element and the second point defines the length of the perpendicular.

Item Selection

On sidebar menu pick: Constr   Perp   From

For window, point to: Palettes   Main   Lines

On paper menu see: LINES

Command Window Prompts:

1. Select the item
   ```
   Identify element
   ```
2. Pick an element
   ```
   Enter endpoint
   ```

To exit item: **Make next selection**

See Also: CONSTRUCT PERPENDICULAR TO, CONSTRUCT TANGENT PERPENDICULAR

CONSTRUCT PERPENDICULAR TO　　　　　CON PE T

Drawing Tool:　Construct Perpendicular to Element
　　The CONSTRUCT PERPENDICULAR TO tool builds a perpendicular line to an element. The element can be a line, line string, circle, arc or other circular element. This tool requires two input points. The first point identifies the perpendicular's end point on the element and the second point defines the origin of the perpendicular.

Item Selection
　On sidebar menu pick:　Constr　Perp　To
　For window, point to:　Palettes　Main　Lines

On paper menu see:　LINES
Command Window Prompts:
　1. Select the item
　　　Identify element
　2. Pick an element
　　　Enter endpoint
　To exit item:　Make next selection
See Also: CONSTRUCT PERPENDICULAR FROM, CONSTRUCT TANGENT PERPENDICULAR

Construct Point

Definition
　　There are five tools which allow you to construct points along elements. These points could aid in dividing an arc or line into equal segments, or connect other points for joining new elements to the original segmented element.
　　The ALONG option lets you place a series of points along an element. The BETWEEN selection places a group of points between two input points. The DISTANCE choice places a point at some defined distance along an element. The INTERSECTION option puts a point at the intersection of two elements. Lastly, the PROJECT selection plots a point on an element the minimum distance to that element.
　　The point is a zero-length line (at the ACTIVE WEIGHT), a cell (at the ACTIVE SCALE) or text (at the ACTIVE TXSIZE). The ACTIVE POINT setting determines the type of point placed.
See Also: CONSTRUCT POINT tools

CONSTRUCT POINT ALONG　　　　　CON PO A

Drawing Tool:　Construct Active Points Along Element
　　Places points along an element at the ACTIVE ANGLE and ACTIVE SCALE settings. This tool expects a key-in for the number of points to place, as well as two input points to define the end points of the target element.

Item Selection
　On sidebar menu pick:　Constr　Point　Along

For window, point to: Palettes   Main   Points

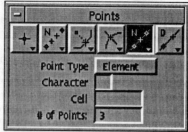

On paper menu see: POINTS
Command Window Prompts:
1. Select the item
   Key in number of points
2. Key in number of points
   Identify element
3. Pick one endpoint
   Enter end point
4. Pick other endpoint
To exit item: Make next selection
See Also: CONSTRUCT POINT tools

CONSTRUCT POINT BETWEEN   CON PO B

Drawing Tool: Construct Active Points Between Data Points
   Places points, at the ACTIVE ANGLE and ACTIVE SCALE, along a distance
defined by two points placed on the display screen. This tool expects a key-in for
the number of points to place, as well as two input points to define the distance
the points will occupy.
   A series of highlighted X's, representing the construction points, will appear
until the second point is given. The constructed points then replace the X's.
Item Selection
On sidebar menu pick: Constr   Point   Betwn
For window, point to: Palettes   Main   Points

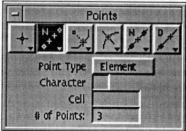

On paper menu see: POINTS
Command Window Prompts:
1. Select the item
   Key in number of points
2. Key in number of points
   Enter first point
3. Pick one point
   Enter end point
4. Pick second point
To exit item: Make next selection
See Also: CONSTRUCT POINT tools

CONSTRUCT POINT DISTANCE   CON PO D

Drawing Tool: Construct Active Point at Distance Along Element

Places a point, at the ACTIVE ANGLE and ACTIVE SCALE, along an element at a defined distance. This tool expects a key-in to determine the distance of the point's placement, as well as two input points to define the endpoints of the target element.

Item Selection

On sidebar menu pick: Constr   Point   Dist

For window, point to: Palettes   Main   Points

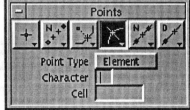

On paper menu see: POINTS

Command Window Prompts:

1. Select the item
   Enter distance
2. Key in distance in working units
   Identify element
3. Pick one endpoint
   Accept/Reject (select next input)
4. Pick other endpoint

To exit item: **Make next selection**

* *Error:* Illegal definition - If this message appears, either the keyed in distance is too large or locate both input points at the element's endpoints.

See Also: CONSTRUCT POINT tools

CONSTRUCT POINT INTERSECTION   CON PO I

Drawing Tool: Construct Active Point at Intersection

Places a point at the intersection of two identified elements. It places the point at the ACTIVE SCALE and ACTIVE ANGLE. This tool expects two input points. The first point identifies the target element and an accept point confirms the placement of the intersecting point.

A highlighted X, representing the intersection, appears until the accept point is given. A point then replaces the X.

Item Selection

On sidebar menu pick: Constr   Point   Intsct

For window, point to: Palettes   Main   Points

On paper menu see: POINTS

Command Window Prompts:

1. Select the item
   Select element for intersection

2. Pick an element
   `Select element for intersection`
3. Pick other element
   `Accept - Initiate intersection`
4. Give an accept point
To exit item: Make next selection
See Also: CONSTRUCT POINT tools

CONSTRUCT POINT PROJECT   CON PO P

Drawing Tool: Project Active Point Onto Element
    Places a point along an element. It places the point at the ACTIVE SCALE and ACTIVE ANGLE. This item expects two input points. One point identifies the target element and an accept point confirms the constructed-point's location.

    A highlighted X, representing the point's location, appears until the accept point is given. A construct point then replaces the X.

Item Selection
On sidebar menu pick: Constr Point Proj
For window, point to: Palettes Main Points

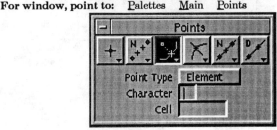

On paper menu see: POINTS
Command Window Prompts:
1. Select the item
   `Identify element`
2. Pick an element
   `Accept/Reject (select next input)`
3. Give an accept point
To exit item: Make next selection
See Also: CONSTRUCT POINT tools

Construct Tangent

Definition
    MicroStation provides six ways to construct tangent lines. The BETWEEN option constructs a tangent line between two circles or arcs. The CIRCLE1 tool places a circle tangent to another element. The CIRCLE3 selection creates a circle tangent to three elements. The FROM choice builds a perpendicular line from a circular element. The PERPENDICULAR option places a line tangent to one element and normal to another. The TO selection creates a perpendicular line to a circular element.
See Also: CONSTRUCT TANGENT tools

CONSTRUCT TANGENT ARC 1   CON T A

Drawing Tool: Construct Tangent Arc by Keyed-in Radius
    Constructs an arc tangent to a line, line string, multi-line or shape created with the CONSTRUCT TANGENT ARC 1 drawing tool. This item requires two input points. The first point identifies the point of tangency to the target element and the second point defines the arc's direction. Enter the length of the arc and its radius in the pop-down field when you pick the tool.

Item Selection
On sidebar menu pick: n/a

For window, point to: Palettes   Main   Arcs

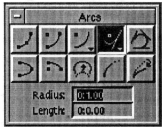

On paper menu see: ARCS
Command Window Prompts:
1. Select the item
   Identify element
2. Pick element
   Accept, define direction/Reject
3. Pick arc's direction
To exit item: Make next selection
See Also: CONSTRUCT TANGENT tools, PLACE ARC tools

CONSTRUCT TANGENT ARC 3   CON T A 3

Drawing Tool: Construct Arc Tangent to Three Elements
Constructs an arc tangent to three lines or three segments of a line string, multi-line or shape created with the CONSTRUCT TANGENT ARC 3 drawing tool. This item requires three input points to identify the target lines or element segments. A fourth point is needed to accept the arc's placement.
Item Selection
On sidebar menu pick: n/a
For window, point to: Palettes   Main   Arcs

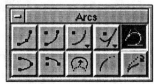

On paper menu see: ARCS
Command Window Prompts:
1. Select the item
   Select first segment
2. Pick first line or segment
   Select second segment
3. Pick second line or segment
   Select third segment
4. Pick third line or segment
   ⇨   Prompt does not change
5. Accept arc with an input point
To exit item: Make next selection
See Also: CONSTRUCT TANGENT tools, PLACE ARC tools

CONSTRUCT TANGENT BETWEEN   CON TAN B

Drawing Tool: Construct Tangent to Two Elements
Constructs a tangent line between two circles or arcs. This tool requires three input points. The first two points identify the elements involved in the tangency and the third point confirms the operation.
Item Selection
On sidebar menu pick: | Constr | | Tan | | Btwn |

For window, point to: Palettes   Main   Lines
On paper menu see: LINES
Command Window Prompts:
1. Select the item
   Identify circle or circular arc
2. Pick an element
   Accept/Reject (select next input)
3. Identify other circular element
   Accept/Reject (select next input)
   ⇨   A tangent line appears.
4. Give an accept point
To exit item: **Make next selection**
See Also: CONSTRUCT TANGENT tools, CONSTRUCT LINE tools

CONSTRUCT TANGENT CIRCLE 1   CON TAN C

Drawing Tool: Construct Circle Tangent to Element
   Constructs a circle tangent to an element. This tool requires two input points.
The first point identifies the element involved in the tangency and the second point
determines the circle's center.

Item Selection
On sidebar menu pick: | Constr | | Tan | | Crc1 |
For window, point to: Palettes   Main   Circles/Ellipses

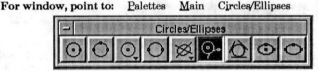

Circles/Ellipses

On paper menu see: CIRCLES
Command Window Prompts:
1. Select the item
   Identify element
2. Pick an element
   Accept/Reject (select next input)
3. The accept point determines the circles center
4. Give an accept point
To exit item: **Make next selection**
See Also: CONSTRUCT TANGENT CIRCLE 3, PLACE CIRCLE tools

CONSTRUCT TANGENT CIRCLE 3   CON TAN C 3

Drawing Tool: Construct Circle Tangent to Three Elements
   Constructs a circle tangent to three lines or two lines and a circle or arc. This
tool requires four input points. The first three points identify the tangent points
of the elements involved. A fourth point makes the circle permanent.

Item Selection
On sidebar menu pick: | Constr | | Tan | | Crc3 |
For window, point to: Palettes   Main   Circles/Ellipses

Circles/Ellipses

On paper menu see: CIRCLES
Command Window Prompts:
1. Select the item
   Select first segment
2. Pick an element
   Select second segment
3. Pick second element
   Select third segment
4. Pick third element

⇨ A highlighted circle appears.
5. Give an accept point
To exit item: Make next selection
See Also: CONSTRUCT TANGENT CIRCLE 1, PLACE CIRCLE tools

CONSTRUCT TANGENT FROM   CON T

Drawing Tool: Construct Tangent from Element
Constructs a tangent line from a circular element. This tool requires two
input points. The first point identifies the element involved in the tangency and
the second point defines the distance of the tangent line.

Item Selection
On sidebar menu pick: | Constr | Tan | From |
For window, point to: <u>P</u>alettes <u>M</u>ain Lines

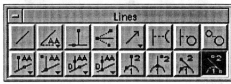

On paper menu see: LINES
Tool Window Prompts:
1. Select the item
   Identify circle, ellipse or arc
2. Pick an element
   Enter endpoint
3. Pick a point
To exit item: Make next selection
See Also: CONSTRUCT TANGENT tools

CONSTRUCT TANGENT PERPENDICULAR   CON T P

Drawing Tool: Construct Tangent to Circular Element and Perpendicular to
Linear Element
Constructs a line tangent to an arc and perpendicular to another line. This
tool requires three input points. The first point identifies the circular element
involved in the tangency. The second point identifies the line and the third point
accepts the placement of the tangent line.

Item Selection
On sidebar menu pick: | Constr | Tan | Perp |
For window, point to: <u>P</u>alettes <u>M</u>ain Lines

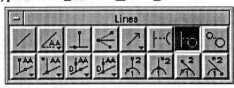

On paper menu see: LINES
Command Window Prompts:
1. Select the item
   Identify circle, ellipse or arc
2. Pick an element
   Accept/Reject (select next input)
3. Select the line
   ⇨ A tangent line appears.
4. Give an accept point
To exit item: Make next selection
See Also: CONSTRUCT TANGENT tools

CONSTRUCT TANGENT TO

Drawing Tool: Construct Tangent to Element

Constructs a tangent line to a circular element. This tool requires two input points. The first point identifies the tangent's endpoint on the element involved in the tangency. The second point defines the length of the tangent line.

Item Selection

On sidebar menu pick: Constr   Tan   To

For window, point to: Palettes   Main   Lines

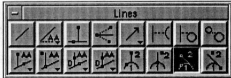

On paper menu see: LINES

Command Window Prompts:
1. Select the item
   Identify circle, circular arc
2. Pick an element
   Enter endpoint
3. Pick a point

To exit item: Make next selection

See Also: CONSTRUCT TANGENT tools

COPY ED

Drawing Tool: Copy Enter Data Field

Copies the contents of one Enter Data Field to another Enter Data Field location. This tool requires two input points. The first point defines the field you wish to copy and the second point picks the destination Enter Data Field.

Item Selection

On sidebar menu pick: Text   ED Fld   Copy   EDFs

For window, point to: Palattes   Main   Enter Data Fields

On paper menu see: ENTER DATA

Command Window Prompts:
1. Select the item
   Select enter data field to copy
2. Pick an enter data field
   Select destination enter data field
3. Pick destination enter data field

To exit item: Click a Reset

See Also: EDIT AUTO, EDIT SINGLE

COPY ELEMENT

Drawing Tool: Copy Element

Duplicates any element or shape. This tool requires two input points. One point identifies the element you wish to copy and a second point determines the copy's final resting place.

Item Selection

On sidebar menu pick: Manip   Copy   Elem

For window, point to: Palettes   Main   Copy Element

On paper menu see: COPY ELEMENT

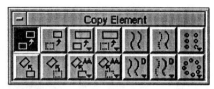

Command Window Prompts:
1. Select the item
   Identify element
2. Pick an element
   Accept/Reject (select next input)
3. Give an accept point
To exit item: Make next selection
See Also: COPY PARALLEL tools, FENCE COPY

COPY PARALLEL DISTANCE   COP P

Drawing Tool: Copy Parallel by Distance
    Copies a new element parallel to an existing element. An input point
graphically defines the distance between the elements. Lines, line strings and arcs
are candidates for the COPY PARALLEL DISTANCE tool. This tool requires two
input points. The first point highlights the element to copy. The second point
defines the distance to copy the new element.

Item Selection
On sidebar menu pick: | Manip | | Copy | | Para d |
For window, point to: Palettes   Main   Copy Element

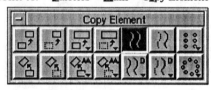

On paper menu see: COPY ELEMENT
Command Window Prompts:
1. Select the item
   Identify element
2. Pick an element
   Accept/Reject (select next input)
3. Give an accept point
To exit item: Make next selection
See Also: COPY PARALLEL KEYIN, COPY ELEMENT

COPY PARALLEL KEYIN   COP K

Drawing Tool: Copy Parallel by Key-in
    Copies a new element parallel to an existing element. A key-in defines the
distance between the elements. Lines, line strings and arcs are candidates for the
COPY PARALLEL KEYIN tool. This tool requires one key-in and two input points.
The first point highlights the element to copy. The second point accepts the element
to copy and determines the direction to copy the element.

Item Selection
On sidebar menu pick: | Manip | | Copy | | Para k |
For window, point to: Palettes   Main   Copy Element
On paper menu see: COPY ELEMENT
Command Window Prompts:
1. Select the item
   Enter distance
2. Enter a distance in working units
   Identify element
3. Pick an element

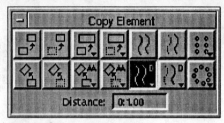

```
Accept/Reject (select next input)
```
4. Give an accept point
To exit item: Make next selection
See Also: COPY PARALLEL DISTANCE, COPY ELEMENT

COPY VIEW   COP VI

View Control: Copy View

Copies the contents of one view into another active view. The destination view assumes the characteristics of the source view such as grid display, weight display etc. This control expects two input points: one in the source view and one in the destination view(s).

Item Selection

On sidebar menu pick: | Window | View | Copy |

For window, point to: View Copy

On paper menu see: VIEW

Command Window Prompts:
1. Select the item
   ```
   Select source view
   ```
2. Pick view you wish to copy
   ```
   Select destination view(s)
   ```
3. Pick the view(s) to copy to

To exit item: Click a Reset, Reset

See Also: SET OVERVIEW, SWAP SCREEN, SWAP VIEWS

Corner Joint

Drawing Tool: Corner Joint

Takes two multi-line elements and makes a corner joint. This MDL application tool requires two input points to identify the multi-lines and a third point to accept the joint process.

Item Selection

On sidebar menu pick: n/a

For window, point to: Palettes Multi-line Joints

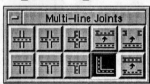

On paper menu see: n/a

Command Window Prompts:
1. Select the item
   ```
   Identify element
   ```
2. Pick first multi-line element
   ⇨   First element highlights, no new prompt
3. Pick second multi-line element
   ⇨   Second element highlights, no new prompt
4. Give an accept point

To exit item: Make next selection

🔊 *Note:* This MDL application is loaded automatically when you start the design session.

See Also: Closed, Open and Merged Cross Joint tools, Closed, Open and Merged Tee Joint tools, PLACE MLINE

CR=

Alternate Key-in
Alternate key-in for RENAME CELL. Use it to rename a cell in the current cell library attached to a drawing file.
See Also: RENAME CELL

CREATE CELL   CC=

Drawing Tool: Create Cell
Takes the elements contained within a fence and files them in a library as a cell.
To create a cell, follow these steps: 1. Create the graphics. 2.Place a fence around the graphics. 3. Use the DEFINE ORIGIN tool to choose the origin point of the cell for placement later. 4. Use the CREATE CELL tool to file the figure away in a cell library. Before adding cells to the library, attach it to a drawing file.
When adding a cell to the cell library, keep three things in mind: 1. The length of the cell name is 6 alpha-numeric characters in length. The cell name must be unique within the cell library. 2. A description up to 27 characters long can accompany the cell name. 3. Cells can be filed in the library with different type designators. The default designation is graphics. A CM after the description stands for a tool menu cell. MM means the cell is a matrix menu cell. A point cell is a cell which is neither view orientation nor level dependent.

Item Selection
On sidebar menu pick: | Cells | | Create |
For window, point to: Settings   Cells   Create...
On paper menu see: n/a
Command Window Prompts:
1. Select the item
    CC=uuu
        Where uuu is a six character user defined cell name.
To exit item: **Exits by itself**

Example(s):
    CC=BOLT,hex top - no threads,point
    CC=ELBOW,,

✳ *Error:* No cell origin defined - If this message appears, use the DEFINE ORIGIN tool to pick a cell origin.
See Also: ATTACH LIBRARY, PLACE CELL tools

CREATE CHAIN AUTOMATIC   CRE CH A

Drawing Tool: Automatic Create Complex Chain
Provides an easy way to associate a group of individual elements into a complex string of elements. This tool is quicker and easier to use than the CREATE CHAIN MANUAL tool.
The tool asks for a tolerance key-in. The tolerance is the distance between the line segments. If no tolerance is given, this tool searches for elements that are touching. On the other hand, if a tolerance is given and the elements in the chain do not actually touch, MicroStation still treats the elements as a chain. After the tolerance key-in, place the cursor on the first element in the chain. Continue to click input points at the same location. Watch the CREATE CHAIN tool hunt for the elements that touch each other to form the complex string of elements.

Item Selection
On sidebar menu pick: | Constr | | Chain | | Auto |

For window, point to: Palettes   Main   Chain

On paper menu see: CHAIN
Command Window Prompts:
1. Select the item
   Enter chain tolerance
   ⇨   No prompt appears.
2. Pick an element
   Accept/Reject (select next input)
3. Give an accept point
   ⇨   Continue picking the elements.
To exit item: Make next selection
See Also: CREATE CHAIN MANUAL, CREATE SHAPE tools

CREATE CHAIN EOF

Drawing Tool: Create Chain EOF
   Creates an open chain of elements. For example, you place a collection of
   elements in the drawing file. Next set the working window with WWSECT and
   WWBYTE at the element where the chain is to begin. Then execute the CREATE
   CHAIN EOF tool from a user command. A chain will be created from that element
   to the end of the file (EOF).
See Also: CREATE CHAIN tools, USERCOMMAND

CREATE CHAIN MANUAL   CRE CH

Drawing Tool: Create Complex Chain
   Builds a chain by prompting you to select each element you wish included in
   the chain. Identify each element in the complex string with an input point.
Item Selection
On sidebar menu pick: Constr   Chain   Man
For window, point to: Palettes   Main   Chain

On paper menu see: CHAIN
Command Window Prompts:
1. Select the item
   Identify element
2. Pick an element
   Accept/Reject (select next input)
3. Give an accept point
   ⇨   Continue picking the elements.
To exit item: Make next selection

✍ *Note:* Use this option when CREATE CHAIN AUTOMATIC has trouble identifying
   the target elements.
See Also: CREATE CHAIN AUTOMATIC, CREATE SHAPE tools

CREATE DRAWING
Ctrl+N

Command: Create Drawing
   Displays the Create Design File dialog. From this dialog box, another drawing can be called up, or new drawing files can be created.

Item Selection
   On sidebar menu pick: n/a
   For window, point to: File New... Name:
   On paper menu see: n/a
Command Window Prompts:
   1. Select the item
   2. Click in the Name window
   3. Enter the file name
      ⇨ Or click the appropriate directory and file name.
   4. Click on OK
   To exit item: Click on OK or Cancel
See Also: CREATE LIBRARY

CREATE ENTITY
CRE E

DOS Only
Command: Create Entity
   Adds new records to the database attached to the drawing file.

Item Selection
   On sidebar menu pick: dBase ActEnt Create

   For window, point to: n/a
   On paper menu see: n/a
See Also: DEFINE AE, FIND, SHOW AE, UPDATE AE

CREATE LIBRARY
CRE L

Command: Create Library
   This item displays the Create Cell Library dialog box. From this dialog box, you can create a new cell library.

Item Selection
   On sidebar menu pick: n/a
   For window, point to: File Cell Library New... Name:
   On paper menu see: n/a
Command Window Prompts:
   1. Select the item
   2. Click in the Name window
   3. Enter the file name
   4. Click on OK
   To exit item: Click on OK or Cancel
See Also: ATTACH LIBRARY

CREATE SHAPE AUTOMATIC
CRE S A

Drawing Tool: Automatic Create Complex Shape
   Provides an easy way to associate a group of individual elements into a complex, closed shape. The tool asks for a chain tolerance. This tolerance is the distance between elements making the shape. If no tolerance is given, this tool searches for elements that are touching.
   After the tolerance key-in, place the cursor on the first element to be identified. Continue to click input points at the same location. Watch the CREATE SHAPE tool hunt for the elements that touch each other to form the closed shape.

Item Selection
   On sidebar menu pick: Constr Shape Auto

For window, point to: P̲alettes   M̲ain   C̲hain

On paper menu see: CHAIN
Command Window Prompts:
1. Select the item
   `Enter chain tolerance`
2. Enter tolerance in working units
   ⇨   No prompt appears yet.
3. Pick an element
   `Accept/Reject (select next input)`
4. Give an accept point
   ⇨   Continue picking points until the shape closes.
   `Shape Closed`
To exit item: **Make next selection**

✍ *Note:* This option is quicker and easier to use than the CREATE SHAPE MANUAL selection.
See Also: CREATE CHAIN tools, CREATE SHAPE MANUAL

CREATE SHAPE EOF

Drawing Tool: Create Shape EOF
   The CREATE SHAPE EOF tool creates a closed shape. For example, you place a collection of elements in the drawing file. Next set the working window with WWSECT and WWBYTE at the element where the chain is to begin. Then execute the CREATE SHAPE EOF tool from inside a user command. A chain will be created from that element to the end of the file (EOF).
See Also: CREATE SHAPE tools, USERCOMMAND

CREATE SHAPE MANUAL   CRE S

Drawing Tool: Create Complex Shape
   Builds a closed shape by prompting you to select each element you want to include in the shape. Identify each element in the complex shape with an input point. To close the shape, the first element must be identified a second time.
Item Selection
On sidebar menu pick: | Constr | Shape | Man |
For window, point to: P̲alettes   M̲ain   C̲hain

On paper menu see: CHAIN
Command Window Prompts:
1. Select the item
   `Identify element`
2. Pick an element
   `Accept/Reject (select next input)`
3. Give an accept point
   ⇨   Continue identifying elements of the complex shape.
   `Shape Closed`
To exit item: **Make next selection**

✍ *Note:* This option is helpful when CREATE SHAPE AUTOMATIC has trouble identifying the right elements needed to form the shape.
See Also: CREATE SHAPE AUTOMATIC, CREATE CHAIN tools

CROSSHATCH CRO

Drawing Tool: Crosshatch Element Area

The CROSSHATCH tool creates a crosshatch pattern inside a closed shape. This tool expects two input points. The first point identifies the shape to pattern and the second point starts the pattern process.

If you want to adjust the spacing between the crosshatch lines, change the ACTIVE PATTERN DELTA (PS=). You can also set the angle of the pattern with the ACTIVE PATTERN ANGLE (PA=) key-in.

Item Selection
On sidebar menu pick: `Pattrn`
For window, point to: Palettes Patterning

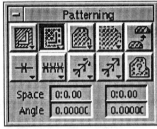

On paper menu see: PATTERNING
Command Window Prompts:
1. Select the item
   Identify element
   Accept/Reject (select next input)
2. Give an accept point
   Patterning in progress...
To exit item: Make next selection
See Also: ACTIVE PATTERN settings, HATCH, PATTERN AREA tools

CT=

Alternate Key-in

Alternate key-in for ATTACH COLORTABLE. The attached colortable determines the color scheme displayed on the screen.
See Also: ATTACH COLORTABLE

Cut All Component Lines

Drawing Tool: Cut All Component Lines

Breaks apart all the lines of a multi-line element. This MDL application tool requires an input point to identify the multi-line and a second point to break the element to two parts.

Item Selection
On sidebar menu pick: n/a
For window, point to: Palettes Multi-line Joints

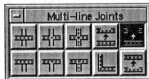

On paper menu see: n/a
Command Window Prompts:
1. Select the item
   Identify element
2. Pick first multi-line element
   ⇨ First element highlights, no new prompt

3. Pick second multi-line element
  ⇨ Second element highlights, no new prompt
To exit item: Make next selection

✍ *Note:* This MDL application is loaded automatically when you start the design session.
See Also: Cut Single Component Line, PLACE MLINE

Cut Single Component Line

Drawing Tool: Cut Single Component Line
  The Cut Single Component Line tool breaks a single line of a multi-line element apart. This MDL application tool requires an input point to identify the multi-line and a second point to bring the element to two parts.

Item Selection
  On sidebar menu pick: n/a
  For window, point to: Palettes   Multi-line Joints

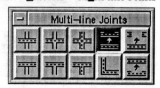

On paper menu see: n/a
Command Window Prompts:
1. Select the item
  Identify element
2. Pick first multi-line element
  ⇨ First element highlights, no new prompt
3. Pick second multi-line element
  ⇨ Second element highlights, no new prompt
To exit item: Make next selection

✍ *Note:* This MDL application is loaded automatically when you start the design session.
See Also: Cut All Component Lines, PLACE MLINE

Cutter Tools

MDL Application
  Use the Cutter tools to modify multi-line elements. This MDL application is loaded automatically when you start the design session.
Choices:

Closed Cross Joint	closes a cross joint between two multi-lines.
Closed Tee Joint	closes a tee joint between two multi-lines.
Corner Joint	creates a corner joint between two multi-lines.
Cut All Component Lines	severs all multi-line members.
Cut Single Component Line	cuts a single multi-line member.
Merged Cross Joint	merges a cross joint between two multi-lines.
Merged Tee Joint	merges a tee joint between two multi-lines.
Open Cross Joint	opens a cross joint between two multi-lines.
Open Tee Joint	opens a tee joint between two multi-lines.
Uncut Component Lines	mends all members of a multi-line.

✍ *Note:* This MDL application is loaded automatically when you start the design session.
See Also: PLACE MLINE

DA=

Alternate Key-in
Alternate key-in for ACTIVE DATYPE. It allows you to associate database
information to text nodes.
See Also: ACTIVE DATYPE

DB=

Alternate Key-in
Alternate key-in for ACTIVE DATABASE. Use to attach a database file to
the active drawing file.
See Also: ACTIVE DATABASE

DD=

Alternate Key-in
Alternate key-in for SET DDEPTH RELATIVE. Alters the display depth in
a three dimensional drawing file. The format of the key-in is DD=[MU:SU:PU,
MU:SU:PU].
See Also: SET DDEPTH RELATIVE

DEFINE ACS ELEMENT   DEF ACS E

Drawing Tool: Define ACS (Aligned with Element)
Conforms the Auxiliary Coordinate System (ACS) with a planar element in
the drawing file. This item requires two input points. The first point identifies the
target element and the second point accepts that element as the candidate.
Item Selection
On sidebar menu pick: Params   ACS   Tools
For window, point to: Palettes   Auxiliary Coordinates

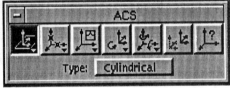

On paper menu see: ACS
Command Window Prompts:
1. Select the item
   Identify element
   Pick an element
   Accept/Reject (select next input)
2. Give an accept point
To exit item: Exits by itself
See Also: DEFINE ACS settings

DEFINE ACS ELEMENT CYLINDRICAL   DEF ACS E C

Drawing Tool: Define Cylindrical ACS (Aligned with Element)
Conforms the Auxiliary Coordinate System (ACS) with a planar element in
the drawing file. This item requires two input points. The first point identifies the
target element, and the second accepts that element as the candidate.
Item Selection
On sidebar menu pick: Params   ACS   Tools

For window, point to: Palettes   Auxiliary Coordinates

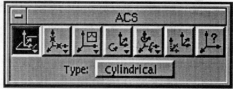

On paper menu see: ACS
Command Window Prompts:
1. Select the item
   Identify element
   Pick an element
   Accept/Reject (select next input)
2. Give an accept point
To exit item: Exits by itself
See Also: DEFINE ACS settings

DEFINE
ACS ELEMENT RECTANGULAR   DEF ACS E R

Drawing Tool: Define Rectangular ACS (Aligned with Element)
   Conforms the Auxiliary Coordinate System (ACS) with a planar element in
the drawing file. This item requires two input points. The first point identifies the
target element, and the second accepts that element as the candidate.
Item Selection
On sidebar menu pick: Params   ACS   Tools
For window, point to: Palettes   Auxiliary Coordinates

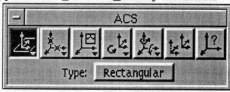

On paper menu see: ACS
Command Window Prompts:
1. Select the item
   Identify element
   Pick an element
   Accept/Reject (select next input)
2. Give an accept point
To exit item: Exits by itself
See Also: DEFINE ACS settings

DEFINE ACS ELEMENT SPHERICAL   DEF ACS E S

Drawing Tool: Define Spherical ACS (Aligned with Element)
   Conforms the Auxiliary Coordinate System (ACS) with a planar element in
the drawing file. This item requires two input points. The first point identifies the
target element, and the second accepts that element as the candidate.
Item Selection
On sidebar menu pick: Params   ACS   Tools

For window, point to: Palettes   Auxiliary Coordinates

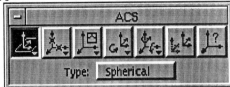

On paper menu see: ACS
Command Window Prompts:
1. Select the item
   Identify element
   Pick an element
   Accept/Reject (select next input)
2. Give an accept point
To exit item: Exits by itself
See Also: DEFINE ACS settings

DEFINE ACS POINTS   DEF ACS P C

Drawing Tool: Define ACS (By Points)
Adjusts the Auxiliary Coordinate System (ACS) with a set of inputs placed in the drawing file. This item requires three input points. The first point identifies the system origin and the second defines the direction of the positive X-axis. A third point is needed to define the positive Y-axis direction in three dimensional files.

Item Selection
On sidebar menu pick: Params | ACS | Tools
For window, point to: Palettes   Auxiliary Coordinates

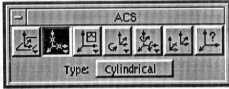

On paper menu see: ACS
Command Window Prompts:
1. Select the item
   Enter first point @X-axis origin
2. Enter first input point
   Enter second point on X-axis
3. Enter second input point
   Enter point to define Y-axis
4. Enter a point on the Y-axis
To exit item: Make next selection
See Also: DEFINE ACS settings

DEFINE ACS POINTS CYLINDRICAL   DEF ACS P C

Drawing Tool: Define Cylindrical ACS (By Points)
Adjusts the Auxiliary Coordinate System (ACS) with a set of inputs placed in the drawing file. This item requires three input points. The first point identifies the system origin, and the second defines the direction of the positive X-axis. A third point is needed to define the positive Y-axis direction in three dimensional files.

Item Selection
On sidebar menu pick: Params | ACS | Tools

For window, point to: <u>P</u>alettes <u>A</u>uxiliary Coordinates

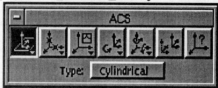

On paper menu see: ACS
Command Window Prompts:
1. Select the item
   First point on X-axis origin
2. Enter first input point
   Second point on X-axis
3. Enter second input point
   Point to define Y-axis
4. Enter a point on the Y-axis
To exit item: **Make next selection**
See Also: DEFINE ACS settings

DEFINE ACS POINTS RECTANGULAR   DEF ACS P R

Drawing Tool: Define Rectangular ACS (By Points)
   Adjusts the Auxiliary Coordinate System (ACS) with a set of inputs placed
in the drawing file. This item requires three input points. The first point identifies
the system origin, and the second defines the direction of the positive X-axis. A
third point is needed to define the positive Y-axis direction in three dimensional
files.

Item Selection
   On sidebar menu pick: Params | ACS | Tools
   For window, point to: <u>P</u>alettes   <u>A</u>uxiliary Coordinates

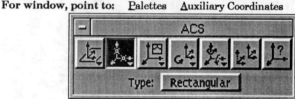

On paper menu see: ACS
Command Window Prompts:
1. Select the item
   First point on X-axis origin
2. Enter first input point
   Second point on X-axis
3. Enter second input point
   Point to define Y-axis
4. Enter a point on the Y-axis
To exit item: **Make next selection**
See Also: DEFINE ACS settings

DEFINE ACS POINTS SPHERICAL   DEF ACS P S

Drawing Tool: Define Spherical ACS (By Points)
   Adjusts the Auxiliary Coordinate System (ACS) with a set of inputs placed
in the drawing file. This item requires three input points. The first point identifies
the system origin and the second defines the direction of the positive X-axis. A
third point is needed to define the positive Y-axis direction in three dimensional
files.

Item Selection
   On sidebar menu pick: Params | ACS | Tools

For window, point to: Palettes   Auxiliary Coordinates

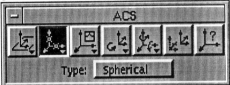

On paper menu see: ACS
Command Window Prompts:
  1. Select the item
     First point on X-axis origin
  2. Enter first input point
     Second point on X-axis
  3. Enter second input point
     Point to define Y-axis
  4. Enter a point on the Y-axis
To exit item: Make next selection
See Also: DEFINE ACS settings

DEFINE ACS VIEW   DEF ACS V C

Drawing Tool: Define ACS (Aligned with View)
     Adjusts the Auxiliary Coordinate System (ACS) with an existing view. This
item requires one input point to identify the target view and set the ACS origin
point.

Item Selection
  On sidebar menu pick: Params   ACS   Tools
  For window, point to: Palettes   Auxiliary Coordinates

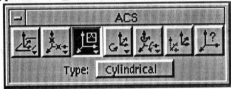

On paper menu see: ACS
Command Window Prompts:
  1. Select the item
     Select source view
  To exit item: Make next selection
See Also: DEFINE ACS settings

DEFINE ACS VIEW CYLINDRICAL   DEF ACS V C

Drawing Tool: Define Cylindrical ACS (Aligned with View)
     Adjusts the Auxiliary Coordinate System (ACS) with an existing view. This
item requires one input point to identify the target view and set the ACS origin
point.

Item Selection
  On sidebar menu pick: Params   ACS   Tools
  For window, point to: Palettes   Auxiliary Coordinates

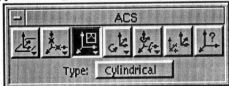

On paper menu see: ACS

Command Window Prompts:
1. Select the item
   Select source view
To exit item: Make next selection
See Also: DEFINE ACS settings

DEFINE ACS VIEW RECTANGULAR   DEF ACS V R

Drawing Tool: Define Rectangular ACS (Aligned with View)

Adjusts the Auxiliary Coordinate System (ACS) with an existing view. This item requires one input point to identify the target view and set the ACS origin point.

Item Selection

On sidebar menu pick: | Params | | ACS | | Tools |

For window, point to: Palettes   Auxiliary Coordinates

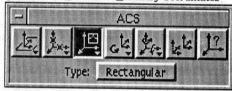

On paper menu see: ACS
Command Window Prompts:
1. Select the item
   Select source view
To exit item: Make next selection
See Also: DEFINE ACS settings

DEFINE ACS VIEW SPHERICAL   DEF ACS V S

Drawing Tool: Define Spherical ACS (Aligned with View)

Adjusts the Auxiliary Coordinate System (ACS) with an existing view. This item requires one input point to identify the target view and set the ACS origin point.

Item Selection

On sidebar menu pick: | Params | | ACS | | Tools |

For window, point to: Palettes   Auxiliary Coordinates

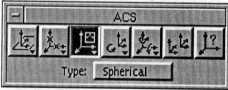

On paper menu see: ACS
Command Window Prompts:
1. Select the item
   Select source view
To exit item: Make next selection
See Also: DEFINE ACS settings

DEFINE AE   DEF AE

Drawing Tool: Define Active Entity Graphically

The DEFINE AE tool is a database equivalent to the FIND setting. This item's duty is to define an entity in a database, after which you attach it to a graphic element, and you are ready to edit.

Item Selection

On sidebar menu pick: | dBase | | ActEnt | | Define |

For window, point to: Palettes   Database

On paper menu see: DATABASE
Command Window Prompts:
1. Select the item
   Identify element
2. Pick an element
   Accept/Reject (select next input)
3. Give an accept point
To exit item: Make next selection
See Also: CREATE ENTITY, FIND, SHOW AE

DEFINE CELL ORIGIN   DEF C

Drawing Tool: Define Cell Origin
    Defines a cell's origin point for placement. When defining the origin of a cell with an input point, a small letter O appears on the display screen. After the cell is filed in the library, the letter O disappears.
    A cell can be any standard symbol or feature that repeats itself throughout a drawing or set of drawings. Define the cell's origin after creating the graphics. The cell origin is a point that becomes the center of the cell placement.

Item Selection
  On sidebar menu pick: Cells   Origin
  For window, point to: Palettes   Main   Cells

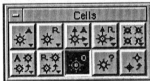

On paper menu see: CELLS
Command Window Prompts:
1. Select the item
   Define origin
2. Pick a point
   ⇨   Places a capitol O at the origin.
To exit item: Make next selection
See Also: ACTIVE CELL, CREATE CELL, PLACE CELL tools

DEFINE LEVELS   DEF LE

Setting: Define Levels
    Setting this item opens the Level Names setting box, through which you can identify levels by names instead of numbers. You can group levels together for manipulation or display purposes. This window also gives you the ability to save or load level structures between drawing files.

Item Selection
  On sidebar menu pick: n/a
  For window, point to: Settings   Level Names
  On paper menu see: n/a
Command Window Prompts:
1. Select the item
2. Make your selections
To exit item: Dismiss the window

See Also: ACTIVE LEVEL, SET LEVELS

DEFINE LIGHTS   DEF LI

Command: Define Lights
    Scan the drawing file for light cells. Once found, MicroStation redefines the
lighting according to the position of the light cells. If you place light cells after a
view is rendered, you must key in DEFINE LIGHTS for the rendering software to
recognize the new light definition.

Item Selection
   On sidebar menu pick: n/a
   For window, point to: n/a
   On paper menu see: n/a
See Also: DEFINE MATERIALS

DEFINE MATERIALS   DEF M

Command: Define Materials
    Rescans the material table to locate changes since the last rendering.
MicroStation uses the material finishes when rendering objects. For example, a
material finish of 0.4 might be used to show a flat or dull surface.

Item Selection
   On sidebar menu pick: n/a
   For window, point to: n/a
   On paper menu see: n/a
See Also: DEFINE LIGHTS

DEFINE NORTH   DEF N

Setting: Define True North
    Allows you to choose a true north direction other than the Y-axis. The north
direction can be any X, Y direction. This item requires two input points, and a
key-in. The first point starts a north line, and the second point defines the line's
end point. If this line defines north, enter a 0 at the prompt, or enter the angle
relative to the line.

Item Selection
   On sidebar menu pick: n/a
   For window, point to: n/a
   On paper menu see: UTILITIES
Command Window Prompts:
   1. Select the item
      Enter true north line start point
   2. Pick a point
      Enter true north end point
   3. Pick line's endpoint
      Enter angle relative to true north
   4. Enter the north value
   To exit item: **Exits by itself**
See Also: DIGITIZING tools

DEFINE SEARCH   DS=

Setting: Define Search
    DEFINE SEARCH is a database related setting that either restricts or
establishes the search criteria for elements being considered during a fence
operation. DEFINE SEARCH is also known as a "Fence Filter".

Item Selection
   On sidebar menu pick: n/a
   For window, point to: n/a
   On paper menu see: n/a
   To disable filter file, key in: **DEFINE SEARCH <database:>**
      Fence Filter : OFF
   To enable filter file, key in: **DEFINE SEARCH <database:filterfile>**

Where database is a valid database file and filterfile an active filter.

Example(s):
```
DEFINE SEARCH USCITIES:LARGE.FIL
```
See Also: FENCE REPORT

DELETE 66ELEMENTS ALL — DEL 66 A

Command: Remove All 66Elements
Strips out certain records placed in a drawing file by MicroStation and other applications (IGDS for example); i.e. 2D reference files attached to 3D drawing files, level names, view orientations and dimension data. DELETE 66ELEMENTS ALL strips out these elements.

Item Selection
On sidebar menu pick: n/a
For window, point to: n/a
On paper menu see: n/a

✍ *Note:* Set the MS_NO66 environment variable and MicroStation will not create any new Type 66 elements. It will also ignore any existing Type 66 elements already in the drawing file.
See Also: DELETE 66ELEMENTS commands

DELETE 66ELEMENTS LEVELNAME — DEL 66 L

Command: Remove Levelname 66Elements
Strips out names assigned to levels in the drawing file by MicroStation and other applications (IGDS for example); i.e. 2D reference files attached to 3D drawing files, level names, view orientations and dimension data.

Item Selection
On sidebar menu pick: n/a
For window, point to: n/a
On paper menu see: n/a

✍ *Note:* Set the MS_NO66 environment variable and MicroStation will not create any new Type 66 elements. It will also ignore any existing Type 66 elements already in the drawing file.
See Also: DELETE 66ELEMENTS commands

DELETE 66ELEMENTS MS — DEL 66 A

Command: Remove MicroStation 66Elements
Strips out certain control elements placed in a drawing file by MicroStation and other applications (IGDS for example); i.e. 2D reference files attached to 3D drawing files, level names, view orientations and dimension data.

Item Selection
On sidebar menu pick: n/a
For window, point to: n/a
On paper menu see: n/a

✍ *Note:* Set the MS_NO66 environment variable and MicroStation will not create any new Type 66 elements. It will also ignore any existing Type 66 elements already in the drawing file.
See Also: DELETE 66ELEMENTS commands

DELETE 66ELEMENTS REFERENCE — DEL 66 R

Command: Remove Reference 66Elements
Strips out certain reference file attachments above 32 and 2D files attached to 3D drawings by MicroStation and other applications (IGDS for example).

Item Selection
On sidebar menu pick: n/a
For window, point to: n/a
On paper menu see: n/a

✍ *Note:* Set the MS_NO66 environment variable and MicroStation will not create any new Type 66 elements. It will also ignore any existing Type 66 elements already in the drawing file.

See Also: DELETE 66ELEMENTS commands

DELETE 66ELEMENTS START   DEL 66 S

Command: Remove Start 66Elements
        Start records can be placed in the drawing file to activate an application program when the drawing file is opened. This command strips out the start records.

Item Selection
        On sidebar menu pick: n/a
        For window, point to: n/a
        On paper menu see: n/a

✍ *Note:* Set the MS_NO66 environment variable and MicroStation will not create any new Type 66 elements. It will also ignore any existing Type 66 elements already in the drawing file.

See Also: DELETE 66ELEMENTS commands

DELETE 66ELEMENTS TCB   DEL 66 T

Command: Delete 66Elements TCB
        Strips out certain dimension and digitizing data and extended Terminal Control Block (TCB) information placed in the drawing file by MicroStation and other applications (IGDS for example).

Item Selection
        On sidebar menu pick: n/a
        For window, point to: n/a
        On paper menu see: n/a

✍ *Note:* Set the MS_NO66 environment variable and MicroStation will not create any new Type 66 elements. It will also ignore any existing Type 66 elements already in the drawing file.

See Also: DELETE 66ELEMENTS commands

DELETE 66ELEMENTS VIEW   DEL 66 V

Command: Delete 66Elements View
        Strips out certain view orientation elements placed in the drawing file by MicroStation and other applications (IGDS for example).

Item Selection
        On sidebar menu pick: n/a
        For window, point to: n/a
        On paper menu see: n/a

✍ *Note:* Set the MS_NO66 environment variable and MicroStation will not create any new Type 66 elements. It will also ignore any exist Type 66 elements already in the drawing file.

See Also: DELETE 66ELEMENTS commands

DELETE ACS   PX=

Command: Delete ACS
        Deletes a previously stored Auxiliary Coordinate System.

Item Selection
        On sidebar menu pick: | Params | | ACS | | Set |
        For window, point to: Settings   Auxiliary Coordinates   Delete
        On paper menu see: n/a

See Also: ATTACH ACS, DEFINE ACS settings

DELETE CELL   CD=

Command: Delete Cell
    Removes cells from a cell library attached to the active drawing file.

Item Selection
  On sidebar menu pick: `Cells` `Delete`
  For window, point to: Settings Cells Delete
  On paper menu see: n/a
Command Window Prompts:
  1. Select the item
    `<- Cell to Delete ?`
  2. Enter cell name to delete
    `Cell uuu deleted`
      Where uuu is a user defined cell name.

  To exit item: Make next selection
See Also: CREATE CELL, PLACE CELL tools, RENAME CELL, REPLACE CELL

DELETE ELEMENT   DEL

Drawing Tool: Delete Element
    Removes elements from the drawing file. This tool deletes single elements, graphic groups or complex strings of elements. It requires two input points. The first point highlights the element to delete, and the second confirms the delete selection.

Item Selection
  On sidebar menu pick: `Manip` `Delete` `Elem`
  For window, point to: Palettes Main

  On paper menu see: MIRROR ELEMENT
Command Window Prompts:
  1. Select the item
    `Identify element`
  2. Pick an element
    `Accept/Reject (select next input)`
  3. Give an accept point
  To exit item: Make next selection

☞ *Tip:* Normally, deleting four elements would take eight input points – four points to identify the elements and four more to confirm their deletion. But the DELETE ELEMENT tool is a cyclical tool. This means you may use the first point to select one element and the second point can be placed on a new element. This second point acts as the confirm point for the first element and the selection point for the new element. This way, you only need to place five points to delete four elements.

See Also: DELETE PARTIAL

DELETE PARTIAL

DEL P

Drawing Tool: Delete Part of Element

Takes a bite out of an element. If you take a bite out of a shape, it turns into an open element; circles and ellipses turn into arcs when partially deleted. Lines and line strings remain in their original element types. This item requires two input points. The first point highlights the element to modify, and the second determines the size of the bite.

Item Selection

On sidebar menu pick: | Manip | | Delete | | Partl |

For window, point to: Palettes Main Modify Element

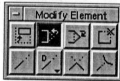

On paper menu see: MODIFY

Command Window Prompts:

1. Select the item
   Select start point for partial
   delete
2. Pick a point
   Select end pnt of partial delete
   ⇨ If the element is a closed shape, the tool asks for direction.
3. Pick a second point

To exit item: Make next selection

☞ *Tip:* It is easier to place a circle and take a chomp out of it with the DELETE PARTIAL tool than placing an arc or partial ellipse.

See Also: DELETE ELEMENT

DELETE VERTEX

DEL VER

Drawing Tool: Delete Vertex

Removes a vertex from a line string or shape. This item requires an input point to identify the target element's vertex, and a second point to confirm the removal of the vertex.

Item Selection

On sidebar menu pick: | Modify | | Vertex | | Del |

For window, point to: Palettes Main Modify Element

On paper menu see: MODIFY

Command Window Prompts:

1. Select the item
   Identify element
2. Pick an element
   Accept/Reject (select next input)
3. Give an accept point

To exit item: Make next selection

See Also: INSERT VERTEX

DELETE VIEW

View Control: Delete View
> Removes named views saved with the drawing file.

Item Selection
On sidebar menu pick: `Window` `View` `Name` `Del`
For window, point to: Y̲iew   Sa̲ved   Delete
On paper menu see: n/a
Command Window Prompts:
1. Select the item
   `<- Name ?`
2. Enter view name uuu to delete
   Where uuu is a user defined view name.
   `Named view deleted`
To exit item: Make next selection
See Also: SAVE VIEW, SHOW VIEWS

DEPTH ACTIVE
DEP A

3D Setting: Set Active Depth
> Changes the active depth in your 3D drawing. This setting asks for one input point to pick the view for change. A second input point defines the depth for that view.

Item Selection
On sidebar menu pick: `Place` `3-D`
For window, point to: P̲alettes   3̲D

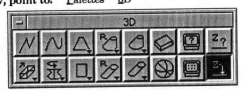

On paper menu see: 3D
Command Window Prompts:
1. Select the item
   `Select view`
2. Pick the view
   `Enter active depth point`
3. Pick the active depth
To exit item: Make next selection
See Also: ACTIVE ZDEPTH settings, SHOW DEPTH ACTIVE

DEPTH ACTIVE PRIMITIVE

Obsolete
> Controls the active depth of a view along the Z-axis in a 3D drawing file. This control expects two input points. The first point selects the depth and the second point declares the view.

See Also: DEPTH ACTIVE

DEPTH DISPLAY
DEP D

3D Setting: Set Display Depth
> Changes the displayed depth in your 3D drawing. This setting asks you for one input point to pick the view for change, and two more to define the front and back clipping planes for that view.

Item Selection
On sidebar menu pick: `Place` `3-D`

For window, point to: Palettes   3D

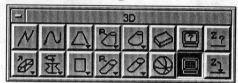

On paper menu see: 3D
Command Window Prompts:
1. Select the item
   Select view for display depth
2. Pick the view
   Define front clipping plane
3. Pick the front plane
   Define back clipping plane
4. Pick back plane
To exit item   **Make next selection**
See Also: SET DDEPTH settings, SHOW DEPTH DISPLAY

DEPTH DISPLAY PRIMITIVE

Obsolete
   Selects a portion of the three dimensional object for display.
See Also: DEPTH DISPLAY, SET DDEPTH settings

DESIGN

Obsolete
   Opens the Design Options window.
See Also: TUTORIAL

DETACH   DET

Drawing Tool: Detach Database Linkage
   Removes database linkages from elements. This drawing tool requires two
input points. The first point identifies the target element, and the second serves
as an accept point.
Item Selection
On sidebar menu pick: dBase   Tools
For window, point to: Palettes   Database

On paper menu see: DATABASE
See Also: ATTACH AE, FENCE DETACH

DF=

Alternate Key-in
   Alternate key-in for SHOW FONT. Use it to display the fonts stored in the
current font library.
See Also: SHOW FONT

DI=

Alternate Key-in
Alternate key-in for POINT DISTANCE. It places an input point some known distance and direction. The command calculates the distance and measures the direction in degrees counterclockwise from the horizontal position of the last point.
See Also: POINT DISTANCE

DIALOG

Command: Dialog
DIALOG choice opens the various dialog and settings boxes. These windows allow you to set, view or select different controls. For example, keying in DIALOG LOCKS displays the LOCKS settings box.

Choices:

ABOUTUSTN	Displays Software version/Drawing file info
ACTIVEANGLE	Opens Active Angle settings box
ACTIVESCALE	Opens Active Scale settings box
ANALYZE	Opens Element Information settings box
ANIMATE	Opens Animation settings box
ATTRIBUTES	Opens Element Attributes settings box
CAMERA	Opens Camera Lens settings box
CELLMAINTENANCE	Opens Cells settings box
CLIPBOARD	Shows clipboard (Mac only)
CMDBROWSE	Opens Key-in Browser window
COLOR	Opens Color Palette settings box
COMMAND	Opens MicroStation Command Window
COORDSYS	Opens Auxiliary Coordinate Systems (ACS) settings box
COPY	Opens Paste/copy settings box (Mac only)
DATABASE	Opens Database settings box
DGNOPTS	Opens Design Options settings box (obsolete)
DIGITIZING	Opens Digitizing settings box
DIMATTRIB	Opens Dimension Attributes settings box
DIMENSIONINGFRAME	Opens Dimensioning tool palette
DIMGEOM	Opens Dimension Geometry settings box
DIMPLACE	Opens Dimension Placement settings box
DIMSTRING	Opens Dimension Text Editor Window
DIMSYMBOL	Opens Dimension Custom Symbols settings box
DIMTEMPLATE	Opens Dimension Tool settings box
DIMTERMINATORS	Opens Dimension Custom Terminators settings box
DIMTEXT	Opens Dimension Text Format settings box
DIMTOLERANCE	Opens Dimension Tolerance settings box
DISPLAY	Opens Display Image File dialog box
ENVIRONMENT	Opens Environment Variables settings box
FENCEFRAME	Opens Fence tool palette
FONT	Opens Fonts settings box
FUNCKEYS	Opens Function keys settings box
GRID	Opens Grid settings box
IMAGE	Opens window for displaying previous saved images
IMAGESAVE	Opens Save Image dialog box
KEY	Opens Network Access Key dialog box
LEVELSYMB	Opens Level Symbology dialog box
LOCKS	Opens Locks settings box
LOCKTOGGLES	Opens Lock Toggles settings box
MAINFRAME	Opens Main tool palette

MDL	Opens MDL settings box
MULTILINE	Opens Multi-lines settings box
NAMEDLEVELS	Opens Level Names settings box
NAMEDVIEWS	Opens Saved Views settings box
NEW	Opens Create Design File dialog box
OPEN	Opens Open Design File dialog box
PALETTE	Palette Opens that sub-palette
ACS	Opens Auxiliary Coordinate Systems tool palette
ANGULARDIMEN	Opens Angular Dimensions tool palette
ARC	Opens Arc tool palette
CELL	Opens Cells tool palette
CHAIN	Opens Chain tool palette
CHANGE	Opens Change Element tool palette
CIRCLE	Opens Circle/Ellipse tool palette
COPYELEMENT	Opens Copy Element tool palette
DATABASE	Opens Database tool palette
DROP	Opens Drop Element tool palette
ENTERDATA	Opens Enter Data Fields tool palette
FENCECHANGE	Opens Fence Change tool palette
FENCECOPY	Opens Fence Copy tool palette
FENCEDELETE	Opens Fence Delete tool palette
FENCEMIRROR	Opens Fence Mirror tool palette
FILLET	Opens Fillets tool palette
LINE	Opens Lines tool palette
LINEARDIMEN	Opens Linear Dimensions tool palette
LINESTRING	Opens Line Strings tool palette
MEASURE	Opens Measure tool palette
MIRRORELEMENT	Opens Mirror Element tool palette
MISCDIMEN	Opens Miscellaneous Dimensions tool palette
MODIFY	Opens Modify Element tool palette
PATTERNING	Opens Patterning tool palette
PLACEFENCE	Opens Place Fence tool palette
POINT	Opens Points tool palette
POLYGON	Opens Polygons tool palette
RADIALDIMEN	Opens Radial Dimensions tool palette
REFERENCEFILE	Opens Reference Files tool palette
TEXT	Opens Text tool palette
THREED	Opens 3D tool palette
VIEWCONTROL	Opens View Control tool palette
PASTE	Opens Paste/copy settings box (Mac only)
PATTERN	Opens Patterning settings box
PLOT	Opens Preview Plot dialog box
PRECISION	Opens Precision Input settings box
READOUT	Opens Coordinate Readout settings box
REFERENCE	Opens Reference Files settings box
REFERENCEATTACH	Opens Reference File Attachment dialog box
REFERENCEROTATE	Opens Rotate Reference File dialog box
REFERENCESCALE	Opens Scale Reference File dialog box
RENDER	Opens Rendering settings box
ROTATEACS	Opens Rotate Active ACS dialog box
SAVEAS	Opens save Design File dialog box
SHOWPLOTTER	Opens Show Plotter settings box (obsolete)
STANDARDALERT	Opens standard alert box
SPLINES	Opens B-splines setting box
TEXT	Opens Text settings box
TEXTEDIT	Opens Text Editor window

TEXTFILEDISPLAY	Opens text file window
TOOLSETTINGS	Opens Tool Settings window
UNITS	Opens Working Units settings box
USERPREF	Opens Preferences dialog box
VIEWLEVELS	Opens View Levels settings box
VIEWROTATION	Opens View Rotation settings box
VIEWSETTINGS	Opens View Attributes settings box

Digitize

Definition

Digitizing is the process of tracing existing paper drawings into the CAD system. This is done by registering the drawing on a digitizing tablet with the aid of the DIGITIZER setup commands. The input device is a cursor instead of a mouse. The graphics on the paper drawing are then traced into an empty drawing file.

See Also: ACTIVE STREAM settings, DEFINE NORTH, DIGITIZER commands and settings

DIGITIZER BUTTONS   SET BU

Setting: Remap Digitizer Buttons

Changes the functions of the Data, Reset and Tentative buttons on puck or mouse.

Item Selection

On sidebar menu pick: `Params` `Set`   `Cursor`

For window, point to: Settings Digitizing Tablet   Button Assignment

On paper menu see: n/a

Command Window Prompts:

1. Select the item
   `Press Data button`
2. Pick the button you want assigned
   `Press Reset button`
3. Pick the button you want assigned
   `Press Tentative button`
4. Pick the button you want assigned
   `When all button are assigned, hit Return`
5. Hit the Enter key

To exit item: Exits by itself

See Also: SET BUTTON

DIGITIZER PARTITION   DIG P

Setting: Partition Tablet Surface

Divides the digitizing table area into two divisions or partitions. One partition, or area, maps the screen partition. For example, if you had a large digitizer, say 36" by 48", you would have to stretch quite a distance to activate the commands on the sidebar menu. If, on the other hand, you setup a digitizer partition 15" by 15", the cursor would track much faster across the display screen. Use the remaining area for digitizing.

By default, the lower left corner of the tablet is for the screen mapping partition.

Item Selection

On sidebar menu pick: `Utils`   `Digitz`

For window, point to: Settings Digitizing Tablet   Partition

On paper menu see: UTILITIES

See Also: DIGITIZER SETUP

DIGITIZER SETUP

Setting: Define Monument Points

Orients the paper drawing you are digitizing to the drawing file on the screen. During the execution of this setting, identify a series of points on the paper drawing and key them into the drawing file. A minimum of two registration points can align the drawing; use more points to gain greater accuracy. After the drawing and design files are aligned, move the cursor around on the digitizing tablet and check the corresponding locations on the display screen.

Item Selection

On sidebar menu pick: Utils   Digitz

For window, point to: Settings Digitizing Tablet Setup

On paper menu see: UTILITIES

See Also: DIGITIZER PARTITION

DIMENSION ANGLE LINES   DIM AN LI

Drawing Tool: Dimension Angle Between Lines

The DIMENSION ANGLE LINES tool places an angular dimension between two lines. Express the dimension of the arc in length or degrees. (See the DIMENSION UNITS settings.) This tool expects three input points. The first two points highlight the target angle legs. The third point determines the placement of the dimension data. If the dimensioned distance needs changing, place the dimension, then hit Enter. The Dimension Text Editor window will appear. Enter the new dimension in the TEXT area, and click OK.

Item Selection

On sidebar menu pick: n/a

For window, point to: Palettes Dimensioning Angular Dimensions

On paper menu see: ANGULAR DIMENSION

☞ *Tip:* Dimensions are grouped together as single dimension elements. If you need to modify part of the dimension, use a MODIFY tool or DROP DIMENSION first. This turns the dimension data into a graphic group. LOCK GGROUP must be off to affect one part of the dimension. Turn LOCK GGROUP on th have the dimension act as one element. Freezing dimensions or setting compatibility mode to 3X also changes dimension elements into graphic groups.

See Also: DIMENSION ANGLE tools

DIMENSION ANGLE LOCATION   DIM AN L

Drawing Tool: Dimension Angle Location

Dimensions an angle in angular degrees or length. (See the DIMENSION UNITS settings.) This tool expects four input points. The first point determines one leg of the angle. The second point picks the length of the dimension witness lines. The third defines the angle's vertex or center, and the last point selects the second leg of the angle.

Place additional points, and the angular distance accumulates to form a new dimension. If the dimensioned distance needs changing, place the dimension, then hit Enter. The Dimension Text Editor window will appear. Enter the new dimension in the TEXT area, and click OK.

Item Selection

On sidebar menu pick: Dimens   Angle   Locate

For window, point to: Palettes Dimensioning Angular Dimensions

On paper menu see: ANGULAR DIMENSION

Command Window Prompts:

1. Select the item

   Select start of dimension

2. Pick an angle le~

```
        Define length of witness line
```
3. Pick a point
```
   Enter center point
```
4. Pick angle vertex
```
   Select dimension endpoint
```
5. Pick other angle leg
To exit item: Make next selection

☞ *Tip:* Dimensions are grouped together as single dimension elements. If you need to modify part of the dimension, use a MODIFY tool or DROP DIMENSION first. This turns the dimension data into a graphic group. LOCK GGROUP must be off to affect one part of the dimension. Turn LOCK GGROUP on th have the dimension act as one element. Freezing dimensions or setting compatibility mode to 3X also changes dimension elements into graphic groups.

See Also: DIMENSION ANGLE tools

DIMENSION ANGLE SIZE   DIM AN S

Drawing Tool: Dimension Angle Size
    Scales, or dimensions, an angle in angular degrees or length. (See the DIMENSION UNITS settings.) This tool expects four input points. The first point determines one leg of the angle. The second point picks the length of the dimension witness lines. Place the third point at the angle's vertex or center. The last point selects the second leg of the angle.

    Place additional points and the angular distance resets before placing new dimensions. If the dimensioned distance needs changing, place the dimension, then hit Enter. The Dimension Text Editor window will appear. Enter the new dimension in the TEXT area, and click OK.

Item Selection
On sidebar menu pick: Dimens   Angle   Size
For window, point to: Palettes   Dimensioning   Angular Dimensions

On paper menu see: ANGULAR DIMENSION
Command Window Prompts:
1. Select the item
```
   Select start of dimension
```
2. Pick an angle leg
```
   Define length of witness line
```
3. Pick a point
```
   Enter center point
```
4. Pick angle vertex
```
   Select dimension endpoint
```
5. Pick other angle leg
To exit item: Make next selection

☞ *Tip:* Dimensions are grouped together as single dimension elements. If you need to modify part of the dimension, use a MODIFY tool or DROP DIMENSION first. This turns the dimension data into a graphic group. LOCK GGROUP must be off to affect one part of the dimension. Turn LOCK GGROUP on th have the dimension act as one element. Freezing dimensions or setting compatibility mode to 3X also changes dimension elements into graphic groups.

See Also: DIMENSION ANGLE tools

DIMENSION ANGLE X   DIM AN X

Drawing Tool: Dimension Angle from X-Axis
    Dimensions the angle between a line segment and the X-axis. Express the dimension of the angle in length or degrees. (See the DIMENSION UNITS settings.) This tool expects two input points. The first point identifies the target line, and the second accepts the line for processing.

If the dimensioned distance needs changing, place the dimension, then hit Enter. The Dimension Text Editor window will appear. Enter the new dimension in the TEXT area, and click OK.

Item Selection
On sidebar menu pick: n/a
For window, point to: Palettes   Dimensioning   Angular Dimensions

On paper menu see: ANGULAR DIMENSION

☞ *Tip:* Dimensions are grouped together as single dimension elements. If you need to modify part of the dimension, use a MODIFY tool or DROP DIMENSION first. This turns the dimension data into a graphic group. LOCK GGROUP must be off to affect one part of the dimension. Turn LOCK GGROUP on th have the dimension act as one element. Freezing dimensions or setting compatibility mode to 3X also changes dimension elements into graphic groups.

See Also: DIMENSION ANGLE tools

DIMENSION ANGLE Y   DIM AN Y

Drawing Tool: Dimension Angle from Y-Axis
Dimensions the angle between a line segment and the Y-axis. Express the dimension of the angle in length or degrees. (See the DIMENSION UNITS settings.) This tool expects two input points. The first point identifies the target line and the second accepts the line for processing.

If the dimensioned distance needs changing, place the dimension, then hit Enter. The Dimension Text Editor window will appear. Enter the new dimension in the TEXT area, and click OK.

Item Selection
On sidebar menu pick: n/a
For window, point to: Palettes   Dimensioning   Angular Dimensions

On paper menu see: ANGULAR DIMENSION

☞ *Tip:* Dimensions are grouped together as single dimension elements. If you need to modify part of the dimension, use a MODIFY tool or DROP DIMENSION first. This turns the dimension data into a graphic group. LOCK GGROUP must be off to affect one part of the dimension. Turn LOCK GGROUP on th have the dimension act as one element. Freezing dimensions or setting compatibility mode to 3X also changes dimension elements into graphic groups.

See Also: DIMENSION ANGLE tools

DIMENSION ARCLENGTH   DIM ARCL

Setting: Dimension Arclength
Allows an arc length accent to display above the arc dimension text. This setting works in conjunction with the DIMENSION ARC tools.

Item Selection
On sidebar menu pick: n/a
For window, point to: n/a
On paper menu see: n/a
To show, key in: `DIMENSION ARCLENGTH`
    Length of Arc Accent : ON
To set, key in: `DIMENSION ARCLENGTH on/off`
    Where on/off is either ON or OFF.

✍ *Note:* A dimension tool must be active to use this setting.

See Also: DIMENSION ARC tools

DIMENSION ARC LOCATION   DIM AR L

Drawing Tool: Dimension Arc Location
  Dimensions circles or other circular elements. Express the dimension of the
arc in length or degrees. (See the DIMENSION UNITS settings.) This tool expects
three input points. The first point defines one of the arc's endpoints. The second
point determines the length of the dimension witness lines. The third point defines
the arc's other endpoint.
  Place additional points and the angular distance accumulates before placing
new dimensions. If the dimensioned distance needs changing, place the dimension,
then hit Enter. The Dimension Text Editor window will appear. Enter the new
dimension in the TEXT area, and click OK.

Item Selection
On sidebar menu pick: Dimens Arc Locate
For window, point to: Palettes Dimensioning Angular Dimensions

  On paper menu see: ANGULAR DIMENSION
Command Window Prompts:
 1. Select the item
   Identify element
 2. Pick an arc endpoint
   Define length of witness line
   Select dimension endpoint
 3. Pick other arc endpoint
 To exit item: Make next selection

☞ *Tip:* Dimensions are grouped together as single dimension elements. If you need to
modify part of the dimension, use a MODIFY tool or DROP DIMENSION first.
This turns the dimension data into a graphic group. LOCK GGROUP must be off
to affect one part of the dimension. Turn LOCK GGROUP on th have the dimension
act as one element. Freezing dimensions or setting compatibility mode to 3X also
changes dimension elements into graphic groups.
See Also: DIMENSION ARC SIZE, DIMENSION RADIUS

DIMENSION ARC SIZE   DIM AR S

Drawing Tool: Dimension Arc Size
  Dimensions circles or other circular elements. Express the dimension of the
arc in length or degrees. (See the DIMENSION UNITS settings.) This tool expects
three input points. The first point defines one of the arc's endpoints. The second
point determines the length of the dimension witness lines. The third point defines
the arc's other endpoint.
  Place additional points and the angular distance resets before placing new
dimensions. If the dimensioned distance needs changing, place the dimension,
then hit Enter. The Dimension Text Editor window will appear. Enter the new
dimension in the TEXT area, and click OK.

Item Selection
On sidebar menu pick: Dimens Arc Size
For window, point to: Palettes Dimensioning Angular Dimensions

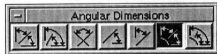

  On paper menu see: ANGULAR DIMENSION
Command Window Prompts:
 1. Select the item

>   Identify element
> 2. Pick an arc endpoint
>   Define length of witness line
>   Select dimension endpoint
> 3. Pick other arc endpoint
> **To exit item: Make next selection**

☞ *Tip:* Dimensions are grouped together as single dimension elements. If you need to modify part of the dimension, use a MODIFY tool or DROP DIMENSION first. This turns the dimension data into a graphic group. LOCK GGROUP must be off to affect one part of the dimension. Turn LOCK GGROUP on th have the dimension act as one element. Freezing dimensions or setting compatibility mode to 3X also changes dimension elements into graphic groups.

See Also: DIMENSION ARC LOCATION, DIMENSION
       RADIUS

DIMENSION AXIS ARBITRARY   DIM AX A

Setting: Dimension Axis Drawing
       Lets you define the angle between the witness line and the dimension line. Normally this angle is 90 degrees. Use this setting to dimension isometric objects. Use this setting with 2D size and location dimension tools.
Item Selection
On sidebar menu pick: n/a
For window, point to: Element Dimensions Placement Alignment:
On paper menu see: n/a
To show, open: Dimension Placement settings box
To set, key in: DIMENSION AXIS DRAWING

☞ *Tip:* LOCK AXIS on works in conjunction with this setting.
See Also: DIMENSION AXIS settings

DIMENSION AXIS DRAWING   DIM AX D

Setting: Dimension Axis Drawing
       Measures linear dimensions parallel to the drawing axis. Rotating the drawing causes the dimension axis to follow one of the drawing axes.
Item Selection
On sidebar menu pick: Dimens   Axis   Draw
For window, point to: Element Dimensions Placement Alignment:
On paper menu see: n/a
To show, open: Dimension Placement settings box
To set, key in: DIMENSION AXIS DRAWING
See Also: DIMENSION AXIS settings

DIMENSION AXIS TRUE   DIM AX T

Setting: Dimension Axis True
       Measures dimensions parallel to the element line being dimensioned.
Item Selection
On sidebar menu pick: Dimens   Axis   True
For window, point to: Element Dimensions Placement Alignment:
On paper menu see: n/a
To show, open: Dimension Placement settings box
To set, key in: DIMENSION AXIS TRUE
See Also: DIMENSION AXIS settings

DIMENSION AXIS VIEW   DIM AX V

Setting: Dimension Axis View
       Measures dimension lines parallel to the X or Y-axis of an active view.
Item Selection
On sidebar menu pick: Dimens   Axis   View

For window, point to: Element Dimensions Placement Alignment:
On paper menu see: n/a
To show, open: **Dimension Placement settings box**
To set, key in: **DIMENSION AXIS VIEW**
See Also: DIMENSION AXIS settings

DIMENSION CENTER   DIM CE

Setting: Dimension Center Mark
Toggles automatic creation of center marks during dimensioning.
Item Selection
On sidebar menu pick: n/a
For window, point to: Element Dimensions Tool Settings
On paper menu see: n/a
See Also: DIMENSION CENTER settings

DIMENSION CENTER MARK   DIM CE M

Drawing Tool: Place Center Mark
Creates a center mark inside circles and circular arcs, with or without center
lines. This tool requires two input points. The first point highlights the target circle
and the second accepts the mark operation. DIMENSION CENTER SIZE controls
the size of the center mark.
Item Selection
On sidebar menu pick: n/a
For window, point to: Palettes Dimensioning Radial

On paper menu see: ANGULAR DIMENSION
See Also: DIMENSION CENTER settings

DIMENSION CENTER SIZE   DIM CE S

Setting: Dimension Center Size
Controls the size of the center mark. Enter the size of the center mark in
working units.
Item Selection
On sidebar menu pick: n/a
For window, point to: Element Dimensions Geometry Center Size:
On paper menu see: n/a
Example(s):
DIM CE S .5   sets the size of the center mark to .5
See Also: DIMENSION CENTER, DIMENSION CENTER MARK

DIMENSION COLOR   DIM CO

Setting: Dimension Color
Controls the color of dimension lines generated in the drawing file. You can
key in a color number, like 1 for blue dimensions, or you can key in the color name,
such as blue. If you key in DIMENSION COLOR ACTIVE, the dimension color
defaults to the active color.
Item Selection
On sidebar menu pick: n/a
For window, point to: Element Dimensions Attributes Color:
On paper menu see: n/a
Example(s):
DIM CO 1   sets the dimension color to blue
See Also: ACTIVE COLOR, ATTACH COLORTABLE

DIMENSION DIAMETER EXTENDED   DIM D E

Drawing Tool: Dimension Diameter (Extended Leader)

Dimensions the diameter of circles with an extended leader line. To express the diameter dimension in length or degrees, see the DIMENSION UNITS settings.

This tool expects two input points. The first point identifies the target circular element. The second point determines the length of a single leader line and the orientation of the dimension. If the dimensioned distance needs changing, place the dimension, then hit Enter. The Dimension Text Editor window will appear. Enter the new dimension in the TEXT area, and click OK.

Item Selection

On sidebar menu pick: n/a

For window, point to: Palettes   Dimensioning   Radial

On paper menu see: ANGULAR DIMENSION

Command Window Prompts:

1. Select the item
   `Identify element`
2. Pick a circle
   `Define length of witness line`
3. Pick witness line endpoint
   ⇨   This second point also controls the orientation of the leader line.

To exit item: `Make next selection`

☞ *Tip:* Dimensions are grouped together as single dimension elements. If you need to modify part of the dimension, use a MODIFY tool or DROP DIMENSION first. This turns the dimension data into a graphic group. LOCK GGROUP must be off to affect one part of the dimension. Turn LOCK GGROUP on th have the dimension act as one element. Freezing dimensions or setting compatibility mode to 3X also changes dimension elements into graphic groups.

See Also: DIMENSION DIAMETER tools

DIMENSION DIAMETER PARALLEL   DIM D PA

Drawing Tool: Dimension Diameter Parallel

Dimensions the diameter of circles, expressed in length or degrees. (See the DIMENSION UNITS settings.) This tool expects two input points. The first point identifies the target circular element. The second point determines the length of the dimension witness lines and the orientation of the dimension.

The DIMENSION AXIS setting also controls the dimension's orientation. If the dimensioned distance needs changing, place the dimension, then hit Enter. The Dimension Text Editor window will appear. Enter the new dimension in the TEXT area, and click OK.

Item Selection

On sidebar menu pick: | Dimens | | Diam | | Para |

For window, point to: Palettes   Dimensioning   Radial

On paper menu see: RADIAL DIMENSION

Command Window Prompts:

1. Select the item
   `Identify element`
2. Pick a circle
   `Select dimension endpoint`
3. Pick dimension endpoint

⇨   This second point also controls the length of the witness lines.
To exit item: Make next selection

☞ *Tip:* Dimensions are grouped together as single dimension elements. If you need to modify part of the dimension, use a MODIFY tool or DROP DIMENSION first. This turns the dimension data into a graphic group. LOCK GGROUP must be off to affect one part of the dimension. Turn LOCK GGROUP on th have the dimension act as one element. Freezing dimensions or setting compatibility mode to 3X also changes dimension elements into graphic groups.
See Also: DIMENSION DIAMETER tools

DIMENSION DIAMETER PERPENDICULAR   DIM D PE

Drawing Tool: Dimension Diameter Perpendicular
Dimensions the diameter of circles in three dimensional files, expressed in length or degrees. (See the DIMENSION UNITS settings.) This tool expects two input points. The first point identifies the target circular element. The second point accepts the circle for dimensioning. If the dimensioned distance needs changing, place the dimension, then hit Enter. The Dimension Text Editor window will appear. Enter the new dimension in the TEXT area, and click OK.

Item Selection
On sidebar menu pick: | Dimens | | Diam | | Perp |
For window, point to: Palettes   Dimensioning   Radial

On paper menu see: RADIAL DIMENSION
Command Window Prompts:
1. Select the item
   Identify element
2. Pick a circle
   Select dimension endpoint
3. Pick dimension endpoint
   ⇨   This second point also controls the length of the leader line.
To exit item: Make next selection

☞ *Tip:* Dimensions are grouped together as single dimension elements. If you need to modify part of the dimension, use a MODIFY tool or DROP DIMENSION first. This turns the dimension data into a graphic group. LOCK GGROUP must be off to affect one part of the dimension. Turn LOCK GGROUP on th have the dimension act as one element. Freezing dimensions or setting compatibility mode to 3X also changes dimension elements into graphic groups.
See Also: DIMENSION DIAMETER tools

DIMENSION DIAMETER POINT   DIM D PO

Drawing Tool: Dimension Diameter
Dimensions the diameter of circles, expressed in length or degrees. (See the DIMENSION UNITS settings.) This tool expects two input points. The first point identifies the target circular element. The second point determines the length of a single leader line and the orientation of the dimension. If the dimensioned distance needs changing, place the dimension, then hit Enter. The Dimension Text Editor window will appear. Enter the new dimension in the TEXT area, and click OK.

Item Selection
On sidebar menu pick: | Dimens | | Diam | | Pnt |
For window, point to: Palettes   Dimensioning   Radial

On paper menu see: RADIAL DIMENSION
Command Window Prompts:
1. Select the item
   Identify element
2. Pick a circle
   Select dimension endpoint
3. Pick dimension endpoint
   ⇨   This second point also controls the length of the leader line.
To exit item: Make next selection

☞ *Tip:* Dimensions are grouped together as single dimension elements. If you need to modify part of the dimension, use a MODIFY tool or DROP DIMENSION first. This turns the dimension data into a graphic group. LOCK GGROUP must be off to affect one part of the dimension. Turn LOCK GGROUP on th have the dimension act as one element. Freezing dimensions or setting compatibility mode to 3X also changes dimension elements into graphic groups.

See Also: DIMENSION DIAMETER tools

DIMENSION ELEMENT   DIM E

Drawing Tool: Dimension Element
  Dimensions a line, line string, multi-line, shape, arc, or circle. This tool asks for an input point to identify the element for dimensioning. The dimension tool name appears in the Command Window. Continue to hit the Enter key until the tool you want to use appears in the window. If you have LOCK ASSOCIATION on, all possible associations are created automatically. If the dimensioned distance needs changing, place the dimension, then hit Enter. The Dimension Text Editor window will appear. Enter the new dimension in the TEXT area, and click OK.

Item Selection
  On sidebar menu pick: n/a
  For window, point to: Palettes   Dimensioning   Miscellaneous

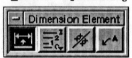

On paper menu see: LINEAR DIMENSION
Command Window Prompts:
1. Select the item
   Select element to dimension
   Accept (Press Return to switch command)
   ⇨   Prompts will vary depending upon the tool selected.
To exit item: Make next selection
See Also: DIMENSION tools and settings

DIMENSION FILE ACTIVE   DIM F A

Setting: Dimension File Active
  Measures objects in the active drawing file.
Item Selection
  On sidebar menu pick: Dimens   Set   File   Act
  For window, point to: Element   Dimensions   Placement
  On paper menu see: n/a
  To show, key in: DIMENSION FILE
   AutoDimensioning File: ACTIVE
  To set, key in: DIMENSION FILE ACTIVE
See Also: DIMENSION FILE REFERENCE

DIMENSION FILE REFERENCE   DIM F R

Setting: Dimension File Reference
  Dimensions objects in the reference file can be dimensioned. The dimension data generated by the objects resides in the active drawing file.

Item Selection

On sidebar menu pick: | Dimens | | Set | | File | | Ref |

For window, point to: E̲lement   D̲imensions   P̲lacement   Reference file units

On paper menu see: n/a

To show, key in: DIMENSION FILE
    AutoDimensioning File: REFERENCE

To set, key in: DIMENSION FILE REFERENCE
See Also: DIMENSION FILE ACTIVE

DIMENSION FONT   DIM F

Setting: Dimension Font
    Controls the style of the text font during dimensioning.

Item Selection

On sidebar menu pick: n/a

For window, point to: E̲lement   D̲imensions   A̲ttributes   Font:

On paper menu see: n/a

Example(s):

DIM F 1   sets the dimension font to 1

See Also: ACTIVE FONT

DIMENSION FONT ACTIVE   DIM F A

Setting: Dimension Font Active
    Changes the style of the dimensioning text font to the active font setting.

Item Selection

On sidebar menu pick: n/a

For window, point to: E̲lement   D̲imensions   A̲ttributes   Font:

On paper menu see: n/a

See Also: ACTIVE FONT, DIMENSION FONT

DIMENSION JUSTIFICATION CENTER   DIM F C

Setting: Dimension Justification Center
    Centers text between the dimension leader lines. This setting affects NEW text placements, NOT the existing dimension text in the drawing file.

Item Selection

On sidebar menu pick: | Dimens | | Just | | Cntr |

For window, point to: E̲lement   D̲imensions   P̲lacement   Justification:

On paper menu see: n/a

To show, key in: DIMENSION JUSTIFICATION
    AutoDimensioning Justification CENTER

To set, key in: DIMENSION JUSTIFICATION CENTER
See Also: DIMENSION JUSTIFICATION settings

DIMENSION JUSTIFICATION LEFT   DIM F L

Setting: Dimension Justification Left
    Justifies text between the dimension leader lines. This setting affects NEW text placements, NOT the existing dimension text in the drawing file.

Item Selection

On sidebar menu pick: | Dimens | | Just | | Left |

For window, point to: E̲lement   D̲imensions   P̲lacement   Justification:

On paper menu see: n/a

To show, key in: DIMENSION JUSTIFICATION
    AutoDimensioning Justification LEFT

To set, key in: DIMENSION JUSTIFICATION LEFT
See Also: DIMENSION JUSTIFICATION settings

DIMENSION JUSTIFICATION RIGHT   DIM F R

Setting: Dimension Justification Right
        Justifies text between the dimension leader lines. This setting affects NEW
text placements, NOT the existing dimension text in the drawing file.
Item Selection
    On sidebar menu pick: | Dimens | | Just | | Rght |
    For window, point to: E̲lement D̲imensions P̲lacement Justification:
    On paper menu see: n/a
    To show, key in: **DIMENSION JUSTIFICATION**
        AutoDimensioning Justification RIGHT
    To set, key in: **DIMENSION JUSTIFICATION RIGHT**
See Also: DIMENSION JUSTIFICATION settings

DIMENSION LEVEL   LD=

Setting: Dimension Level
        Determines the drawing level of the dimension data.
        BE CAREFUL! If the level selected for the dimensions is off, you will not see
the dimension data.
Item Selection
    On sidebar menu pick: n/a
    For window, point to: E̲lement D̲imensions A̲ttributes Level:
    On paper menu see: n/a
    To show, key in: **DIMENSION LEVEL**
        AutoDimensioning Level: ACTIVE
    To set, key in: **DIMENSION LEVEL nn**
        Where nn is a whole number between 1 and 63.

Example(s):
    DIM LE 15   sets the dimension level to 15
See Also: ACTIVE LEVEL, SET LEVEL, CHANGE LEVEL

DIMENSION LEVEL ACTIVE   DIM LE A

Setting: Dimension Level Active
        Changes the dimension level setting to the active level.
Item Selection
    On sidebar menu pick: n/a
    For window, point to: E̲lement D̲imensions A̲ttributes Level:
    On paper menu see: n/a
See Also: ACTIVE LEVEL, DIMENSION LEVEL

DIMENSION LINEAR   DIM LI

Drawing Tool: Dimension Size (Custom)
        Provides three dimension size tools, each with different terminators. This
means you don't have to go back to the settings box and reselect terminators as
often.
Item Selection
    On sidebar menu pick: n/a
    For window, point to: P̲alettes D̲imensioning Linear

    On paper menu see: LINEAR DIMENSION
Command Window Prompts:
    1. Select the item

```
Select start of dimension
```
2. Pick begin point
```
Define length of witness line
```
3. Pick length of witness lines
```
Select dimension endpoint
```
4. Pick endpoint
   To exit item: **Make next selection**
See Also: DIMENSION SIZE tools, DIMENSION LOCATION tools

DIMENSION LOCATION SINGLE   DIM LO SI

Drawing Tool: Dimension Location
    Dimensions a linear distance on the basis of three input points. The first point identifies the starting point of the measured distance. The second point determines the length of the dimension witness lines. The third point identifies the endpoint.

    Place additional points and the linear distance resets before placing new dimensions. If the dimensioned distance needs changing, place the dimension, then hit Enter. The Dimension Text Editor window will appear. Enter the new dimension in the TEXT area, and click OK.

Item Selection
  On sidebar menu pick: Dimens   Locatn   Single
  For window, point to: P̲alettes   Di̲mensioning   Li̲near   Justification:

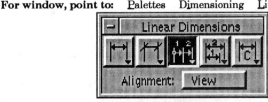

  On paper menu see: LINEAR DIMENSION
Command Window Prompts:
1. Select the item
```
Select start of dimension
```
2. Pick an endpoint
```
Define length of witness line
```
3. Pick a point
```
Select dimension endpoint
```
4. Pick endpoint of dimension
   To exit item: **Make next selection**

☞ *Tip:* Dimensions are grouped together as single dimension elements. If you need to modify part of the dimension, use a MODIFY tool or DROP DIMENSION first. This turns the dimension data into a graphic group. LOCK GGROUP must be off to affect one part of the dimension. Turn LOCK GGROUP on to have the dimension act as one element. Freezing dimensions or setting compatibility mode to 3X also changes dimension elements into graphic groups.

✍ *Note:* An element keypoint can be used to define the base point of a witness line.

✍ *Note:* A second RESET restarts the dimension tool.
    See Also: DIMENSION LOCATION STACKED

DIMENSION LOCATION STACKED   DIM LO ST

Drawing Tool: Dimension Location (Stacked)
    Dimensions several linear distances based on three input points. The first point identifies the starting point of the measured distance. The second point determines the length of the dimension witness lines. The third point identifies the dimension's endpoint.

    Placing additional points causes the command to measure the linear distance from the original endpoint, stacking the new dimensions on top of the first dimension placed. If the dimensioned distance needs changing, place the dimension, then hit Enter. The Dimension Text Editor window will appear. Enter the new dimension in the TEXT area, and click OK.

Item Selection
On sidebar menu pick: | Dimens | Locatn | Stack |
For window, point to: Palettes Dimensioning Linear

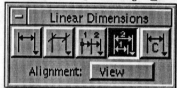

On paper menu see: LINEAR DIMENSION
Command Window Prompts:
1. Select the item
   `Select start of dimension`
2. Pick an endpoint
   `Define length of witness line`
3. Pick a point
   `Select dimension endpoint`
4. Pick endpoint of dimension
To exit item: Make next selection

☞ *Tip:* Dimensions are grouped together as single dimension elements. If you need to modify part of the dimension, use a MODIFY tool or DROP DIMENSION first. This turns the dimension data into a graphic group. LOCK GGROUP must be off to affect one part of the dimension. Turn LOCK GGROUP on to th have the dimension act as one element. Freezing dimensions or setting compatibility mode to 3X also changes dimension elements into graphic groups.

✍ *Note:* An element keypoint can be used to define the base point of a witness line.

✍ *Note:* A second RESET restarts the dimension tool.
See Also: DIMENSION LOCATION SINGLE

DIMENSION ORDINATE   DIM O

Drawing Tool: Dimension Ordinates
Labels a distance along a specified angle or axis. This drawing tool requires five input points. The first three points define the ordinate origin, direction and angle. The subsequent points define the beginning and endpoints of dimensions from the ordinate point.
Item Selection
On sidebar menu pick: | Dimens | Ord |
For window, point to: Palettes Dimensioning Miscellaneous

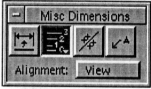

On paper menu see: LINEAR DIMENSION
Command Window Prompts:
1. Select the item
   `Select ordinate origin`
2. Pick ordinate origin
   `Select ordinate direction`
3. Pick ordinate angle
   `Select ordinate endpoint`
4. Pick ordinate endpoint
   `Select start of dimension`
5. Pick begin point
   `Select dimension endpoint.`
To exit item: Make next selection

✍ *Note:* If you use Automatic to place the dimensions, MicroStation automatically aligns each ordinate line with the first ordinate placed. If you use the Manual option to place the dimensions, MicroStation requires an input point to locate the end of the ordinate line.

✍ *Note:* If you're using Left justification, the alignment point of the text is outside dimension text. With Right or Center justification, the inside of the text becomes the alignment point.

DIMENSION PLACEMENT AUTO   DIM P

Setting: Dimension Placement Automatic
     Places dimension text automatically when turned on. This mode of operation is the fastest way to place dimension text.

Item Selection
On sidebar menu pick: Dimens | Mode | Auto
For window, point to: Element Dimensions Placement Location:
On paper menu see: n/a
To show, key in: DIMENSION PLACEMENT
     AutoDimensioning Placement AUTO
To set, key in: DIMENSION PLACEMENT AUTO
See Also: DIMENSION PLACEMENT settings

DIMENSION PLACEMENT MANUAL   DIM P M

Setting: Dimension Placement Manual
     Gives you the opportunity to pick the location for the dimension text. The text characters appear attached to the cursor on the screen, allowing you to place the text anywhere along the dimension leader line.

Item Selection
On sidebar menu pick: Dimens | Mode | Man
For window, point to: Element Dimensions Placement Location:
On paper menu see: n/a
To show, key in: DIMENSION PLACEMENT
     AutoDimensioning Placement MANUAL
To set, key in: DIMENSION PLACEMENT MANUAL
See Also: DIMENSION PLACEMENT settings

DIMENSION PLACEMENT SEMIAUTO   DIM P S

Setting: Dimension Placement Semi-automatic
     Constructs dimension lines and places the text along the leader line, but the text is not permanently placed. You can adjust its position if needed.

Item Selection
On sidebar menu pick: n/a
For window, point to: Element Dimensions Placement Location:
On paper menu see: n/a
To show, key in: DIMENSION PLACEMENT
     AutoDimensioning Placement MANUAL
To set, key in: DIMENSION PLACEMENT SEMIAUTO

✍ *Note:* Same as DIMENSION PLACEMENT MANUAL if the text does not fit in between the witness lines.
See Also: DIMENSION PLACEMENT settings

DIMENSION POST DIAMETER   DIM PO D

Setting: Dimension Post Diameter
     Places a diameter symbol at the end of a dimensioned radius or diameter. This item can be viewed or changed with a key-in or through the Dimension Tool settings box.
Item Selection
On sidebar menu pick: n/a

For window, point to: Element   Dimensions   Tool Settings   Tool:
On paper menu see: n/a

✍ *Note:* A dimension tool must be active before keying in this setting.
See Also: DIMENSION POST settings, DIMENSION PRE settings

DIMENSION POST OFF   DIM PO O

Setting: Dimension Post Off
        Stops the creation of the diameter symbol at the end of a dimensioned radius
        or diameter. This item can be viewed or changed with a key-in or through the
        Dimension Tool settings box.

Item Selection
On sidebar menu pick: n/a
For window, point to: Element   Dimensions   Tool Settings   Tool:
On paper menu see: n/a

✍ *Note:* A dimension tool must be active before keying in this setting.
See Also: DIMENSION POST settings, DIMENSION PRE settings

DIMENSION POST RADIUS   DIM PO R

Setting: Dimension Post Radius
        Places a radius symbol at the end of a dimensioned radius or diameter. This
        item can be viewed or changed with a key-in or through the Dimension Tool
        settings box.

Item Selection
On sidebar menu pick: n/a
For window, point to: Element   Dimensions   Tool Settings   Tool:
On paper menu see: n/a

✍ *Note:* A dimension tool must be active before keying in this setting.
See Also: DIMENSION POST settings, DIMENSION PRE settings

DIMENSION POST SQUARE   DIM PO S

Setting: Dimension Post Square
        Places a square symbol at the end of a dimensioned radius or diameter. This
        item can be viewed or changed with a key-in or through the Dimension Tool
        settings box.

Item Selection
On sidebar menu pick: n/a
For window, point to: Element   Dimensions   Tool Settings   Tool:
On paper menu see: n/a

✍ *Note:* A dimension tool must be active before keying in this setting.
See Also: DIMENSION POST settings, DIMENSION PRE settings

DIMENSION PRE DIAMETER   DIM PR D

Setting: Dimension Pre Diameter
        Places a diameter symbol at the beginning of a dimensioned radius or
        diameter. This item can be viewed or changed with a key-in or through the
        Dimension Tool settings box.

Item Selection
On sidebar menu pick: n/a
For window, point to: Element   Dimensions   Tool Settings   Tool:
On paper menu see: n/a

✍ *Note:* A dimension tool must be active before keying in this setting.
See Also: DIMENSION PRE settings, DIMENSION POST settings

DIMENSION PRE OFF

DIM PR O

Setting: Dimension Pre Off
Stops the creation of the diameter symbol at the beginning of a dimensioned radius or diameter. This item can be viewed or changed with a key-in or through the Dimension Tool settings box.

Item Selection
On sidebar menu pick: n/a
For window, point to: Element   Dimensions   Tool Settings   Tool:
On paper menu see: n/a

✍ *Note:* A dimension tool must be active before keying in this setting.
See Also: DIMENSION POST settings, DIMENSION PRE settings

DIMENSION PRE RADIUS

DIM PR R

Setting: Dimension Pre Radius
Places a radius symbol at the beginning of a dimensioned radius or diameter. This item can be viewed or changed with a key-in or through the Dimension Tool settings box.

Item Selection
On sidebar menu pick: n/a
For window, point to: Element   Dimensions   Tool Settings   Tool:
On paper menu see: n/a

✍ *Note:* A dimension tool must be active before keying in this setting.
See Also: DIMENSION PRE settings, DIMENSION POST settings

DIMENSION PRE SQUARE

DIM PR S

Setting: Dimension Pre Square
Places a square symbol at the beginning of a dimensioned radius or diameter. This item can be viewed or changed with a key-in or through the Dimension Tool settings box.

Item Selection
On sidebar menu pick: n/a
For window, point to: Element   Dimensions   Tool Settings   Tool:
On paper menu see: n/a

✍ *Note:* A dimension tool must be active before keying in this setting.
See Also: DIMENSION PRE settings, DIMENSION POST settings, DIMENSION RADIUS EXTENDED

DIMENSION RADIUS EXTENDED

DIM R E

Drawing Tool: Dimension Radius (Extended Leader)
Dimensions the radius of a circle or circular element with an extended leader line. This item expects two input points. The first point identifies the target element. The second point determines the length of a single leader line and the orientation of the dimension data.

Item Selection
On sidebar menu pick: n/a
For window, point to: Palettes   Dimensioning   Radial

On paper menu see: RADIAL DIMENSION
Command Window Prompts:
1. Select the item
   Identify element
2. Pick a circle
   Define length of witness line

3. Pick witness line endpoint
 ⇨ This second point also controls the orientation of the leader line.
To exit item: **Make next selection**

☞ *Tip:* Dimensions are grouped together as single dimension elements. If you need to modify part of the dimension, use a MODIFY tool or DROP DIMENSION first. This turns the dimension data into a graphic group. LOCK GGROUP must be off to affect one part of the dimension. Turn LOCK GGROUP on th have the dimension act as one element. Freezing dimensions or setting compatibility mode to 3X also changes dimension elements into graphic groups.

See Also: DIMENSION RADIUS

DIMENSION RADIUS POINT   DIM R

Drawing Tool: Dimension Radius
 Dimensions the radius of a circle or circular element. This command expects two input points. The first point identifies the target element. The second determines the length of a single leader line and the orientation of the dimension data.

Item Selection
On sidebar menu pick: Dimens Radius
For window, point to: Palettes Dimensioning Radial

On paper menu see: RADIAL DIMENSION
Command Window Prompts:
 1. Select the item
  Identify element
 2. Pick a circle
  Select dimension endpoint
 3. Pick dimension endpoint
  ⇨ This second point also controls the length of the leader line.
To exit item: **Make next selection**

☞ *Tip:* Dimensions are grouped together as single dimension elements. If you need to modify part of the dimension, use a MODIFY tool or DROP DIMENSION first. This turns the dimension data into a graphic group. LOCK GGROUP must be off to affect one part of the dimension. Turn LOCK GGROUP on th have the dimension act as one element. Freezing dimensions or setting compatibility mode to 3X also changes dimension elements into graphic groups.

See Also: DIMENSION ARC commands

DIMENSION SCALE   DIM SC

Setting: Dimension Scale
 Controls the scale factor of the dimensions. Leave this setting at zero to calculate the actual scale of the object. Dimension a half scale prototype of an existing object by changing the DIMENSION SCALE to .5. If you want to dimension an existing part twice its normal size, set the DIMENSION SCALE to 2. This setting can be viewed or changed with a key-in or through the Dimension Text Format settings box.

Item Selection
On sidebar menu pick: n/a
For window, point to: Element Dimensions Text Format Scale Factor:
On paper menu see: n/a
To show, open: **Dimension Text Format settings box**
To set, key in: **DIMENSION SCALE nnn**
 Where nnn is a number representing the scale factor.

Example(s):
 DIM SC .5   sets the dimension scale factor to .5
See Also: ACTIVE SCALE, DIMENSION SCALE RESET

◼ DIMENSION SCALE RESET   DIM SC R

Setting: Reset Dimension Scale Factor
      Changes the dimensioning scale factor to 1.0. This setting can be viewed or
changed with a key-in or through the Dimension Text Format settings box.

Item Selection
      On sidebar menu pick: n/a
      For window, point to: Element   Dimensions   Text Format   Scale Fac-
tor:
      On paper menu see: n/a
      To show, open: **Dimension text Format settings box**
      See Also: ACTIVE SCALE, DIMENSION SCALE

◼ DIMENSION SIZE ARROW   DIM SI A

Drawing Tool: Dimension Size with Arrows
      Measures linear distances and then expresses those distances between arrow
symbols. This item expects three input points. The first point defines one endpoint
of the distance. The second point controls the length of the dimension witness lines.
The last point determines the dimension endpoint.
      Place additional points and the origin point resets before placing new dimen-
sions. If the dimensioned distance needs changing, place the dimension, then hit
Enter. The Dimension Text Editor window will appear. Enter the new dimension
in the TEXT area, and click OK.

Item Selection
      On sidebar menu pick: [Dimens] [Size] [Arrow]

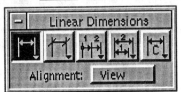

      For window, point to: Palettes   Dimensioning   Linear
      On paper menu see: LINEAR DIMENSION
Command Window Prompts:
      1. Select the item
          Select start of dimension
      2. Pick an endpoint
          Define length of witness line
      3. Pick a point
          Select dimension endpoint
      4. Pick other endpoint
      To exit item: **Make next selection**

☞ *Tip:* Dimensions are grouped together as single dimension elements. If you need to
     modify part of the dimension, use a MODIFY tool or DROP DIMENSION first.
     This turns the dimension data into a graphic group. LOCK GGROUP must be off
     to affect one part of the dimension. Turn LOCK GGROUP on th have the dimension
     act as one element. Freezing dimensions or setting compatibility mode to 3X also
     changes dimension elements into graphic groups.

✐ *Note:* An element keypoint can be used to define the base point of a witness line.

✐ *Note:* A second RESET restarts the dimension tool.
     See Also: DIMENSION LOCATION commands, DIMENSION SIZE STROKE

◼ DIMENSION SIZE STROKE   DIM SI S

Drawing Tool: Dimension Size with Strokes
      Measures linear distances and then expresses those distances between stroke
symbols. This item expects three input points. The first point defines one endpoint
of the distance. The second point controls the length of the dimension witness lines.
The last point determines the dimension endpoint.

Place additional points and the origin point resets before placing new dimensions. If the dimensioned distance needs changing, place the dimension, then hit Enter. The Dimension Text Editor window will appear. Enter the new dimension in the TEXT area, and click OK.

Item Selection
On sidebar menu pick: Dimens | Size | Stroke
For window, point to: Palettes Dimensioning Linear

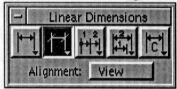

On paper menu see: LINEAR DIMENSION
Command Window Prompts:
1. Select the item
   Select start of dimension
2. Pick an endpoint
   Define length of witness line
3. Pick a point
   Select dimension endpoint
4. Pick other endpoint
To exit item: Make next selection

☞ *Tip:* Dimensions are grouped together as single dimension elements. If you need to modify part of the dimension, use a MODIFY tool or DROP DIMENSION first. This turns the dimension data into a graphic group. LOCK GGROUP must be off to affect one part of the dimension. Turn LOCK GGROUP on th have the dimension act as one element. Freezing dimensions or setting compatibility mode to 3X also changes dimension elements into graphic groups.

✍ *Note:* An element keypoint can be used to define the base point of a witness line.

✍ *Note:* A second RESET restarts the dimension tool.
See Also: DIMENSION LOCATION commands, DIMENSION SIZE ARROW

DIMENSION STACKED   DIM ST

Setting: Dimension Stacked
Controls whether dimensions are standard (not stacked) or stacked by using the DIMENSION STACKED settings. This setting toggles on or off.

Item Selection
On sidebar menu pick: n/a
For window, point to: Element Dimenions Tool Settings Stack dimensions
On paper menu see: n/a
To show, key in: DIM ST
To set, key in: DIM ST on/off
Where on/off is either ON or OFF.

DIMENSION
TERMINATOR FIRST ARROW   DIM TE F A

Setting: Dimension Terminator First Arrow
Controls the creation of the first terminator in a dimension string. Keying in DIMENSION TERMINATOR FIRST ARROW produces an arrow terminating symbol. Keying in DIMENSION TERMINATOR FIRST OFF stops the creation of the first terminator symbol. You can also modify the terminating symbol.

Item Selection
On sidebar menu pick: Dimens | Set | Term | First
For window, point to: Element Dimensions Tool Settings Terminators
On paper menu see: n/a

To show, key in: **DIMENSION TERMINATOR**
AutoDimensioning Terminator ON
To set, key in: **DIMENSION TERMINATOR FIRST on/off**
Where on/off is either ON or OFF.
See Also: DIMENSION TERMINATOR settings

DIMENSION TERMINATOR FIRST ORIGIN   DIM TE F O

Setting: Dimension Terminator First Origin
Controls the creation of the first terminator in a dimension string. Keying in DIMENSION TERMINATOR FIRST ORIGIN produces an origin terminating symbol. Keying in DIMENSION TERMINATOR FIRST OFF stops the creation of the first terminator symbol. You can also modify the terminating symbol.

Item Selection
On sidebar menu pick: | Dimens | | Set | | Term | | First |
For window, point to: Element Dimensions Tool Settings Terminators
On paper menu see: n/a
To show, key in: **DIMENSION TERMINATOR**
AutoDimensioning Terminator ON
To set, key in: **DIMENSION TERMINATOR FIRST on/off**
Where on/off is either ON or OFF.
See Also: DIMENSION TERMINATOR settings

DIMENSION TERMINATOR FIRST STROKE   DIM TE F S

Setting: Dimension Terminator First Stroke
Controls the creation of the first terminator in a dimension string. Keying in DIMENSION TERMINATOR FIRST STROKE produces a stroke terminating symbol. Keying in DIMENSION TERMINATOR FIRST OFF stops the creation of the first terminator symbol. You can also modify the terminating symbol.

Item Selection
On sidebar menu pick: | Dimens | | Set | | Term | | First |
For window, point to: Element Dimensions Tool Settings Terminators
On paper menu see: n/a
To show, key in: **DIMENSION TERMINATOR**
AutoDimensioning Terminator ON
To set, key in: **DIMENSION TERMINATOR FIRST on/off**
Where on/off is either ON or OFF.
See Also: DIMENSION TERMINATOR settings

DIMENSION TERMINATOR LEFT ARROW   DIM TE L A

Setting: Dimension Terminator Left Arrow
Controls the creation of the left terminator in a dimension string. Keying in DIMENSION TERMINATOR LEFT ARROW produces an arrow terminating symbol. Keying in DIMENSION TERMINATOR LEFT OFF stops the creation of the left terminator symbol. You can also modify the terminating symbol.

Item Selection
On sidebar menu pick: | Dimens | | Set | | Term | | Left |
For window, point to: Element Dimensions Tool Settings Terminators
On paper menu see: n/a
To show, key in: **DIMENSION TERMINATOR**
AutoDimensioning Terminator ON
To set, key in: **DIMENSION TERMINATOR LEFT on/off**
Where on/off is either ON or OFF.
See Also: DIMENSION TERMINATOR settings

DIMENSION
TERMINATOR LEFT ORIGIN   DIM TE L O

Setting: Dimension Terminator Left Origin
    Controls the creation of the left terminator in a dimension string. Keying in
DIMENSION TERMINATOR LEFT ORIGIN produces an origin terminating
symbol. Keying in DIMENSION TERMINATOR LEFT OFF stops the creation of
the left terminator symbol. You can also modify the terminating symbol.

Item Selection
On sidebar menu pick: | Dimens | Set | Term | Left |

For window, point to: Element Dimensions Tool Settings Terminators
On paper menu see: n/a
To show, key in: DIMENSION TERMINATOR
    AutoDimensioning Terminator ON
To set, key in: DIMENSION TERMINATOR LEFT on/off
    Where on/off is either ON or OFF.
See Also: DIMENSION TERMINATOR settings

DIMENSION
TERMINATOR LEFT STROKE   DIM TE L S

Setting: Dimension Terminator Left Stroke
    Controls the creation of the left terminator in a dimension string. Keying in
DIMENSION TERMINATOR LEFT STROKE produces a stroke terminating
symbol. Keying in DIMENSION TERMINATOR LEFT OFF stops the creation of
the left terminator symbol. You can also modify the terminating symbol.

Item Selection
On sidebar menu pick: | Dimens | Set | Term | Left |

For window, point to: Element Dimensions Tool Settings Terminators
On paper menu see: n/a
To show, key in: DIMENSION TERMINATOR
    AutoDimensioning Terminator ON
To set, key in: DIMENSION TERMINATOR LEFT on/off
    Where on/off is either ON or OFF.
See Also: DIMENSION TERMINATOR settings

DIMENSION
TERMINATOR RIGHT ARROW   DIM TE R A

Setting: Dimension Terminator Right Arrow
    Controls the creation of the right terminator in a dimension string. Keying
in DIMENSION TERMINATOR RIGHT ARROW produces an arrow terminating
symbol. Keying in DIMENSION TERMINATOR RIGHT OFF stops the creation
of the right terminator symbol. You can also modify the terminating symbol.

Item Selection
On sidebar menu pick: | Dimens | Set | Term | Right |

For window, point to: Element Dimensions Tool Settings Terminators
On paper menu see: n/a
To show, key in: DIMENSION TERMINATOR
    AutoDimensioning Terminator ON
To set, key in: DIMENSION TERMINATOR RIGHT on/off
    Where on/off is either ON or OFF.
See Also: DIMENSION TERMINATOR settings

DIMENSION
TERMINATOR RIGHT ORIGIN   DIM TE R O

Setting: Dimension Terminator Right Origin
    Controls the creation of the right terminator in a dimension string. Keying
in DIMENSION TERMINATOR RIGHT ORIGIN produces an origin terminating

symbol. Keying in DIMENSION TERMINATOR RIGHT OFF stops the creation
of the right terminator symbol. You can also modify the terminating symbol.

Item Selection
On sidebar menu pick: | Dimens | Set | Term | Right |
For window, point to: Element Dimensions Tool Settings Terminators
On paper menu see: n/a
To show, key in: DIMENSION TERMINATOR
   AutoDimensioning Terminator ON
To set, key in: DIMENSION TERMINATOR RIGHT on/off
   Where on/off is either ON or OFF.
See Also: DIMENSION TERMINATOR settings

DIMENSION TERMINATOR RIGHT STROKE   DIM TE R S

Setting: Dimension Terminator Right Stroke
   Controls the creation of the right terminator in a dimension string. Keying
in DIMENSION TERMINATOR RIGHT STROKE produces a stroke terminating
symbol. Keying in DIMENSION TERMINATOR RIGHT OFF stops the creation
of the right terminator symbol. You can also modify the terminating symbol.

Item Selection
On sidebar menu pick: | Dimens | Set | Term | Right |
For window, point to: Element Dimensions Tool Settings Terminators
On paper menu see: n/a
To show, key in: DIMENSION TERMINATOR
   AutoDimensioning Terminator ON
To set, key in: DIMENSION TERMINATOR RIGHT on/off
   Where on/off is either ON or OFF.
See Also: DIMENSION TERMINATOR settings

DIMENSION TEXT BOX   DIM TE B

Setting: Dimension Text Box
   Places a box around dimension text. This setting toggles on or off.

Item Selection
On sidebar menu pick: n/a
For window, point to: Element Dimensions Placement Text frame:
On paper menu see: n/a
To show, key in: DIMENSION TEXT BOX
   Dimension Text BOX : ON
To set, key in: DIMENSION TEXT BOX on/off
   Where on/off is either ON or OFF.
See Also: DIMENSION TEXT settings

DIMENSION TEXT CAPSULE   DIM TE CA

Setting: Dimension Text Capsule
   Places a capsule around the dimension text. This setting toggles on or off.

Item Selection
On sidebar menu pick: n/a
For window, point to: Element Dimensions Placement Text frame:
On paper menu see: n/a
To show, key in: DIMENSION TEXT CAPSULE
   Dimension Text CAPSULE : ON
To set, key in: DIMENSION TEXT CAPSULE on/off
   Where on/off is either ON or OFF.
See Also: DIMENSION TEXT settings

DIMENSION TEXT COLOR   DIM TE CO

Setting: Dimension Text Color
    Displays the current color setting. When followed by a color number or name,
it resets the color of the dimension text.

Item Selection
On sidebar menu pick: n/a
For window, point to: Element   Dimensions   Attributes   Color:
On paper menu see: n/a
To show, key in: DIMENSION TEXT COLOR
    Dimension Text Color : 2
To set, key in: DIMENSION TEXT COLOR opt
    Where opt is either a COLOR NUMBER or NAME.
See Also: DIMENSION TEXT settings

DIMENSION TEXT COLOR ACTIVE   DIM TE CO A

Setting: Dimension Text Color Active
    Entering DIMENSION TEXT COLOR ACTIVE sets the color of the dimen-
sion text to the active color.

Item Selection
On sidebar menu pick: n/a
For window, point to: Element   Dimensions   Attributes   Color:
On paper menu see: n/a
See Also: DIMENSION TEXT settings

DIMENSION TEXT WEIGHT   DIM TE W

Setting: Dimension Text Weight
    Keying in DIMENSION TEXT WEIGHT followed by some weight number
adjusts the weight of the dimension text. Entering DIMENSION TEXT WEIGHT
by itself, displays the current text weight setting.

Item Selection
On sidebar menu pick: n/a
For window, point to: Element   Dimensions   Attributes   Weight:
On paper menu see: n/a
To show, key in: DIMENSION TEXT WEIGHT
    Active Text Weight : 3
To set, key in: DIMENSION TEXT COLOR opt
    Where opt is either a COLOR NUMBER or NAME.
See Also: DIMENSION TEXT settings

DIMENSION TEXT WEIGHT ACTIVE   DIM TE W A

Setting: Dimension Text Weight Active
    Adjusts the weight of the dimension text to the active weight settings.
Item Selection
On sidebar menu pick: n/a
For window, point to: Element   Dimensions   Attributes   Weight:
On paper menu see: n/a
See Also: DIMENSION TEXT WEIGHT

DIMENSION TOLERANCE   TV=

Setting: Dimension Tolerance
    Displays the upper and lower dimension tolerances. This setting can be
viewed or changed with a key-in or through the Dimension Tolerance settings box.
Item Selection
On sidebar menu pick: n/a
For window, point to: Element   Dimensions   Tolerance   Upper:/Lower:
On paper menu see: n/a
To show, open: Dimension Tolerance settings box

To set, key in: TV=nnn
Where nnn is any whole or decimal number.

Example(s):
TV=.005   sets the upper/lower tolerance scale factor to
                                      .005
See Also: DIMENSION TOLERANCE settings

DIMENSION TOLERANCE LOWER   DIM TO L

Setting: Active Tolerance Lower
Controls creation of lower dimension tolerance symbol. Leave the lower value
at zero, and no symbol will appear with the dimension.

Item Selection
On sidebar menu pick: n/a
For window, point to: Element   Dimensions   Tolerance   Lower:
On paper menu see: n/a
To show, open: Dimension Tolerance settings box
To set, key in: DIMENSION TOLERANCE -nnn,+nnn
Where -nnn is a number which represents the lower tolerance.

Example(s):
DIM TO -.5   sets the tolerance to -.5
See Also: DIMENSION TOLERANCE settings

DIMENSION TOLERANCE SCALE   DIM TO S

Setting: Dimension Tolerance Scale
Controls the size of the dimension tolerance text.

Item Selection
On sidebar menu pick: n/a
For window, point to: Element   Dimensions   Tolerance   Text Size:
On paper menu see: n/a
To show, open: Dimension Tolerance settings box
To set, key in: DIMENSION TOLERANCE SCALE nnn
Where nnn is any whole or decimal number.

Example(s):
DIM TO S .005   sets the tolerance scale factor to .005
See Also: DIMENSION TOLERANCE settings

DIMENSION TOLERANCE UPPER   DIM TO U

Setting: Dimension Tolerance Upper
Controls creation of upper dimension tolerance symbol. Leave the upper
value at zero, and no symbol will appear with the dimension.

Item Selection
On sidebar menu pick: n/a
For window, point to: Element   Dimensions   Tolerance   Upper:
On paper menu see: n/a
To show, open: Dimension Tolerance settings box
To set, key in: DIM TO ,+nnn
Where +nnn is a number which represents the upper tolerance.

Example(s):
DIM TO ,+.5   sets the tolerance to +.5
See Also: DIMENSION TOLERANCE settings

DIMENSION TUTORIAL   DIM TU

Command: Dimension Tutorial
Opens the Dimension Placement settings box. Through this window you can
select either orientation, justification, location and text frame characteristics. You
can also pick view alignment, reference file units, and turn witness line generation
on or off.

Item Selection
  On sidebar menu pick: | Dimens | | Opts |
  For window, point to: Element   Dimensions   Text Format   Text frame:
  On paper menu see: n/a
Command Window Prompts:
  1. Select the item
  To show, key in: DIMENSION TUTORIAL
See Also: DESIGN

DIMENSION UNITS DEGREES   DIM U D

Setting: Dimension Units Degrees
    Expresses the angle of an arc in degrees.
Item Selection
  On sidebar menu pick: | Dimens | | Unit | | Deg |
  For window, point to: Element   Dimensions   Text Format   Angle measure:
  On paper menu see: n/a
  To show, key in: DIMENSION UNITS
    AutoDim Angle: DEGREES
  To set, key in: DIMENSION UNITS DEGREES
See Also: DIMENSION UNITS LENGTH

DIMENSION UNITS LENGTH   DIM U L

Setting: Dimension Units Length
    Reflects the distance, or length, of the arc rather than it's degree reading.
Item Selection
  On sidebar menu pick: | Dimens | | Unit | | Len |
  For window, point to: Element   Dimensions   Text Format   Angle measure:
  On paper menu see: n/a
  To show, key in: DIMENSION UNITS
    AutoDim Angle: DEGREES
  To set, key in: DIMENSION UNITS LENGTH
See Also: DIMENSION UNITS DEGREES

DIMENSION UPDATE

Obsolete
    Changes existing MicroStation (not IGDS) dimension data to the current
dimension settings.
See Also: DIMENSION settings

DIMENSION VERTICAL MIXED   DIM V M

Setting: Dimension Vertical Mixed
    Places the dimensions vertically, if they won't fit horizontally.
Item Selection
  On sidebar menu pick: n/a
  For window, point to: Element   Dimensions   Tool Settings   Text:
  On paper menu see: n/a
See Also: DIMENSION VERTICAL settings

DIMENSION VERTICAL OFF   DIM V OF

Setting: Dimension Vertical Off
    Orients the dimensions horizontally.
Item Selection
  On sidebar menu pick: n/a
  For window, point to: Element   Dimensions   Tool Settings   Text:
  On paper menu see: n/a
See Also: DIMENSION VERTICAL settings

DIMENSION VERTICAL ON   DIM V ON

Setting: Dimension Vertical On
        Places dimensions vertically.
Item Selection
    On sidebar menu pick: n/a
    For window, point to: Element   Dimensions   Tool Settings   Text:
    On paper menu see: n/a
See Also: DIMENSION VERTICAL settings

DIMENSION WEIGHT   DIM WE

Setting: Dimension Weight
        Controls the thickness, or weight, of the dimension lines.
Item Selection
    On sidebar menu pick: n/a
    For window, point to: Element   Dimensions   Attributes   Weight:
    On paper menu see: n/a
Example(s):
    DIM WE 5   sets the dimension weight to 5
See Also: ACTIVE WEIGHT, DIMENSION WEIGHT ACTIVE

DIMENSION WEIGHT ACTIVE   DIM WE A

Setting: Dimension Weight Active
        Changes the weight of the dimension lines to the active weight setting.
Item Selection
    On sidebar menu pick: n/a
    For window, point to: Element   Dimensions   Attributes   Weight:
    On paper menu see: n/a
See Also: ACTIVE WEIGHT, DIMENSION WEIGHT

DIMENSION WITNESS BOTTOM   DIM W B

Setting: Dimension Witness Bottom
        Controls the generation of bottom witness lines during the creation of center
    marks. Works in conjunction with the PLACE CENTER MARK tool.
Item Selection
    On sidebar menu pick: n/a
    For window, point to: n/a
    On paper menu see: n/a
    To show, key in: DIMENSION WITNESS BOTTOM
        BOTTOM WITNESS LINE : OFF
    To set, key in: DIMENSION WITNESS BOTTOM on/off
        Where on/off is either ON or OFF.

✍ ***Note:*** DIMENSION WITNESS OFF stops the creation of witness lines.

✳ ***Error:*** No Dimension Tool Active - **Select PLACE CENTER MARK tool and
    identify circle first.**
    See Also: DIMENSION WITNESS settings

DIMENSION WITNESS LEFT   DIM W L

Setting: Dimension Witness Left
        Controls the generation of left witness lines during the creation of the
    dimension data. DIMENSION WITNESS LEFT OFF stops the creation of the left
    witness line.
Item Selection
    On sidebar menu pick: Dimens   Set   Witnes
    For window, point to: Element   Dimension   Tool Settings   Left Witness
    On paper menu see: n/a
    To show, key in: DIMENSION WITNESS

AutoDimensioning Witness lines OFF
To set, key in: DIMENSION WITNESS LEFT on/off
Where on/off is either ON or OFF.
See Also: DIMENSION WITNESS settings

DIMENSION WITNESS OFF   DIM W OF

Setting: Dimension Witness Off
Prevents both witness lines from being generated.
Item Selection
On sidebar menu pick: | Dimens | Set | Witnes |
For window, point to: Element Dimension Tool Settings L&R Witness
On paper menu see: n/a
To show, key in: DIMENSION WITNESS
AutoDimensioning Witness lines OFF
To set, key in: DIMENSION WITNESS on/off
Where on/off is either ON or OFF.
See Also: DIMENSION WITNESS settings

DIMENSION WITNESS RIGHT   DIM W R

Setting: Dimension Witness Right
Controls the generation of right witness lines during the creation of the
dimension data. DIMENSION WITNESS RIGHT OFF stops the creation of the
right witness line.
Item Selection
On sidebar menu pick: | Dimens | Set | Witnes |
For window, point to: Element Dimension Tool Settings Right Witness
On paper menu see: n/a
To show, key in: DIMENSION WITNESS
AutoDimensioning Witness lines OFF
To set, key in: DIMENSION WITNESS RIGHT on/off
Where on/off is either ON or OFF.
See Also: DIMENSION WITNESS settings

DIMENSION WITNESS TOP   DIM W T

Setting: Dimension Witness Top
Controls the generation of top witness lines during the creation of center
marks. Works in conjunction with the PLACE CENTER MARK tool.
Item Selection
On sidebar menu pick: n/a
For window, point to: n/a
On paper menu see: n/a
To show, key in: DIMENSION WITNESS TOP
TOP WITNESS LINE : OFF
To set, key in: DIMENSION WITNESS TOP on/off
Where on/off is either ON or OFF.

📖 *Note:* DIMENSION WITNESS OFF stops the creation of witness lines.

✳ *Error:* No Dimension Tool Active - **Select PLACE CENTER MARK tool and
identify circle first.**
See Also: DIMENSION WITNESS settings

Dimensioning

Definition
Dimensions are witness lines, leader lines, arrows and text, expressing the
distance the dimension represents. MicroStation adds dimensions to the drawing
file in three modes: automatically, semi-automatically or manually.

The dimension tools are responsible for building the dimension data in the drawing file. With MicroStation's dimensioning commands you are able to perform radial dimensioning, angular dimensioning and linear dimensioning.

Control the display and creation of the dimension data with the dimension control settings. For example, the DIMENSION COLOR setting determines the color of the dimension lines and text. The DIMENSION LEVEL tells MicroStation where to place the dimension data in the drawing file.

See Also: DIMENSION tools and settings

|DISCONNECT |DISCONNECT

Command: Disconnect

Log off of an Oracle database with the |DISCONNECT command. After signing off, you can perform any database functions necessary outside MicroStation.

Item Selection
On sidebar menu pick:	n/a
For window, point to:	n/a
On paper menu see:	n/a
To log off, key in:	|DISCONNECT

Command Window Prompts:

✍ *Note:* Oracle's implementation on the PC is for a single user system only. If you log on to your database in MicroStation, remember to log off, or you will not be able to access the database from DOS.

See Also: |CONNECT

DISPLAY ERASE

Command: Display Erase

Execute the DISPLAY ERASE command from a user command. Store the element in DGNBUF, and then erase it using the ERASE option.

See Also: DISPLAY commands, FENCE LOCATE, LOCELE

DISPLAY HILITE

Command: Display Hilite

Highlights displayed elements stored in DGNBUF. Execute DISPLAY HILITE from a user command.

See Also: DISPLAY commands

DISPLAY SET

Command: Display Set

Returns an element to original display symbology after storage in DGNBUF. Execute DISPLAY SET from a user command.

t use:

See Also: DISPLAY commands

DL=

Alternate Key-in

Alternate key-in for POINT DELTA. Use to place a point at some known distance and direction relative to the design plane, not the view.

See Also: POINT DELTA

DOS   DOS

DOS Only
Command: DOS

Suspends the design session and places you in the DOS operating system environment. This command is handy if you need to jump out of the design session to execute a DOS command.

For example, to check on a filename, key in DOS DIR <filename> for the file in question. Then hit Enter to return to MicroStation.

Item Selection

On sidebar menu pick: | Utils | | DOS |

For window, point to: n/a

On paper menu see: UTILITIES

Command Window Prompts:

1. Select the item

   [USTN] C:\CGSI>

   ⇨ You may now execute your operating system command.

To return to graphics: Key in EXIT

See Also: !, %

DP=

Alternate Key-in

Alternate key-in for SET DDEPTH ABSOLUTE. Use it to set an absolute display depth in a three dimensional drawing file.

See Also: SET DDEPTH ABSOLUTE

DR=

Alternate Key-in

Alternate key-in for TYPE. Use it to display a text file while remaining in the drawing file.

See Also: TYPE

DROP ASSOCIATION   DR A

Drawing Tool: Drop Association

Drops a point's association with a connected point on another element, or a multi-line. Dropping the point's association gives the point its own X,Y,Z coordinates. This tool requires two input points. One point to identify the target vertex and a second point to confirm the change.

Item Selection

On sidebar menu pick: n/a

For window, point to: Palettes   Main   Drop Element

On paper menu see: LINEAR DIMENSION

Command Window Prompts:

1. Select the item

   Identify element

2. Pick an element

   Accept/Reject (select next input)

3. Give an accept point

To exit item: Make next selection

See Also: DROP tools

DROP COMPLEX   DR

Drawing Tool: Drop Complex Status

Drops the complex status of a cell or text node so that you can manipulate the single elements making up the cell or text nodes individually. This item requires two input points. One point to identify the complex group of elements and a second point to accept the drop operation.

Item Selection

On sidebar menu pick: | Modify | | Drop | | Cmplex |

For window, point to: Palettes   Main   Drop Element

On paper menu see: DROP
Command Window Prompts:
   1. Select the item
      Identify Element
   2. Pick a complex element
      Accept / Reject (select next input)
   3. Give an accept point
   To exit item: Make next selection
See Also: DROP STRING, FENCE DROP COMPLEX

DROP DIMENSION   DR A

Drawing Tool: Drop Dimension Element
      Converts dimension data to their base elements. This tool requires an input
point to identify the dimension and a second point to convert the dimension data
to its base elements.

Item Selection
   On sidebar menu pick: n/a
   For window, point to: Palettes   Main   Drop Element

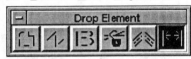

On paper menu see: n/a
Command Window Prompts:
   1. Select the item
      Identify element
   2. Pick an element
      Accept/Reject (select next input)
   3. Give an accept point
   To exit item: Make next selection
See Also: DROP tools

DROP MLINE   DR M

Drawing Tool: Drop Multi-line
      Converts multi-line elements to their base elements. This tool requires an
input to identify the multi-line and a second point to convert the multi-line to its
base elements.

Item Selection
   On sidebar menu pick: n/a
   For window, point to: Palettes   Main   Drop Element

On paper menu see: n/a
Command Window Prompts:
   1. Select the item
      Identify element
   2. Pick an element
      Accept/Reject (select next input)
   3. Give an accept point
   To exit item: Make next selection

See Also: DROP tools

DROP SHARECELL

DR SH

Drawing Tool: Drop Shared Cell
>   Makes a shared cell unshared. An unshared cell is no longer affected by shared cell modifications.

Item Selection
>   **On sidebar menu pick:**   n/a
>   **For window, point to:**   n/a
>   **On paper menu see:**   CELLS

Command Window Prompts:
>   1. Select the item
>   Identify element
>   2. Pick an element
>   Accept/Reject (select next input)
>   3. Give an accept point

To exit item: Make next selection

See Also: DROP tools

DROP STRING

DR S

Drawing Tool: Drop Line String/Shape Status
>   Reduces a line string, or shape, to its individual line segments. This item requires two input points. One point to identify the target line string and a second point to accept the drop operation.

Item Selection
>   **On sidebar menu pick:**   Modify   Drop   LnStr
>   **For window, point to:**   Palettes   Main   Drop Element

>   **On paper menu see:**   DROP

Command Window Prompts:
>   1. Select the item
>   Identify Element
>   2. Pick a line string
>   Accept / Reject (select next input)
>   3. Give an accept point

To exit item: Make next selection

See Also: DROP COMPLEX

DROP TEXT

DR T

Drawing Tool: Drop Text
>   Converts text characters into lines, arcs, circles, etc.

Item Selection
>   **On sidebar menu pick:**   n/a
>   **For window, point to:**   Palettes   Main   Drop Element

>   **On paper menu see:**   DROP

Command Window Prompts:
>   1. Select the item
>   Identify element
>   2. Pick text
>   Accept/Reject (select next input)

3. Give an accept point
To exit item: **Make next selection**
See Also: PLACE TEXT tools

DS=

Alternate Key-in
    Alternate key-in for DEFINE SEARCH. Use it to control the elements considered during a fence operation.
See Also: DEFINE SEARCH

DUPLICATE   Ctrl+D

Drawing Tool: Duplicate
    Reproduces a selection set. A selection set is any group of elements that have been placed in a selection set.
Item Selection
   On sidebar menu pick: n/a
   For window, point to: Edit   Duplicate
   On paper menu see: n/a
Command Window Prompts:
  1. Select the item
    Identify element
  2. Pick an element
    Accept/reject (select next input)
  3. Give an accept point
To exit item: **Make next selection**
See Also: COPY tools

DV=

Alternate Key-in
    Alternate key-in for DELETE VIEW. Use it to delete named views stored in the drawing file.
See Also: DELETE VIEW

DX=

Alternate Key-in
    Alternate key-in for POINT VDELTA. Use it to place points along the view axis relative to the last data or tentative point placed.
See Also: POINT VDELTA

DZ=

Alternate Key-in
    Alternate key-in for ACTIVE ZDEPTH RELATIVE. Use it to adjust the active depth in a three dimensional drawing file by a keyed-in displacement.
See Also: ACTIVE ZDEPTH RELATIVE

ECHO   EC

Setting: Echo
    Turns on the message and prompt display in MicroStation's message fields. This is a default setting. Keying ECHO toggles the setting on or off.
Item Selection
   On sidebar menu pick: n/a
   For window, point to: n/a
   On paper menu see: n/a

Note: This setting cannot be saved with the FILE DESIGN command.
See Also: NOECHO

Edges

Obsolete

The word EDGES cannot be typed in to execute the edges hidden line removal process. Perform this process with the EDGES.UCM user command called EDGES.UCM.

See Also: RENDER tools

EDIT AE   ED AE

DOS Only
Drawing Tool: Edit AE

Interactively edits database records. Keying in EDIT AE clears the screen and opens the active screen format file. Setup the record you wish to change with DEFINE AE or FIND first.

Item Selection
On sidebar menu pick: `dBase` `ActEnt` `Edit`

For window, point to: n/a
On paper menu see: DATABASE
See Also: DEFINE AE, FIND, SHOW AE

EDIT AUTO   ED AU

Drawing Tool: Automatic Fill in Enter Data Fields

Automatically enters text into Enter Data Fields automatically. This item asks you to select a view with a data point. MicroStation then searches for the first Enter Data Field it finds in the view. A box appears around the field representing the size of the text to place. Enter the text at the prompt.

Item Selection
On sidebar menu pick: `Text` `ED Fld` `Fill` `Auto`

For window, point to: n/a
On paper menu see: ENTER DATA
Command Window Prompts:
1. Select the item
   Select view
2. Pick a point
   <Return> or DATA for next field
3. Enter the text
To exit item: **Make next selection**
See Also: EDIT AUTO DIALOG, EDIT SINGLE tools, Enter Data Fields

EDIT AUTO DIALOG   ED AU D

Drawing Tool: Automatic Fill in Enter Data Fields

Automatically enters text into Enter Data Fields through the Text Editor window. This item asks you to select a view with a data point. MicroStation searches for the first Enter Data Field it finds in the view. A box appears around the field representing the size of the text to place. Enter the text at the prompt. The text automatically appears in the Enter Data Field area.

Item Selection
On sidebar menu pick: `Text` `ED Fld` `Fill` `Auto`

For window, point to: Palettes   Main   Enter Data Fields

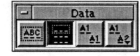

On paper menu see: n/a
Command Window Prompts:
1. Select the item
   Select view
2. Pick a point
   <Return> or DATA for next field

3. Enter the text
To exit item: Make next selection
See Also: EDIT SINGLE, Enter Data Fields

EDIT SINGLE   ED S

Drawing Tool: Fill in Single Enter Data Field
Enters text to a specified Enter Data Field when placing text. This item asks
you to identify the Enter Data Field with an input point. A box appears around
the field representing the size of the text to place. Enter the text at the prompt.
The text then appears in the Enter Data Field area.

Item Selection
On sidebar menu pick: | Text | | ED Fld | | Fill | | Single |

For window, point to: n/a
On paper menu see: ENTER DATA
Command Window Prompts:
1. Select the item
   Identify element
2. Pick a point
   ED FIELD:
3. Enter the text
To exit item: Make next selection
See Also: EDIT AUTO tools, ENTER SINGLE DIALOG, Enter Data Fields

EDIT SINGLE DIALOG   ED S

Drawing Tool: Fill in Single Enter Data Field
Enters text to a specific Enter Data Field using the Text Editor window. This
item asks you to identify the Enter Data Field with an input point. A box appears
around the field representing the size of the text to place. Enter the text at the
prompt. The text then appears in the Enter Data Field area.

Item Selection
On sidebar menu pick: | Text | | ED Fld | | Fill | | Single |

For window, point to: Palettes   Main   Enter Data Fields

On paper menu see: n/a
Command Window Prompts:
1. Select the item
   Identify element
2. Pick a point
   ED FIELD:
3. Enter the text
To exit item: Make next selection
See Also: EDIT AUTO, Enter Data Fields

EDIT TEXT   ED

Drawing Tool: Edit Text
Allows you to change, or correct, text in a drawing file. This item expects an
input point to define the text you want to change. The text highlights. A second
input point loads the text into a buffer. Change the text. Hit the Enter key. The
new text appears in the drawing.

Item Selection
On sidebar menu pick: | Text | | Edit |

For window, point to: P̲alettes   M̲ain   T̲ext

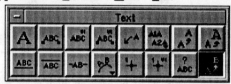

On paper menu see: n/a
Command Window Prompts:
1. Select the item
   Identify element
2. Pick a point
   Accept/Reject (select next input)
3. Give an accept point
   [> This is a test   <]
   Press Return to continue
To exit item: **Make next selection**

☞ *Tip:* Use the arrow keys on the keyboard to move right and left along the buffered text string.
See Also: PLACE TEXT tools

EL=

Alternate Key-in
    Alternate key-in for ELEMENT LIST. Use it to create an element list file.
See Also: ELEMENT LIST

ELEMENT LIST   EL=

Command: Element List
    Creates an element list file. Place a fence around elements in the drawing file to create a list file containing design file block and byte pointers for the elements. This list is useful in communicating information about the elements to application programs.
    The fence contents search criteria determines the element selection.

Item Selection
    On sidebar menu pick: n/a
    For window, point to: n/a
    On paper menu see: n/a

Enter Data Fields

Definition
    Enter Data Fields control and speed up the process of placing text in the drawing file. Let's use the title block of a typical drawing file for our example. The text, which represents the drawing number, the project number and the engineer's name, are usually different text sizes, text fonts and line weights. You must pay close attention to these details to avoid mistakes. Changing the text size, text font and line weight each time you place text creates a second problem: it is time consuming.
    By entering text through Enter Data Fields, MicroStation helps you with the control aspect of placing text. The text placement process speeds up because you no longer need to make decisions before entering the text.
    To create an Enter Data Field set the text attributes the way you want the text to look. Use a PLACE TEXT tool to place an under bar (_) character to represent each text letter to enter later. Then use the EDIT AUTO or EDIT SINGLE tool to place the text in the Enter Data Fields.
    Think of using Enter Data Fields for part names in or around cells. Use Enter Data Fields when entering text in schedules, revision blocks, bill of material lists or bubbles.
See Also: EDIT AUTO, EDIT SINGLE

EXCHANGEFILE   XD=

Command: Exchange Drawing File
   Save time! Use EXCHANGEFILE when you want to open another drawing
file. Even though the drawing files are exchanged, the view configuration on the
display screen remains the same.

Item Selection
   On sidebar menu pick: n/a
   For window, point to: n/a
   On paper menu see: n/a
See Also: NEWFILE

EXIT   Ctrl+Q

Command: Exit
   Ends the design session. MicroStation writes the last few elements to the
drawing file on disk, then closes the file and returns control to the computer's
operating system.

Item Selection
   On sidebar menu pick: Exit
   For window, point to: File   Exit
   On paper menu see: n/a
See Also: EXIT commands, QUIT

EXIT NOCLEAR

Command: Exit NOClear
   Ends the design session. MicroStation writes the last few elements to the
drawing file on disk, but does not clear the display screen upon exiting. This option
is run from a user command. EXIT NOCLEAR works on MicroStation PC only.

See Also: EXIT commands, QUIT

EXIT NOUC

Command: Exit NOUnClear
   Ends the design session. MicroStation writes the last few elements to the
drawing file on disk, but does not run the EXITUC step of a user command upon
exiting. This option is run from a user command.

See Also: EXIT commands, QUIT

EXTEND ELEMENT 2   EXT E 2

Drawing Tool: Extend Two Elements to Intersection
   Forces the intersection of two lines, line strings or arcs. This item requires
three input points. Two points highlight the lines to act upon. The third point
extends the two elements until they touch.

Item Selection
   On sidebar menu pick: Modify   Extend   2-X
   For window, point to: Palettes   Main   Modify Element

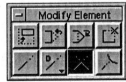

   On paper menu see: EXTEND
See Also: EXTEND ELEMENT INTERSECTION, EXTEND LINE tools

EXTEND ELEMENT INTERSECTION

EXT E

Drawing Tool: Extend Element to Intersection

Forces the intersection of a line, line string or arc with another line or arc. This item requires three input points. The first point highlights the line to act upon. The second point determines the intersecting element. The third point extends the line until it touches the second element.

Item Selection

On sidebar menu pick: Modify   Extend   1-X

For window, point to: Palettes   Main   Modify Element

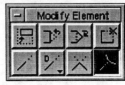

On paper menu see: EXTEND
See Also: EXTEND ELEMENT 2, EXTEND LINE tools

EXTEND LINE INTERSECTION

Same as EXTEND ELEMENT INTERSECTION
See Also: EXTEND ELEMENT INTERSECTION

EXTEND LINE

EXT L

Drawing Tool: Extend Line

Lengthens or shortens a line or line string by a graphically defined distance. This item requires two input points. The first point highlights the line to act upon. The second point determines the distance the line is stretched or shrunk.

Item Selection

On sidebar menu pick: Modify   Extend   Line

For window, point to: Palettes   Main   Modify Element

On paper menu see: EXTEND
Command Window Prompts:

1. Select the item
   Identify element
2. Pick a line
   Accept/Reject (select next input)
3. Give an accept point

To exit item: Make next selection
See Also: EXTEND LINE tools

EXTEND LINE 2

Same as EXTEND ELEMENT 2
See Also: EXTEND ELEMENT 2

EXTEND LINE KEYIN

EXT L

Drawing Tool: Extend Line by Key-in

Lengthens or shortens a line or line string by a keyed in distance. This item requires a key-in and two input points. The key-in determines the distance the line is stretched or shrunk. The first point highlights the line and the second point accepts the extend operation.

Item Selection
   On sidebar menu pick: | Modify | | Extend | | Line |
   For window, point to: Palettes   Main   Modify Element

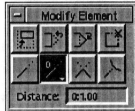

On paper menu see: EXTEND
Command Window Prompts:
  1. Select the item
    Enter distance
  2. Enter key-in distance
    Identify element
  3. Pick an element
    Accept/Reject (select next input)
  4. Give an accept point
  To exit item: Make next selection
See Also: EXTEND LINE tools

EXTRACT
BSPLINE SURFACE BOUNDARY   EXT BS S B

3D Drawing Tool: Extract B-spline Surface Boundary
    Constructs a B-spline curve from a B-spline surface. A specified tolerance can be entered through the pop-down field of the Space Curves palette.
Item Selection
   On sidebar menu pick: | Place | | BSpl | | SpcCur |
   For window, point to: Palettes   3D B-splines   Space Curves

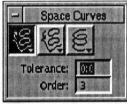

On paper menu see: SPACE CURVES
Command Window Prompts:
  1. Select the item
    Identify element
  2. Pick surface
    Accept/Reject (select next input)
  3. Give an accept point
  To exit item: Make next selection

✍ *Note:* The order number can be entered through the pop-down field in the Space Curve palette.
  See Also: CONSTRUCT B-SPLINE tools, IMPOSE BSPLINE SURFACE BOUNDARY, PLACE B-SPLINE tools

FENCE ARRAY POLAR   FEN AR P

Drawing Tool: Polar Array Fence Contents

Places the contents of a fence in a circular array. For example, if you are drawing a steel plate with eight bolt holes 45 degrees apart, you place the first hole. Then place a fence around the hole and select FENCE ARRAY POLAR to place the other seven.

This item asks you the number of array elements to generate, the sweep of the angle and if the fence should be rotated. Then it expects one input point to identify the origin and an accept point to array the elements.

Item Selection

On sidebar menu pick: Fences Array Polr

For window, point to: Palettes Fence Fence Copy

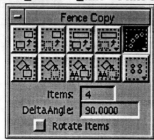

On paper menu see: FENCE COPY

Command Window Prompts:
1. Select the item
   Enter number of array items
2. Enter some whole number
   Enter angle to fill
3. Enter some whole number
   (1) ROTATE (y) :
4. Key in Y to rotate array elements
   Select point & array radius
5. Pick a point
   Accept, select center/Reject
6. Pick a second point
   Processing fence contents
   ⇨ Array elements appear.

To exit item: **Exits by itself**

See Also: ARRAY tools, FENCE ARRAY RECTANGULAR

FENCE ARRAY RECTANGULAR   FEN AR

Drawing Tool: Rectangular Array Fence Contents

Places the contents of a fence in a rectangular array. Two examples of rectangular arrays are building columns on a plan view drawing or tiles in an acoustic ceiling plan.

This item asks you how many rows and columns are in the array and the spacing between those rows and columns. Then it expects one input point to identify the origin and an accept point to array the elements.

Item Selection

On sidebar menu pick: Fences Array Rect

For window, point to: Palettes   Fence   Fence Copy

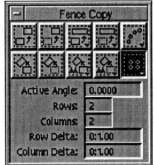

On paper menu see: FENCE COPY
Command Window Prompts:
1. Select the item
   Enter number of array rows
2. Enter some whole number
   ⇨   Rows go horizontally.
   Enter number of array columns
3. Enter some whole number
   ⇨   Columns go vertically.
   Enter distance between horizontal rows
4. Enter some whole number
   Enter distance between vertical columns
5. Enter some whole number
   Accept/Reject Fence Contents
6. Pick a point
   Processing working set
To exit item: Exits by itself
See Also: ARRAY tools, FENCE ARRAY POLAR

FENCE ATTACH   FE AT

Drawing Tool: Attach Active Entity to Fence Contents
   Attaches the elements contained within a fence to an active entity. The item requires one input point to link the elements to a database record. Use the SET DATABASE tool to define the database; then use the FIND tool to signify the database record.
Item Selection
On sidebar menu pick: n/a
For window, point to: Palettes   Database

On paper menu see: DATABASE
Command Window Prompts:
1. Select the item
   Accept/Reject Fence Contents
2. Give an accept point
To exit item: Make next selection

☞ *Tip:* The LOCK FENCE settings determine the fate of elements contained in a fence.
See Also: FENCE DETACH

FENCE
CHANGE CLASS CONSTRUCTION   FEN CH CL P

Drawing Tool: Change Fence Contents to Active Class
   Toggles the class of elements contained in a fence to CONSTRUCTION. Use this class of elements as visual aids in construction drawings.

Item Selection
   On sidebar menu pick: n/a
   For window, point to: n/a
   On paper menu see: FENCE CHANGE
Command Window Prompts:
   1. Select the item
      Accept/Reject Fence Contents
   2. Pick a point in any view
      Processing fence contents
   To exit item: **Exits by itself**

☞ *Tip:* The LOCK FENCE settings determine the fate of elements contained in the fence.
See Also: FENCE CHANGE CLASS PRIMARY, FENCE CHANGE tools

FENCE CHANGE CLASS PRIMARY   FEN CH CL P

Drawing Tool: Change Fence Contents to Active Class
   The FENCE CHANGE CLASS PRIMARY tool toggles the class of elements contained in a fence to PRIMARY. Primary is the default class for elements.

Item Selection
   On sidebar menu pick: n/a
   For window, point to: n/a
   On paper menu see: FENCE CHANGE
Command Window Prompts:
   1. Select the item
      Accept/Reject Fence Contents
   2. Pick a point in any view
      Processing fence contents
   To exit item: **Exits by itself**

☞ *Tip:* The LOCK FENCE settings determine the fate of elements contained in the fence.
See Also: FENCE CHANGE CLASS CONSTRUCTION, FENCE CHANGE tools

FENCE CHANGE COLOR   FEN CH CO

Drawing Tool: Change Fence Contents to Active Color
   Changes the color of elements contained within a fence. First place a fence around the target elements. Then select FENCE CHANGE COLOR and place one input point to start the fence processing.

Item Selection
   On sidebar menu pick: Fences Change Color
   For window, point to: Palettes   Fence   Fence Change

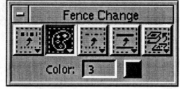

   On paper menu see: FENCE CHANGE
Command Window Prompts:
   1. Select the item
      Accept/Reject Fence Contents
   2. Pick a point in any view
      Processing fence contents

To exit item: **Exits by itself**

☞ *Tip:* The LOCK FENCE settings determine the fate of elements contained in the fence.

See Also: FENCE CHANGE tools

FENCE CHANGE LEVEL   FEN CH LE

Drawing Tool: Change Fence Contents to Active Level

Moves elements between levels. First, place a fence around the target elements and change the ACTIVE LEVEL setting. Then select FENCE CHANGE LEVEL and place one input point to start the fence processing.

Item Selection

On sidebar menu pick: | Fences | | Change | | Level |

For window, point to: <u>P</u>alettes   <u>F</u>ence   Fence <u>C</u>hange

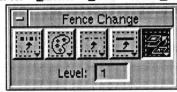

On paper menu see: FENCE CHANGE

Command Window Prompts:

1. Select the item
   Accept/Reject Fence Contents
2. Pick a point in any view
   Processing fence contents

To exit item: **Exits by itself**

☞ *Tip:* The LOCK FENCE settings determine the fate elements contained in the fence.

See Also: FENCE CHANGE tools

FENCE CHANGE LOCK   FEN CH L

Drawing Tool: Lock Fence Contents

Secures, or locks up, a group of elements. Once locked, the elements cannot be manipulated or changed in any way. This item requires one input point to start processing the fence contents.

Item Selection

On sidebar menu pick: n/a

For window, point to: n/a

On paper menu see: n/a

Command Window Prompts:

1. Select the item
   Accept/Reject Fence Contents
2. Pick a point in any view
   Processing fence contents

To exit item: **Exits by itself**

See Also: CHANGE LOCK, CHANGE UNLOCK, FENCE CHANGE UNLOCK

FENCE CHANGE STYLE   FEN CH ST

Drawing Tool: Change Fence Contents to Active Style

Alters the line style of elements. First place a fence around the target elements. Then select FENCE CHANGE STYLE and place one input point to start the fence processing.

Item Selection

On sidebar menu pick: | Fences | | Change | | Style |

For window, point to: Palettes   Fence   Fence Change

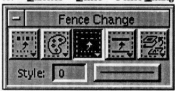

On paper menu see: FENCE CHANGE
Command Window Prompts:
1. Select the item
   Accept/Reject Fence Contents
2. Pick a point in any view
   Processing fence contents
To exit item: Exits by itself

☞ *Tip:* The LOCK FENCE settings determine the fate of elements contained in the fence.
See Also: FENCE CHANGE tools

FENCE CHANGE SYMBOLOGY   FEN CH SY

Drawing Tool: Change Fence Contents to Active Symbology
   Changes the symbology of elements contained in a fence. First place a fence around the target elements. Then select FENCE CHANGE SYMBOLOGY and place one input point to start the fence processing.
Item Selection
On sidebar menu pick: Fences   Change   Symb
For window, point to: Palettes   Fence   Fence Change

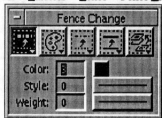

On paper menu see: FENCE CHANGE
Command Window Prompts:
1. Select the item
   Accept/Reject Fence Contents
2. Pick a point in any view
   Processing fence contents
To exit item: Exits by itself

☞ *Tip:* The LOCK FENCE settings determine the fate of elements contained in the fence.
See Also: CHANGE SYMBOLOGY, FENCE CHANGE tools

FENCE CHANGE UNLOCK   FEN CH U

Drawing Tool: Unlock Fence Contents
   Unlocks elements so they can be altered. This item requires one input point to unlock the elements.
Item Selection
On sidebar menu pick: n/a
For window, point to: n/a
On paper menu see: n/a
See Also: CHANGE LOCK, CHANGE UNLOCK, FENCE CHANGE LOCK

FENCE CHANGE WEIGHT

FEN CH W

Drawing Tool: Change Fence Contents to Active Weight

Changes the thickness of elements contained within a fence. First place a fence around the target elements. Then select FENCE CHANGE WEIGHT and place one input point to start the fence processing.

Item Selection

On sidebar menu pick: Fences Change Weight

For window, point to: Palettes Fence Fence Change

On paper menu see: FENCE CHANGE

Command Window Prompts:

1. Select the item
   Accept/Reject Fence Contents
2. Pick a point in any view
   Processing fence contents

To exit item: Exits by itself

☞ *Tip:* The LOCK FENCE settings determine the fate of elements contained in the fence.

See Also: FENCE CHANGE tools

FENCE COPY

FEN CO

Drawing Tool: Copy Fence Contents

Makes a copy of the elements contained within a fence. This item expects two input points. The first point picks the origin in the fence and the second point defines the distance to copy the elements.

Item Selection

On sidebar menu pick: Fences Copy

For window, point to: Palettes Fence Fence Copy

On paper menu see: FENCE COPY

Command Window Prompts:

1. Select the item
   Define origin
2. Pick a point
   Define distance
3. Pick a second point
   Processing fence contents

To exit item: Make next selection

☞ *Tip:* The lock fence settings determine the fate of elements contained in the fence.

See Also: COPY ELEMENT, PLACE FENCE tools

FENCE DELETE

FEN DE

Drawing Tool: Delete Fence Contents

Deletes a fenced group of elements. Place the fence and select FENCE DELETE. One input point placed in any view starts the fence processing.

Item Selection
On sidebar menu pick: Fences Delete
For window, point to: Palettes Fence
On paper menu see: FENCE MIRROR
Command Window Prompts:
1. Select the item
   Accept/Reject Fence Contents
2. Pick a point in any view
   Processing fence contents
To exit item: **Make next selection**

☞ *Tip:* The lock fence settings determine the fate of elements contained in the fence.
See Also: DELETE ELEMENT, PLACE FENCE tools

FENCE DETACH   FEN DET

Drawing Tool: Detach Database Linkage from Fence Contents
Breaks the linkages between the database and the elements contained within a fence. This item expects one input point to start processing the fence contents.
Item Selection
On sidebar menu pick: dBase Tools
For window, point to: Palettes Database

On paper menu see: DATABASE
Command Window Prompts:
1. Select the item
   Accept/Reject Fence Contents
2. Pick a point
   Processing fence contents
To exit item: **Exits by itself**

☞ *Tip:* The LOCK FENCE settings determine the fate of elements contained in the fence.
See Also: DETACH, FENCE ATTACH

FENCE DROP ASSOCIATION   FE DR A

Drawing Tool: Drop Associations in Fence
        Drops any fenced point's association with any other connected points of
elements or multi-lines. The points now have their own X,Y,Z coordinates. This
tool requires one input point to process the fence contents.

Item Selection
    On sidebar menu pick: n/a
    For window, point to: n/a
    On paper menu see: n/a
Command Window Prompts:
    1. Select the item
        Accept/Reject Fence Contents
    2. Place an input in the fence
        Processing fence contents
    To exit item: Make next selection

✍ Note:

☞ *Tip:*

✳ *Error:* message - **comment**
    See Also:

FENCE DROP COMPLEX   FEN DR

Drawing Tool: Drop Complex Status of Fence Contents
        Drops the complex status of elements within a fenced cell or shape. To change
one element contained within the cell or shape, first place a fence around the
figure. Then use the FENCE DROP COMPLEX tool to drop the complex status of
many cells or complex shapes. This item requires one input point to start processing the fence contents.

Item Selection
    On sidebar menu pick: Fences   Drop
    For window, point to: Palettes   Fence   Fence Delete

    On paper menu see: FENCE DROP
Command Window Prompts:
    1. Select the item
        Accept/Reject Fence Contents
    2. Pick a point
        Processing fence contents
    To exit item: Exits by itself

☞ *Tip:* The LOCK FENCE settings determine the fate of elements contained in the
    fence.

☞ *Tip:* Using the COMPRESS DESIGN command after dropping the cells, you can
    pick up free disk space.
    See Also: DROP COMPLEX

FENCE DROP DIMENSION

FE DR D

Drawing Tool: Drop Dimension Elements in Fence

Converts dimension data contained within a fence to their base elements. This tool requires an input to start the fence processing operation.

Item Selection

On sidebar menu pick: n/a

For window, point to: n/a

On paper menu see: n/a

Command Window Prompts:

1. Select the item

   Accept/Reject Fence Contents

2. Give an accept point

To exit item: **Make next selection**

See Also: DROP MLINE

FENCE DROP MLINE

FE DR M

Drawing Tool: Drop Multi-lines in Fence

Multi-line elements contained within a fence to their base elements. This tool requires an input to start the fence operation.

Item Selection

On sidebar menu pick: n/a

For window, point to: n/a

On paper menu see: n/a

Command Window Prompts:

1. Select the item

   Accept/Reject Fence Contents

2. Give an accept point

To exit item: **Make next selection**

See Also: DROP MLINE

FENCE FILE

FF=

Command: Fence File

Copies a portion of a drawing file into a second file. If the output design file does not exist, MicroStation will create it automatically. First place a fence around the elements you want to appear in the second file. Then give an input point to start the fence processing.

Item Selection

On sidebar menu pick: Fences Export FF=:

For window, point to: n/a

On paper menu see: n/a

Command Window Prompts:

1. Select the item

   Define origin, <- Dest File?

2. Key in the output filename

   ⇨ Where filename is the output drawing name.

   Accept/Reject Fence Contents

3. Pick a point

   Processing fence contents

To exit item: **Exits by itself**

✳ *Error:* Clip Lock not supported for this command - If this message appears, select either LOCK FENCE INSIDE or LOCK FENCE OVERLAP.

☞ *Tip:* The LOCK FENCE settings determine the fate of elements contained in the fence.

See Also: FENCE SEPARATE

Fence Filter

Alias
Fence Filter is not a valid setting and cannot be keyed. It is an alias for
FENCE REPORT. Issuing a DS= keyin activates a fence filter. Keying in
DS=NONE deactivates the filter. DS= is short for DEFINE SEARCH.
See Also: DEFINE SEARCH, FENCE REPORT

FENCE FREEZE   FE FR

Command: Freeze Elements in Fence
Freezes multi-line elements, shared cells and dimension data contained
within a fence. Once frozen, these elements can be displayed in Interactive
Graphics Design Software (IGDS).

Item Selection
On sidebar menu pick: n/a
For window, point to: n/a
On paper menu see: n/a
Command Window Prompts:
 1. Select the item
    Accept/reject Fence Contents
 2. Give an accept point
 To exit item: **Make next selection**
See Also: FENCE THAW, FREEZE

FENCE LOAD   FEN LOA

Drawing Tool: Load Displayable Attributes to Fence Contents
Loads attribute data contained within the fence onto an existing text node in
the drawing file.

Item Selection
On sidebar menu pick: n/a
For window, point to: Palettes   Database

On paper menu see: DATABASE
Command Window Prompts:
 1. Select the item
    Identify element
 2. Pick a text node
    Accept/Reject (select next input)
 3. Give an accept point
 To exit item: **Make next selection**
See Also: LOAD DA

FENCE LOCATE
FEN L

Drawing Tool: Fence Locate

Loads a fenced element's characteristics into DGNBUF. You place a fence and key in FENCE LOCATE. Load the element in DGNBUF by placing an input point in a view. Once in DGNBUF, a user command can display or manipulate the element's attributes. Each successive input point loads the next element that satisfies the fence criteria.

Item Selection
On sidebar menu pick: n/a
For window, point to: n/a
On paper menu see: n/a
See Also: USERCOMMAND

FENCE MIRROR COPY HORIZONTAL
FEN MI CO

Drawing Tool: Mirror Fence Contents About Horizontal (Copy)

Copies and mirrors the elements contained within a fence about the horizontal axis. This item takes one input point to define the distance the fenced elements are mirrored.

Item Selection
On sidebar menu pick: Fences   Mirror   Copy-H
For window, point to: Palettes   Fence   Fence Mirror

On paper menu see: FENCE MIRROR
Command Window Prompts:
1. Select the item
   Accept/Reject Fence Contents
2. Pick a point
   Processing fence contents
To exit item: Make next selection

☞ *Tip:* The LOCK FENCE settings determine the fate of elements contained in the fence.

See Also: FENCE MIRROR tools, MIRROR tools

FENCE MIRROR COPY LINE

FEN MI CO L

Drawing Tool: Mirror Fence Contents About Line (Copy)

Copies and mirrors the elements contained within a fence about a swing line. This item takes two input points to define the swing line.

Item Selection

On sidebar menu pick: | Fences | Mirror | Copy-L |

For window, point to: Palettes Fence Fence Mirror

On paper menu see: FENCE MIRROR

Command Window Prompts:
1. Select the item
   Enter 1st pnt on mirror line (or reject)
2. Pick a line endpoint
   Enter 2nd pnt on mirror line
3. Pick other line endpoint
   Processing fence contents

To exit item: Make next selection

☞ *Tip:* The LOCK FENCE settings determine the fate of elements contained in the fence.

See Also: FENCE MIRROR tools, MIRROR tools

FENCE MIRROR COPY VERTICAL

FEN MI CO V

Drawing Tool: Mirror Fence Content About Vertical (Copy)

Copies and mirrors the elements contained within a fence about the vertical axis. This item takes one input point to define the distance the fenced elements are mirrored.

Item Selection

On sidebar menu pick: | Fences | Mirror | Copy-V |

For window, point to: Palettes Fence Fence Mirror

On paper menu see: FENCE MIRROR

Command Window Prompts:
1. Select the item
   Enter 1st pnt on mirror line (or reject)
2. Pick a line endpoint
   Enter 2nd pnt on mirror line
3. Pick other line endpoint
   Processing fence contents

To exit item: Make next selection

☞ *Tip:* The LOCK FENCE settings determine the fate of elements contained in the fence.

See Also: FENCE MIRROR tools, MIRROR tools

FENCE MIRROR ORIGINAL HORIZONTAL

FEN MI

Drawing Tool: Mirror Fence Contents About Horizontal (Original)

Mirrors the elements contained within a fence about the horizontal axis. This item takes one input point to define the distance the fenced elements are mirrored.

Item Selection

On sidebar menu pick: | Fences | Mirror | Orig-H |

For window, point to: Palettes   Fence   Fence Mirror

On paper menu see: FENCE MIRROR
Command Window Prompts:
1. Select the item
   Accept/Reject Fence Contents
2. Pick a point
   Processing fence contents
To exit item: Make next selection

☞ *Tip:* The LOCK FENCE settings determine the fate of elements contained in the fence.

See Also: FENCE MIRROR tools, MIRROR tools

FENCE MIRROR ORIGINAL LINE   FEN MI L

Drawing Tool: Mirror Fence Contents About Line (Original)
   Mirrors the elements contained within a fence about a swing line. This item takes two input points to define the swing line.

Item Selection
On sidebar menu pick: Fences   Mirror   Orig-L
For window, point to: Palettes   Fence   Fence Mirror

On paper menu see: FENCE MIRROR
Command Window Prompts:
1. Select the item
   Enter 1st pnt on mirror line (or reject)
2. Pick a line endpoint
   Enter 2nd pnt on mirror line
3. Pick other line endpoint
   Processing fence contents
To exit item: Make next selection

☞ *Tip:* The LOCK FENCE settings determine the fate of elements contained in the fence.

See Also: FENCE MIRROR tools, MIRROR tools

FENCE MIRROR ORIGINAL VERTICAL   FEN MI V

Drawing Tool: Mirror Fence Contents About Vertical (Original)
   Mirrors the elements contained within a fence about the vertical axis. This item takes one input point to define the distance the fenced elements are mirrored.

Item Selection
On sidebar menu pick: Fences   Mirror   Orig-V
For window, point to: Palettes   Fence   Fence Mirror

On paper menu see: FENCE MIRROR
Command Window Prompts:
1. Select the item
   Enter 1st pnt on mirror line (or reject)
2. Pick a line endpoint

```
Enter 2nd pnt on mirror line
```
3. Pick other line endpoint
```
Processing fence contents
```
To exit item: Make next selection

☞ *Tip:* The LOCK FENCE settings determine the fate of elements contained in the fence.

See Also: FENCE MIRROR tools, MIRROR tools

FENCE MOVE   FEN MO

Drawing Tool: Move Fence Contents

Moves fenced elements. Place a fence around the elements. Select FENCE MOVE and place an input point inside the fence to define the fence origin. Then place a second point to define the distance to move the elements.

Item Selection

On sidebar menu pick: Fences   Move

For window, point to: Palettes   Fence   Fence Copy

On paper menu see: FENCE COPY

Command Window Prompts:
1. Select the item
```
   Define origin
```
2. Pick a point
```
   Define Distance
```
   second point
```
   Processing fence contents
```
To exit item: Make next selection

☞ *Tip:* The LOCK FENCE settings determine the fate of elements contained in the fence.

See Also: MOVE ELEMENT, PLACE FENCE tools

FENCE REPORT   FE RE

Drawing Tool: Generate Report Table

Generates a table report of all the elements contained within the fence. The window invoked by the SET DATABASE item sets the table report. This item requires one input point to start the fence operation.

BE CAREFUL! This item destroys the report file each time you execute this tool.

Item Selection

On sidebar menu pick: dBase   Setup   Search

For window, point to: Palettes   Database

On paper menu see: DATABASE

Command Window Prompts:
1. Select the item
```
   Accept/Reject Fence Contents
```
2. Give an accept point
To exit item: Make next selection

☞ *Tip:* Read the LOCK FENCE setting to determine how elements contained in the fence are affected.

See Also: DEFINE SEARCH

FENCE ROTATE COPY

FEN RO C

Drawing Tool: Rotate Fence Contents by Active Angle (Copy)

Revolves and copies fenced elements. First place a fence around the elements you want rotated. Then place an input to start the rotation process and define the axis of rotation.

Set the ACTIVE ANGLE (AA=) setting to some non-zero degree setting before attempting to rotate the elements.

Item Selection

On sidebar menu pick: Fences Rotate Copy

For window, point to: Palettes Fence Fence Copy

On paper menu see: FENCE COPY

Command Window Prompts:

1. Select the item

   Accept/Reject Fence Contents
2. Pick a point

   Processing fence contents

To exit item: Make next selection

☞ *Tip:* The LOCK FENCE settings determine the fate of elements contained in the fence.

☞ *Tip:* Setting the ACTIVE ANGLE to 0 forces MicroStation to update the element's range blocks. This could fix range errors.

See Also: FENCE ROTATE tools, ROTATE tools, SPIN tools

FENCE ROTATE ORIGINAL

FEN RO

Drawing Tool: Rotate Fence Contents by Active Angle (Original)

Revolves the elements enclosed in a fence. First place a fence around the elements you want rotated. Then place an input to start the rotation process and define the axis of rotation.

Set the ACTIVE ANGLE (AA=) setting to some non-zero degree setting before attempting to rotate the elements.

Item Selection

On sidebar menu pick: Fences Rotate Orig

For window, point to: Palettes Fence Fence Copy

On paper menu see: FENCE COPY

Command Window Prompts:

1. Select the item

   Accept/Reject Fence Contents
2. Pick a point

   Processing fence contents

To exit item: Make next selection

☞ *Tip:* The LOCK FENCE settings determine the fate of elements contained in the fence.

☞ *Tip:* Setting the ACTIVE ANGLE to 0 forces MicroStation to update the element's range blocks. This could fix range errors.

See Also: FENCE ROTATE tools, ROTATE tools, SPIN tools

FENCE SCALE COPY   FEN SC C

Drawing Tool: Scale Fence Contents (Copy)
Shrinks or enlarges fenced elements while copying them. Changing the scale factor of an element is a two step process. First, set the desired ACTIVE SCALE. For instance, an ACTIVE SCALE of .5 would scale the element down to half its original size. Choosing an ACTIVE SCALE factor of 2 would scale the object up to twice its original size.

After setting the scale factor, place a fence around the elements and select the FENCE SCALE COPY tool. This item requires one input point to accept the scaling operation.

Item Selection
On sidebar menu pick: Fences | Scale | Copy
For window, point to: Palettes Fence Fence Copy

On paper menu see: FENCE COPY
Command Window Prompts:
1. Select the item
   Accept/Reject Fence Contents
      ⇨   The fence reflects the newly scaled elements.
2. Pick a point
   Processing fence contents
   To exit item: Make next selection

✐ *Note:* If the ACTIVE SCALE is 1.0, the size of the elements within the fence do not change.

✐ *Note:* Cells and graphic groups can be scaled with this tool.

☞ *Tip:* The LOCK FENCE settings determine the fate of elements contained in the fence.

See Also: FENCE SCALE ORIGINAL

FENCE SCALE ORIGINAL   FEN SC

Drawing Tool: Scale Fence Contents (Original)
Shrinks or enlarges fenced elements. Changing the scale factor of an element is a two step process. First, set the ACTIVE SCALE setting to the scale factor you want. For example, an ACTIVE SCALE of .5 would scale the element down to half its original size. Choosing an ACTIVE SCALE of 2 would scale the object up to twice its original size.

After you have set the ACTIVE SCALE setting, place a fence around the elements and select the FENCE SCALE ORIGINAL tool. This item requires one input point to accept the scaling operation.

Item Selection
On sidebar menu pick: Fences | Scale | Orig

For window, point to: Palettes   Fence   Fence Copy

On paper menu see: FENCE COPY
Command Window Prompts:
1. Select the item
   `Accept/Reject Fence Contents`
      ⇨   The fence reflects the newly scaled elements.
2. Pick a point
   `Processing fence contents`
   To exit item: **Make next selection**

✍ *Note:* If the ACTIVE SCALE is 1.0, the size of the elements within the fence do not change.

✍ *Note:* Cells and graphic groups can be scaled with this tool.

☞ *Tip:* The LOCK FENCE settings determine the fate of elements contained in the fence.

See Also: FENCE SCALE COPY

FENCE SEPARATE   FS=

Command: Fence Separate
   Removes (A.K.A. deletes, eliminates, erases) the elements contained within a fence and moves them into a new drawing file. This item expects an output filename and one input point to start the processing.
   BE CAREFUL! If the name you pick for the output file already exists, the existing file is overwritten when this command executes.

Item Selection
   On sidebar menu pick: Fences   Export   FS=:
   For window, point to: n/a
   On paper menu see: n/a
Command Window Prompts:
1. Select the item
   `<- Dest File?`
2. Key in the output filename
   Where filename is the output drawing name.
   `Accept/Reject Fence Contents`
1. Pick a point
   `Processing fence contents`
   To exit item: **Exits by itself**

✳ *Error:* `Clip Lock not supported for this tool` - If this message appears, select either LOCK FENCE INSIDE or LOCK FENCE OVERLAP.

☞ *Tip:* The LOCK FENCE settings determine the fate of elements contained in the fence.

See Also: FENCE FILE

FENCE SPIN COPY   FEN SP C

Drawing Tool: Spin Fence Contents (Copy)
   Spins and copies a group of fenced elements by defining the angle of swing with an input point. This item requires two input points. The first point sets the pivot point and the second point defines the angle of the spin.
   Cells and graphic groups can be spun with this tool.

Item Selection
   On sidebar menu pick: Fences   Spin   Copy

For window, point to: P̲alettes   F̲ence   Fence Co̲py

On paper menu see: FENCE COPY
Command Window Prompts:
  1. Select the item
    `Accept (define pivot point) /Reject`
  2. Pick a point
    `Define rotation angle`
  3. Pick a point
    `Processing fence contents`
  To exit item: Make next selection

☞ *Tip:* Place the second input point with a precision key-in.

☞ *Tip:* The LOCK FENCE settings determine the fate of elements contained in the fence.
  See Also: FENCE SPIN ORIGINAL, ROTATE tools, SPIN tools

FENCE SPIN ORIGINAL   FEN SP

Drawing Tool: Spin Fence Contents (Original)
    Spins a group of elements contained in a fence. This item requires two input points. The first point identifies the pivot point and the second point defines the angle of the spin.
    Cells and graphic groups can be spun with this tool.

Item Selection
  On sidebar menu pick: `Fences`   `Spin`   `Orig`
  For window, point to: P̲alettes   F̲ence   Fence Co̲py

On paper menu see: FENCE COPY
Command Window Prompts:
  1. Select the item
    `Accept (define pivot point) /Reject`
  2. Pick a point
    `Define rotation angle`
  3. Pick a point
    `Processing fence contents`
  To exit item: Make next selection

☞ *Tip:* Place the second input point with a precision key-in.

☞ *Tip:* The LOCK FENCE settings determine the fate of elements contained in the fence.
  See Also: FENCE SPIN COPY, ROTATE tools, SPIN tools

FENCE STRETCH   FEN ST

Drawing Tool: Fence Stretch
    Stretches fenced elements. Only elements crossing the fence boundaries are stretched. Elements contained within the fence are not stretched. This item expects one point to define the fence origin and one point to define the distance the element(s) are lengthened.

This drawing tool ignores the FENCE LOCK settings.

Item Selection
On sidebar menu pick: | Fences | | Strch |
For window, point to: Palettes Fence

On paper menu see: FENCE COPY
Command Window Prompts:
1. Select the item
   Define origin
2. Pick a point
   Define distance
3. Pick a second point
   Processing fence contents
To exit item: Make next selection
See Also: FENCE MOVE, LOCK CELLSTRETCH, PLACE FENCE

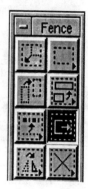

FENCE SURFACE PROJECTION   FEN P

3D Drawing Tool: Project Fence Contents
        Projects the contents of a fence, creating surfaces. This item requires two input points. The first point identifies the point within the fence to project. The second point defines the distance to project the fence contents.

Item Selection
On sidebar menu pick: n/a
For window, point to: n/a
On paper menu see: n/a

☞ *Tip:* If the elements are skewed when projected, try snapping to the element when identifying it and use a precision key-in for the projection distance.
See Also: FENCE SURFACE REVOLUTION, SURFACE tools

FENCE SURFACE REVOLUTION   FEN SU

3D Drawing Tool: Revolve Fence Contents
        Rotates the contents of a fence about a pivot point, creating a surface of revolution. This item requires a rotation angle key-in and two input points. The first point identifies the rotation point within the fence. The second point defines the axis of rotation.

Item Selection
On sidebar menu pick: n/a
For window, point to: n/a
On paper menu see: n/a

☞ *Tip:* If the elements are skewed when rotated, try snapping to the element when identifying it and use a precision key-in for the projection distance.

☞ *Tip:* If you want to revolve the fence contents again and the angle and axis remain the same, use the DX=0,0,0 key-in.
See Also: FENCE SURFACE PROJECTION, SURFACE tools

FENCE THAW   FE TH

Command: Thaw Elements in Fence
        Thaws multi-line elements, shared cells and dimension data that were frozen. Once thawed, these elements can be manipulated.

Item Selection
On sidebar menu pick: n/a
For window, point to: n/a
On paper menu see: n/a

Command Window Prompts:
1. Select the item
   `Accept/Reject Fence Contents`
2. Give an accept point
To exit item: Make next selection
See Also: FENCE FREEZE, THAW

FENCE TRANSFORM

Drawing Tool: Transform Fence Contents
Transforms the contents of a fence by a previously defined transformation matrix. The fence search criteria determines the element selection within the fence. User command and application programs use this drawing tool.

See Also: Tmatrx, TRANSFORM

FENCE WSET ADD   FEN W

Drawing Tool: Add Fence Contents to Working Set
Adds fenced elements into a working set. WSET is short for working set. FENCE WSET ADD is quicker to use than the WSET ADD tool. Place a fence around a group of elements and select FENCE WSET ADD. Place one input point to process the fence contents.
Once in a group, you can manipulate these elements using any FENCE tool.

Item Selection
On sidebar menu pick: | Fences | WrkSet | Add |

For window, point to: n/a
On paper menu see: W-SET
Command Window Prompts:
1. Select the item
   `Accept/Reject Fence Contents`
2. Pick a point
   `Processing fence contents`
To exit item: Make next selection

✍ *Note:* Any elements placed while a working set is active become part of the active working set.
See Also: FENCE WSET tools, Graphic Group, WSET tools

FENCE WSET COPY   FEN W C

Drawing Tool: Add Copy of Fence Contents to Working Set
Adds fenced elements to a working set, and then copies them. WSET is short for working set. FENCE WSET COPY is quicker to use than the WSET COPY tool. Place a fence around a group of elements. Select FENCE WSET COPY. Place one input point to process the fence contents. Manipulate these grouped elements using any FENCE tool.

Item Selection
On sidebar menu pick: | Fences | WrkSet | Copy |

For window, point to: n/a
On paper menu see: W-SET
Command Window Prompts:
1. Select the item
   `Accept/Reject Fence Contents`
2. Pick a point
   `Processing fence contents`
To exit item: Make next selection

✍ *Note:* Any elements placed while a working set is active become part of the active working set.
See Also: FENCE WSET tools, Graphic Group, WSET tools

FENCE WSET DROP

Obsolete
Deactivates, or discontinues, the active working set.
See Also: FENCE WSET tools, Graphic Group, WSET tools

FF=

Alternate Key-in
Alternate key-in for FENCE FILE. Use it to copy the elements within a fence into a new drawing file.
See Also: FENCE FILE

FI=

Alternate Key-in
Alternate key-in for FIND. Use it to define an active entity in a database.
See Also: FIND

FILEDESIGN

Ctrl+F

Command: File Design
Saves a drawing file's active settings between design sessions. Selecting this tool saves the active settings.

Item Selection
On sidebar menu pick: Utils   File
For window, point to: File   Save Settings
On paper menu see: TOP BORDER
Command Window Prompts:
1. Select the item
   Active Parameters Saved
To exit item: Exits by itself
See Also: COMPRESS DESIGN

FILLET MODIFY

FILL M

Drawing Tool: Circular Fillet and Truncate Both
Places an arc tangent to two elements, then trims the original elements. The elements can be linear or circular. This item requires a key-in to define the radius of the fillet and an input point to identify each of the two elements. After the arc appears, the two elements are trimmed back to points tangent to the arc.

Item Selection
On sidebar menu pick: Modify   Fillet   Mod2
For window, point to: Palettes   Main   Fillets

On paper menu see: FILLET
Command Window Prompts:
1. Select the item
   Key in radius
2. Key in a radius
   Select first segment
3. Pick a line
   Select second segment
4. Pick second line
To exit item: Click a Reset, Reset

* **Error:** Illegal definition - If this message appears, the radius of the arc may be too large for both legs of the angle to be tangent to the arc.
See Also: FILLET tools, PLACE ARC tools

FILLET NOMODIFY   FILL

Drawing Tool: Circular Fillet (No Truncation)
Places an arc tangent to two elements, but does not trim the original elements. The elements can be linear or circular. This item requires a key-in to define the radius of the fillet and an input point to identify each of the two elements.

Item Selection
On sidebar menu pick: `Modify` `Fillet` `No Mod`
For window, point to: Palettes   Main   Fillets

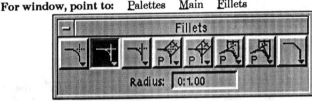

On paper menu see: FILLET
Command Window Prompts:
1. Select the item
   Key in radius
2. Key in a radius
   Select first segment
3. Pick a line
   Select second segment
4. Pick second line
To exit item: Click a Reset, Reset

* **Error:** Illegal definition - If this message appears, the radius of the arc may be too large for both legs of the angle to be tangent to the arc.
See Also: FILLET tools, PLACE ARC tools

FILLET SINGLE   FILL S

Drawing Tool: Circular Fillet and Truncate Single
Places an arc tangent to two elements, then trims one of the original elements. The elements can be linear or circular. This item requires a key-in to define the radius of the fillet and an input point to identify each of the two elements. After the arc appears, the first element identified is trimmed back to a point tangent to the arc.

Item Selection
On sidebar menu pick: `Modify` `Fillet` `Mod1`
For window, point to: Palettes   Main   Fillets

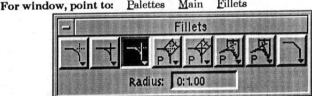

On paper menu see: FILLET
Command Window Prompts:
1. Select the item
   Key in radius
2. Key in a radius
   Select first segment
3. Pick a line
   Select second segment

4. Pick second line
To exit item: Click a Reset, Reset

* *Error:* Illegal definition - If this message appears, the radius of the arc may be too large for both legs of the angle to be tangent to the arc.
See Also: FILLET tools, PLACE ARC tools

FIND
FI=

Command: Find
Defines an active entity in a database. Once defined, the entity can be edited or attached to an element in the drawing file.
Item Selection
On sidebar menu pick: dBase ActEnt Find
For window, point to: n/a
On paper menu see: n/a
See Also: CREATE ENTITY, DEFINE AE

FIT ACTIVE
FIT

View Control: Fit Active Design
Displays an entire design file in one view. This item expects one input point in the view to display the file.
Item Selection
On sidebar menu pick: Window Fit
For window, point to: View Fit Active Design
On paper menu see: VIEW
Command Window Prompts:
1. Select the item
   Select view
To exit item: Make next selection
See Also: UPDATE, WINDOW view controls, ZOOM view controls

FIT ALL
FIT A

View Control: Fit Design and Reference Files
Displays the active file, as well as any attached reference files, in a single view. Selecting this item fits all files to the active drawing file.
Item Selection
On sidebar menu pick: n/a
For window, point to: View Fit All
On paper menu see: VIEW
Command Window Prompts:
1. Select the item
   Select view
To exit item: Make next selection
See Also: FIT view controls, UPDATE, WINDOW view controls, ZOOM view controls

FIT REFERENCE
FIT R

View Control: Fit Reference
Displays in a single view any reference files attached to the active drawing. Selecting this item fits any attached reference files to the drawing file.
Item Selection
On sidebar menu pick: n/a
For window, point to: View Fit Reference
On paper menu see: VIEW
Command Window Prompts:
1. Select the item
   Select view
To exit item: Make next selection
See Also: FIT view controls, UPDATE, WINDOW view controls, ZOOM view controls

FLUSH   FL

Command: Flush
  Causes MicroStation to write the information from memory into the disk file.
MicroStation does not write to the drawing file every time you place an element.
The time between disk writes depends upon the platform and operating system
version. MicroStation also time-stamps the drawing file's directory entry.

Item Selection
  On sidebar menu pick: n/a
  For window, point to: n/a
  On paper menu see: n/a
See Also: NEWFILE

FORMS   FO

DOS Only

Setting: Enable SQL Forms
  Key in FORMS followed by on or off to enable or disable the display of
SQL FORMS in MicroStation. By default the FORMS item is toggled off. This
setting can be changed or viewed in the Database settings box.

Item Selection
  On sidebar menu pick: n/a
  For window, point to: Settings   Database   Forms
  On paper menu see: n/a

Note: The database must be warm started before FORMS can be activated.
See Also: FORMS DISPLAY

FORMS DISPLAY   FO D

Command: Display SQL Forms Screen Form
  Key in FORMS DISPLAY followed by a form name and a list of arguments.
MicroStation will search the directories identified by the MS_DBASE variable
until it finds this form, and then passes the argument list to Oracle. The Oracle
RUNFORM utility then displays the form on the screen.

Item Selection
  On sidebar menu pick: n/a
  For window, point to: n/a
  On paper menu see: n/a
See Also: FORMS

FREE   FR

Command: Free
  Displays the total size of the hard drive in bytes, as well as the remaining
free space on the drive. The information is displayed in the Command Window.

Item Selection
  On sidebar menu pick: | Params | | Show | | Disk |
  For window, point to: n/a
  On paper menu see: n/a
Command Window Prompts:
  1. Select the item
  To show, key in: FREE
    Free space is 1198K out of 2081K
  To exit item: Exits by itself
See Also: COMPRESS DESIGN, COMPRESS LIBRARY

FREEZE   FREEZ

Command: Freeze Element
  Freezes multi-line elements, shared cells and dimension data. Once frozen,
these elements can be displayed in Interactive Graphics Design Software (IGDS).

Item Selection
  On sidebar menu pick: n/a
  For window, point to: n/a
  On paper menu see: LINEAR DIMENSION
Command Window Prompts:
  1. Select the item
     `Identify element`
     `Accept/Reject (select next input)`
  2. Give an accept point
  To exit item: **Make next selection**
See Also: THAW, FENCE FREEZE, PLACE MLINE

FT=

Alternate Key-in
    Alternate key-in for ACTIVE FONT. Use it to control the font of text in the drawing.
See Also: ACTIVE FONT

GO=

Alternate Key-in
    Alternate key-in for ACTIVE ORIGIN. Use it to define a global origin. All input points placed in the drawing file are placed relative to the global origin.
See Also: ACTIVE ORIGIN

Good Things

Expression
    An expression that describes the mood of the CAD environment if adequate training is provided, books studied, instructions adhered to, time allotted for becoming proficient, and the advice of outside consultants followed. If you maintain these conditions at your work place, then "Good Things" are in your the future.
See Also: Bad Things, Bentley Boys, Bingo, Brain Fault

GR=

Alternate Key-in
    Alternate key-in for ACTIVE GRIDREF. Use it to control the number of dots between each reference cross on the display grid.
See Also: ACTIVE GRIDREF

Graphic Group

Definition
    Associates a group of elements in the drawing file on an on-again-off-again basis. When the LOCK GGROUP setting is toggled on, the elements relate to each other. When the lock is off, the elements act as individuals. Toggle the lock back on and they're associated again.
    Use the GROUP ADD tool to add new elements to the existing group. Use the GROUP DROP tool to take a single element out of a group, or to disband the entire cluster of elements.
See Also: GROUP tools

GROUP ADD   GR A

Drawing Tool: Add to Graphic Group
    Creates a new graphic group or adds elements to an existing group. This item expects two input points. The first point identifies the element to add and the second point accepts the element into the group.
Item Selection
  On sidebar menu pick: `Manip` `Groups` `GrGp` `Add`

For window, point to: P̲alettes   M̲ain   C̲hain

On paper menu see: GROUP
Command Window Prompts:
1. Select the item
   Identify element
2. Pick an element
   Add to new group (Accept/Reject)
3. Give an accept point
To exit item: Make next selection

✍ *Note:* See the LOCK GGROUP setting to toggle the graphic group association on and off.

☞ *Tip:* Toggle LOCK GGROUP on before adding elements to the graphic group.
See Also: GROUP DROP, PLACE CELL tools, WSET tools

GROUP DROP   GR D

Drawing Tool: Drop from Graphic Group
   Drops an element from a graphic group, or disbands an entire group. This command expects two input points. One point identifies the element and a second point accepts the element for processing.
   Identifying an element, with the LOCK GGROUP setting off, drops the individual element from the group. Identifying an element with the LOCK GGROUP on highlights all the elements in the group and breaks up the entire group.
Item Selection
On sidebar menu pick: Manip   Groups   GrGp   Drop
For window, point to: P̲alettes   M̲ain   C̲hain

On paper menu see: GROUP
Command Window Prompts:
1. Select the item
   Identify element
2. Pick an element
   Accept/Reject (select next input)
3. Give an accept point
To exit item: Make next selection
See Also: GROUP ADD

GROUP HOLES   GR H

Drawing Tool: Group Holes
   Associates a solid with its related holes. GROUP HOLES then copies the solid and the related hole or holes to an orphan cell, then places the orphan cell at the end of the drawing file.
Item Selection
On sidebar menu pick: Utils   GHole
For window, point to: P̲alettes   M̲ain   C̲hain

On paper menu see: GROUP
Command Window Prompts:
1. Select the item
   `Identify Solid Element`
2. Pick the solid
   `Accept/Identify hole element`
3. Pick a hole element
   ⇨ The prompt repeats.
   ⇨ Click a RESET when complete.
To exit item: Make next selection

✍ *Note:* Holes and solids that reside on different planes cannot be grouped.
See Also: ACTIVE AREA settings

GROUP SELECTION   Ctrl+G

Command: Group Selection
   Places selected elements into a group. Once in a group, elements can be manipulated as one entity. Handles appear at the groups boundary.

On sidebar menu pick: n/a
For window, point to: Edit Group
On paper menu see: TOP BORDER

☞ *Tip:* Store the selected group in a cell library as a cell.
See Also: UNGROUP

GU=

Alternate Key-in
   Alternate key-in for ACTIVE GRIDUNIT. Use it to set the number of working units between each grid unit on the display grid.
See Also: ACTIVE GRIDUNIT

HATCH   HA

Drawing Tool: Hatch Element Area
   Patterns a closed shape, or area, with lines. This item expects two input points. The first point identifies the shape. The second point accepts the element and starts the patterning process.
   The ACTIVE PATTERN DELTA setting controls the space between the pattern lines. The ACTIVE PATTERN ANGLE controls the angle of the pattern lines.

Item Selection
On sidebar menu pick: `Pattrn`
For window, point to: Palettes Patterning

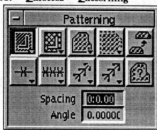

On paper menu see: PATTERNING
Command Window Prompts:
1. Select the item
   `Identify element`
2. Pick the pattern element
   `Patterning in progress...`
To exit item: Make next selection

See Also: ACTIVE PATTERN settings, CROSSHATCH, PATTERN AREA and PATTERN LINEAR tools

HELP   HE

Command: Help
Invokes an online help utility while in a drawing file. Key in HELP followed by the complete tool name.
Entering HELP by itself brings up an index listing of the available MicroStation items. Unix users need to see SET HELP.
HELP CIRCLE brings up a selection list of all the items with the word CIRCLE in the name.
HELP PLACE CIRCLE CENTER produces a help screen to assist in placing a circle by its center.

Item Selection
On sidebar menu pick: | HELP |

For window, point to: Help
On paper menu see: TOP BORDER
See Also: HELP commands, other books by OnWord Press

HELP CONTEXT   HE C

Command: Help on Context
Opens the Help window and activates the context-sensitive help portion of MicroStation. You select any menu item, control or dialog box item, and the Help information appears in the Help window.

Item Selection
On sidebar menu pick: | Help | | Contxt |

For window, point to: Help   On Context
On paper menu see: n/a

✍ *Note:* Close the Help window when you are done using Help.
See Also: HELP commands

HELP HELP   HE

Command: Help on Help
Opens the Help window and activates the article which explains the MicroStation Help features.

Item Selection
On sidebar menu pick: | Help | | Help |

For window, point to: Help   On Help
On paper menu see: n/a

✍ *Note:* Close the Help window when you are done using Help.
See Also: HELP commands

HELP KEYS   HE K

Command: Help on Keys
Opens the Help window and displays a cross-reference list of Tool names and their alternate key-ins.

Item Selection
On sidebar menu pick: | Help | | Keys |

For window, point to: Help   On Keys
On paper menu see: n/a

✍ *Note:* Close the Help window when you are done using Help.
See Also: HELP commands

HELP TOPICS
HE T

Command: Help on Topics

Opens the Help window and displays a list help topics. Choose the topic of interest with the pointer. The information about the tool, control or setting appears in the Help window.

Item Selection

On sidebar menu pick: `Help` `Topic`

For window, point to: Help On Topics

On paper menu see: n/a

🖎 *Note:* Close the Help window when you are done using Help.

See Also: HELP commands

Hline

Obsolete

The word HLINE cannot be typed in to execute the hidden line removal process. Perform this process with the HLINE.UCM user command.

See Also: RENDER ALL HLINE, RENDER FENCE HLINE, RENDER VIEW HLINE

Hlinelem

Obsolete

The word HLINELEM cannot be typed in to execute the hidden line removal process. Perform this process with the HLINELEM.UCM user command.

See Also: RENDER ALL HLINE, RENDER FENCE HLINE, RENDER VIEW HLINE

ICONS
IC

Unix Only
View Control: Icons

Displays a box of four icons on the computer screen. They are the collapse, pop-to-top, pop-to-bottom and repaint icons.

Item Selection

For window, point to: n/a

On sidebar menu pick: n/a

On paper menu see: UTILITIES

IDENTIFY CELL
ID

Drawing Tool: Identify Cell

Displays the name and level of an existing cell in the drawing file. This item takes two input points. The first point identifies the cell and the second point starts the identification process.

Item Selection

On sidebar menu pick: `Cells` `Ident`

For window, point to: Palettes Main Cells

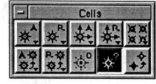

On paper menu see: CELLS

Command Window Prompts:

1. Select the item
   Identify element
2. Pick a cell
   Accept/Reject (select next input)

```
Cell: DESK, Levels: 2
```
To exit item: Make next selection
See Also: IDENTIFY TEXT

IDENTIFY TEXT   ID T

Drawing Tool: Display Attributes of Text Element

Displays the node number, line length, line spacing, level, font and text type of existing text in the drawing file. This item takes one input point for the identification to take place.

Item Selection

On sidebar menu pick: Text   Attr   Disp

For window, point to: Palettes   Main   Text

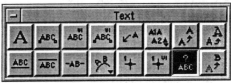

On paper menu see: MODIFY TEXT

Command Window Prompts:
1. Select the item
   ```
   Identify text element
   ```
2. Pick the text
   ```
   Type=TEXT, Level=1
   NN=3, LL=255, LS=0:8.000, LV=1, FT=0
   ```
To exit item: Make next selection
See Also: IDENTIFY CELL

IMPOSE BSPLINE SURFACE BOUNDARY   I B S B

Drawing Tool: Impose B-spline Surface Boundary

Impose a B-spline curve onto a 3D surface. The curve is projected perpendicular to the to the b-spline surface. This tool expects two input points to identify the curve and the surface, and a third point to complete the projection.

Item Selection

On sidebar menu pick: Place   BSpl   SpcCur

For window, point to: Palettes   3D B-splines   Space Curves

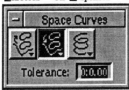

On paper menu see: SPACE CURVE

Command Window Prompts:
1. Select the item
   ```
   Identify element
   ```
2. Pick B-spline curve to project
   ```
   Identify surface
   ```
3. Pick B-spline surface
   ```
   Accept/Reject (select next input)
   ```
4. Give an accept point

To exit item: Make next selection
See Also: EXTRACT BSPLINE SURFACE BOUNDARY

INCLUDE INCL=filename

Command: Import Text File

Inserts an ASCII text file into a drawing file. The characters from the text file become a text node. Text files longer than 40 lines or four hundred characters become graphic groups.

At the "USTN" prompt, key in INCLUDE filename. Where filename is the name of the text file you want placed in the drawing file. A rectangle attaches to the cursor. This box shows the size and justification of the text. You can adjust any active text setting at this time. Place the text in the drawing by clicking an input point.

Item Selection
On sidebar menu pick: Text Incl
For window, point to: File Import Text...
On paper menu see: n/a
Where filename is a user defined text file.

See Also: ACTIVE TAB

INCREMENT ED INCR E

Drawing Tool: Copy and Increment Enter Data Field

Copies and increments numeric characters in Enter Data Fields. For example, copy the text string 001 four times to generate the text strings 002, 003, 004 and 005. This item requires one input point to identify the text string to increment. Each subsequent point will create a newly incremented test string.

See ACTIVE TAG to determine the increment number. If the copied text is a combination of alphabetic and numeric characters the numbers must be at the right end of the text string.

Item Selection
On sidebar menu pick: Text ED Fld Copy CIEd
For window, point to: Palettes Main Enter Data Fields

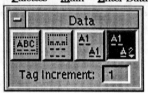

On paper menu see: ENTER DATA
Command Window Prompts:
1. Select the item
   Select enter data field to copy
2. Pick enter data field
   Select destination enter data field
3. Select the new enter data field
To exit item: Make next selection

✍ *Note:* You can only copy text in an Enter Data field to another Enter Data field.

✳ *Error:* Tag overflow - If this message appears, you cannot increment the text any higher. For instance, you cannot increment 998 until it reaches 1000. 999 is as high as you can go.
See Also: INCREMENT TEXT

INCREMENT TEXT INCR

Drawing Tool: Copy and Increment Text

Copies and increments text. For example, you can copy the text 001 four times to generate the text strings 002, 003, 004 and 005. This command requires one input point to identify the text string to increment. Each subsequent point creates a newly incremented test string.

The ACTIVE TAG (TI=) setting determines the increment number for the text. If the text copied is a combination of alphabetic and numeric characters the numbers must be at the end of the text string.

Item Selection
On sidebar menu pick: `Text` `ED Fld` `Copy` `CITx`
For window, point to: Palettes Main Text

On paper menu see: TEXT
Command Window Prompts:
1. Select the item
   `Identify element`
2. Pick the text number
   `Accept/Reject (select next input)`
3. The accept point is also the placement point for the text.
To exit item: Make next selection

✱ *Error:* `Tag overflow` - If this message appears you cannot increment the text any higher. For instance, you cannot increment 998 until it reaches 1000. 999 is as high as you can go.
See Also: INCREMENT ED

INDEX   IND

Command: Index Cell Library
Creates an index or catalog file. The file contains the cell name of every cell in the cell library attached to the drawing file. Indexing a cell library shortens the search time needed to look up a cell name. Selecting this item automatically starts the indexing process. If you indexed CGSI.CEL, the indexing process produces a file in the default directory named CGSI.CDX.

Item Selection
On sidebar menu pick: `Cells` `Index`
For window, point to: n/a
On paper menu see: n/a
See Also: ATTACH LIBRARY, CREATE CELL

INSERT VERTEX   INS

Drawing Tool: Insert Vertex
Adds a vertex to an existing line or line string. Adding a vertex to a line turns it into a line string. This item requires two input points. The first point identifies where you want the vertex added. The second point accepts the operation.

Item Selection
On sidebar menu pick: `Modify` `Vertex` `Ins`

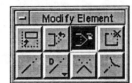

For window, point to: Palettes Main Modify Element
On paper menu see: MODIFY
See Also: DELETE VERTEX

INTERSECTION

Obsolete

The INTERSECTION command allows you to snap to the intersection of two elements. Make sure to set either LOCK SNAP PROJECT or LOCK SNAP KEYPOINT are on. This command requires two tentative input points. First snap to one of the intersecting elements with a tentative point. Then key in the word INTERSECTION. Now identify the other element with a tentative point. The cursor jumps to the intersection of the two elements.

See Also: LOCK SNAP INTERSECTION

IUPDATE   IU opt

View Control: Iupdate

Performs an incremental update of a selected view. Application programs use this update feature. Two variables, WWSECT and WWBYTE, store the current working window location. IUPDATE refreshes the screen from the working window to the end of the drawing file.

Item Selection
On sidebar menu pick: n/a
For window, point to: n/a
On paper menu see: n/a

Where opt is a valid IUPDATE choice:

Choices(s):
IUPDATE
⇨   Asks you to pick the view to update.
IUPDATE [viewnumber]
⇨   Updates a specified view number (1,2,3,etc).
IUPDATE [screen]
⇨   Updates all views on the RIGHT or LEFT screen.
IUPDATE BOTH
⇨   Updates both the primary and secondary screens.
IUPDATE VIEW
⇨   Updates a view based on a "Select view" data point.
IUPDATE GRID
⇨   Updates the display grid for a view.
IUPDATE FENCE [choice]
⇨   Updates a fence area (INSIDE or OUTSIDE).
IUPDATE FILE
⇨   Updates the active drawing file.
IUPDATE TUTORIAL
⇨   Updates the active tutorial.
IUPDATE REFERENCE
⇨   Updates attached reference files.
IUPDATE ALL
⇨   Updates the whole shootin' match.

See Also: UPDATE view control

JUSTIFY CENTER   JU

Setting: Center Justify Enter Data Field

Centers text placed in an Enter Data field area. Set justification before placing the text.

Item Selection
On sidebar menu pick: | Text | | ED Fld | | Just | | Center |
For window, point to: n/a
On paper menu see: ENTER DATA
To show, key in: ᴊᴜ
    LEFT JUSTIFY ED FIELD
To set, key in: ᴊᴜ c
See Also: JUSTIFY settings, PLACE TEXT tools, Enter Data Fields

JUSTIFY LEFT   JU

Setting: Left Justify Enter Data Field
Left justifies text placed in an Enter Data field area. Set justification before placing the text.

Item Selection
   On sidebar menu pick: | Text | | ED Fld | | Just | | Left |

   For window, point to: n/a
   On paper menu see: ENTER DATA
   To show, key in: ʊ
     LEFT JUSTIFY ED FIELD
   To set, key in: ʊ **L**
See Also: JUSTIFY settings, PLACE TEXT tools, Enter Data Fields

JUSTIFY RIGHT   JU

Setting: Right Justify Enter Data Field
Right justifies text placed in an Enter Data Field area. Set justification before placing the text.

Item Selection
   On sidebar menu pick: | Text | | ED Fld | | Just | | Right |

   For window, point to: n/a
   On paper menu see: ENTER DATA
   To show, key in: ʊ
     LEFT JUSTIFY ED FIELD
   To set, key in: ʊ **R**
See Also: JUSTIFY settings, PLACE TEXT tools, Enter Data Fields

KY=

Alternate Key-in
Alternate key-in for ACTIVE KEYPNT. Use it to divide elements into segments for snapping purposes.
See Also: ACTIVE KEYPNT

LABEL LINE   LA

2D Drawing Tool: Label Line
Labels a line's length and direction. This item requires two input points. The first point identifies the line and labels it. The next input point accepts the label.
The Design Options setting determines the text characteristics of the label placed on the line.

Item Selection
   On sidebar menu pick: | Dims | | Line |
   For window, point to: Palettes   Dimensioning

   On paper menu see: LINEAR DIMENSION
Command Window Prompts:
  1. Select the item
    Identify element
  2. Pick an element.
    ⇨   The element highlights and the label text appears.
    Accept/Reject
  3. Give an accept point.
    ⇨   Move the cursor away from the element before you give the accept point.
  To exit item: **Make next selection**

✐ *Note:* Identifying the endpoints of a line clockwise versus counterclockwise gives you different degree readings.

☞ *Tip:* Use the LABEL LINE tool as a quick measurement tool. After you click the first point and see the measurements, hit RESET and the label disappears.

See Also: Dimensioning

LC=

Alternate Key-in
> Alternate key-in for ACTIVE STYLE. Use it to set the line style during element creation.

See Also: ACTIVE STYLE

LD=

Alternate Key-in
> An alternate key-in for DIMENSION LEVEL. Use it to determine the level for the dimension data.

See Also: DIMENSION LEVEL

LISTEN   LI

Setting: Listen
> Plays back a recorded design session. Key in LISTEN followed by the filename of the record session. You may need to use the full path if the file does not reside in the default directory.

Item Selection
On sidebar menu pick: n/a
For window, point to: n/a
On paper menu see: n/a

☞ *Tip:* To halt a recording session early, type Ctrl+C.

☞ *Tip:* To pause a recording session, type Ctrl+S. Press any key to resume the play-back session.

See Also: RECORD

LL=

Alternate Key-in
> Alternate key-in for ACTIVE LINE LENGTH. Use it to control the number of characters in a line of text of a text node.

See Also: ACTIVE LINE LENGTH

LOAD DA   LOA DA

Drawing Tool: Load Displayable Attributes
> Assigns attribute data to a text node attached to a database record. This item expects two input points. The first point placed highlights the target text node and the second point accepts that text node as the candidate for database linkage.

Item Selection
On sidebar menu pick: n/a
For window, point to: Palettes   Database

On paper menu see: DATABASE
See Also: ACTIVE DATYPE, FENCE LOAD

LOCELE

Command: Locate Element

Identifies an element's type and level. (LOCELE is short for LOCATE ELEMENT.) This item requires one input point to identify the element in question. The element's type and level location appear in the status field.

Item Selection
On sidebar menu pick: n/a
For window, point to: n/a
On paper menu see: n/a
Command Window Prompts:
1. Select the item
   Identify element
2. Pick an element
   Type=LINE, Level=1
   ⇨   This message appears in the status field.
To exit item: Make next selection

☞ **Tip:** Use this command if you are unsure of an element's type or the level that it occupies.

LOCK

Setting: Lock

Key in LOCK to see what locks are set. Locks help you control your input points in the drawing file. They make sure you get the same results each time. For example, if you want to work with elements on level 5 only, you turn on the LOCK LEVEL setting. Now your input points lock onto level 5 until you turn the lock off. If you want all of the input points placed on the display grid, use the LOCK GRID setting.

The LOCK settings work just like light switches. If you click the lock setting when it's on, it goes off. When you want the lock back on, you toggle the switch one more time.

Item Selection
On sidebar menu pick: Locks   Show
For window, point to: Settings   Locks   Full
On paper menu see: LOCKS
Command Window Prompts:
To show, open: Locks settings box
See Also: LOCK settings

LOCK ACS

3D Setting: Lock Auxiliary Coordinate System Plane

Locks input points to points on the active Auxiliary Coordinate System (ACS). This lock only works in 3D drawing files. Selecting the setting toggles the lock on and off.

Item Selection
On sidebar menu pick: Locks   ACS
For window, point to: Settings   Locks   Full   ACS Plane
On paper menu see: LOCKS
Command Window Prompts:
To show, key in: LOCK CONSTRUCTION
   ACS Plane Lock : OFF
To set, key in: LOCK CONSTRUCTION on/off
   Where on/off is either ON or OFF.
See Also: DEFINE ACS tools

LOCK ANGLE

Setting: Lock Angle

Rounds Define Angle/Scale values to the ACTIVE ANGLE setting. Selecting the setting toggles the lock on or off. For instance, set the angle round-off 1.0 and

170 LOCK ASSOCIATION

turn the lock on. When you enter AA=30.25 the ACTIVE ANGLE setting rounds
off to 30.

Item Selection
On sidebar menu pick: `Locks` `Angle`
For window, point to: Settings   Active Angle   Angle Lock
On paper menu see: LOCKS
To show, key in: LOCK
  Locks=AN,
To set, key in: LOCK ANGLE nnn
  Where nnn is a positive or negative number.

☞ *Tip:* Use the LOCK ANGLE setting to round off angle entries automatically.
See Also: ACTIVE ANGLE

LOCK ASSOCIATION   LOCK AS

Setting: Lock Association
  Associates related elements. For example, a point along an element might be
  associated with a dimension line or multi-line element. Create association points
  by toggling LOCK ASSOCIATION on.

Item Selection
On sidebar menu pick: `Locks` `Snap` `Assoc.`
For window, point to: Settings   Locks   Full   Association
On paper menu see: LOCKS
To show, key in: LOCK
  Locks=AS,GG,SN
To set, key in: LOCK ASSOCIATION on/off
  Where on/off is either ON or OFF.
See Also: LOCK SNAP ACS

LOCK AXIS   LOCK AX

Setting: Lock Axis
  Forces every input point to fall on the ACTIVE AXIS. Selecting the setting
  toggles the lock on of off. Set the ACTIVE AXIS to 0/90, and all input points lock
  onto the X or Y axis.

Item Selection
On sidebar menu pick: `Locks` `Axis`
For window, point to: Settings   Locks   Full   Axis Lock
On paper menu see: LOCKS
To show, key in: LOCK
  Locks=AX,
To set, key in: LOCK AXIS opt
  Where opt either Key, 30, 45, 0/90 or Orig.

☞ *Tip:* Use the 0/90 axis setting when you need to draw straight lines and 90 degree
  angles. Use the 30 degree setting to draw isometric piping spool drawings in 2D
  files.
See Also: LOCK ANGLE

LOCK BORESITE   LOCK B

3D Setting: Lock Boresite
  Aids location of elements in a 3D file. Normally, it is very difficult to snap to
  elements in a 3D file unless the elements reside close to the active depth. With
  LOCK BORESITE on, any element in the 3D file can be located easily. Selecting
  the setting toggles the lock on or off.

Item Selection
On sidebar menu pick: `Locks` `Bore`
For window, point to: Settings   Locks   Full   Boresite
On paper menu see: LOCKS
To show, key in: LOCK
  Boresite - OFF

To set, key in: `LOCK BORESITE on/off`
Where on/off is either ON or OFF.

✍ *Note:* The status of the boresite lock does not appear on the status line when keying in LOCKS.

✳ *Error:* `3-D cmd in 2-D file` - If this appears in the error field, you are attempting to use the LOCK BORESITE setting in a 2D file. This lock only operates in 3D files.
See Also: LOCK SNAP settings

LOCK CELLSTRETCH   LOCK CE

Setting: Lock Cellstretch
Stops cells from being stretched during a FENCE STRETCH operation. Cells contained totally within the limits of the fence move, but do not stretch. Selecting the setting toggles the lock on or off.

Item Selection
On sidebar menu pick: n/a
For window, point to: n/a
On paper menu see: LOCKS
To show, key in: `LOCK CELLSTRETCH`
`    Cell Stretch Lock : OFF`
To set, key in: `LOCK CELLSTRETCH on/off`
Where on/off is either ON or OFF.
See Also: FENCE STRETCH

LOCK CONSTRUCTION

DOS and Unix Only

This item is the same as LOCK ACS.
See Also: LOCK ACS

LOCK CONSPLANE   LOCK CO

Mac Only
3D Setting: Lock Construction Plane
Locks all input points to the XY axis of an Auxiliary Coordinate System. Selecting the setting toggles the lock on or off.

Item Selection
On sidebar menu pick: n/a
For window, point to: Settings Locks Full   ACS Plane
On paper menu see: n/a
To show, key in: `LOCK CONSPLANE`
`    Construct Plane Lock : OFF`
To set, key in: `LOCK CONSPLANE on/off`
Where on/off is either ON or OFF.
See Also: DEFINE ACS tools

LOCK FENCE CLIP   LOCK F C

Setting: Lock Fence Clip
Breaks apart, or clips, elements crossing the fence boundary. Selecting the setting toggles the lock on.

Item Selection
On sidebar menu pick: `Locks` `Fence` `Clip`
For window, point to: Settings Locks Full   Fence Selection:
On paper menu see: PLACE FENCE
To show, key in: `LOCK FENCE`
`    Fence contents - Inside`
To set, key in: `LOCK FENCE CLIP`

✍ *Note:* The LOCK FENCE setting cannot be toggled on or off. One of the LOCK FENCE settings is always active.

☞ *Tip:* Keep the LOCK FENCE INSIDE active as the default setting. This way you only effect elements within the fence.
See Also: LOCK FENCE settings, FENCE tools

LOCK FENCE INSIDE   LOCK F I

Setting: Lock Fence Inside
        Locks only those elements lying entirely within a fence. Selecting the setting toggles the lock on.

Item Selection
On sidebar menu pick: | Locks | | Fence | | Inside |
For window, point to: Settings   Locks   Full   Fence Selection:
On paper menu see: PLACE FENCE
To show, key in: LOCK FENCE
        Fence contents - Inside
    To set, key in: LOCK FENCE INSIDE

✍ *Note:* The LOCK FENCE setting cannot be toggled on or off. One of the LOCK FENCE settings is always active.

☞ *Tip:* Keep the LOCK FENCE INSIDE active as the default setting. This way you only effect elements within the fence.
See Also: LOCK FENCE settings, FENCE tools

LOCK FENCE OVERLAP   LOCK F O

Setting: Lock Fence Overlap
        Locks elements contained within a fence as well as those passing through it. Selecting the setting toggles the lock on.

Item Selection
On sidebar menu pick: | Locks | | Fence | | Ovrlap |
For window, point to: Settings   Locks   Full   Fence Selection:
On paper menu see: PLACE FENCE
To show, key in: LOCK FENCE
        Fence contents - Inside
    To set, key in: LOCK FENCE OVERLAP

✍ *Note:* The LOCK FENCE setting cannot be toggled on or off. One of the LOCK FENCE settings is always active.

☞ *Tip:* Keep the LOCK FENCE INSIDE active as the default setting. This way you only effect elements within the fence.
See Also: LOCK FENCE settings, FENCE tools

LOCK FENCE VOID CLIP   LOCK F V C

Setting: Lock Fence Void Clip
        Clips out an area of the drawing file that you do not want shown. First, select the LOCK FENCE VOID CLIP setting. Then place a fence around the area you no longer want to see. Now update the screen; the clipped area appears as a void on the display screen.

Item Selection
On sidebar menu pick: | Locks | | Fence | | Void | | Clip |
For window, point to: Settings   Locks   Full   Fence Selection:
On paper menu see: n/a
To show, key in: LOCK FENCE VOID
        Fence contents - Outside Void
    To set, key in: LOCK FENCE VOID CLIP
See Also: LOCK FENCE VOID settings, REFERENCE CLIP VOID

LOCK FENCE VOID OUTSIDE   LOCK F V OU

Setting: Lock Fence Void Outside
Turns off display of elements outside a fenced area. First select the LOCK
FENCE VOID OUTSIDE setting. Then place a fence around the area you want
shown. Now update the screen; the elements outside the fence are hidden from
view.

Item Selection
On sidebar menu pick: | Locks | | Fence | | Void | | Outsde |
For window, point to: Settings   Locks   Full   Fence Selection:
On paper menu see: n/a
To show, key in: LOCK FENCE VOID
   Fence contents - Outside Void
To set, key in: LOCK FENCE VOID OUTSIDE
See Also: LOCK FENCE VOID settings, REFERENCE CLIP VOID

LOCK FENCE VOID OVERLAP   LOCK F V OV

Setting: Lock Fence Void Overlap
Turns off the display of elements outside a fenced area, as well as those which
pass through the fence. First select the LOCK FENCE VOID OVERLAP setting.
Then place a fence around the area you want shown. Now update the screen. The
elements outside the fence are hidden from view.

Item Selection
On sidebar menu pick: | Locks | | Fence | | Void | | Overlp |
For window, point to: Settings   Locks   Full   Fence Selection:
On paper menu see: PLACE FENCE
To show, key in: LOCK FENCE VOID
   Fence contents - Outside Void
To set, key in: LOCK FENCE VOID OVERLAP
See Also: LOCK FENCE VOID settings

LOCK GGROUP   LOCK GG

Setting: Lock Graphic Group
Locks/Unlocks an associated group of graphic elements. Elements in a
graphic group only respond as a group when the LOCK GGROUP setting is on.
Elements have an on-again-off-again association by turning the LOCK GGROUP
on and off. GGROUP is short for Graphic Group; patterns and dimension data are
two examples of graphic groups.

Item Selection
On sidebar menu pick: | Locks | | GrGrp |
For window, point to: Settings   Locks   Full   Graphic Group
On paper menu see: LOCKS
To show, key in: LOCK
   Locks=GG,
To set, key in: LOCK GGROUP on/off
   Where on/off is ON or OFF.

☞ *Tip:* When you want to delete a piece of a pattern, make sure the LOCK GGROUP
setting is off.

☞ *Tip:* If you want to delete the whole pattern, turn the lock on. If you want to modify
part of the pattern, turn the lock off.

☞ *Tip:* When deleting just one witness line in a dimension, turn the lock off. If you
want the whole dimension to disappear, the lock must be on. All dimensions placed
with a single dimension tool, stacked dimensions for instance, are deleted if LOCK
GGROUP is on.

See Also: Graphic Group

LOCK GRID   LOCK GR

Setting: Lock Grid
    Locks all input points onto the display grid. Selecting the setting toggles the
lock on or off.

Item Selection
    On sidebar menu pick: `Locks`   `Grid`
    For window, point to: Settings   Locks   Full   Grid Lock
    On paper menu see: LOCKS
    To show, key in: LOCK
        Locks=GR,
    To set, key in: LOCK GRID

☞ *Tip:* When you are working on non-scaled drawings like forms, charts or flow
    diagrams, turn on the LOCK GRID setting to force input points to the display grid.
    See Also: SET GRID

LOCK ISOMETRIC   LOCK IS

Setting: Lock Isometric
    Locks all input points to the current isometric plane. Selecting the setting
toggles the lock on or off.

Item Selection
    On sidebar menu pick: `Locks`   `Iso`
    For window, point to: Settings   Locks   Full   Isometric Lock
    On paper menu see: LOCKS
    To show, key in: LOCK
        Locks=ISO
    To set, key in: LOCK ISOMETRIC
    See Also: ACTIVE GRIDMODE ISOMETRIC, SET ISOPLANE

LOCK LEVEL   LOCK L

Setting: Lock Level
    Prevents manipulations of elements on levels other than the active level. For
example, if the active level is 3 and there are lines on level 3 and circles on level
4, you can manipulate the lines. You cannot touch the circles until you either turn
off LOCK LEVEL or change the active level to 4. Selecting the setting toggles the
lock on or off.

Item Selection
    On sidebar menu pick: `Locks`   `Level`
    For window, point to: Settings   Locks   Full   Level Lock
    On paper menu see: LOCKS
    To show, key in: LOCK
        Locks=LV,
    To set, key in: LOCK LEVEL

✍ *Note:* Fence operations do not honor the LOCK LEVEL setting. If you have elements
    on different levels and you have the LOCK LEVEL on, a fence tool affects all the
    elements, not just the ones on the active level.

☞ *Tip:* Turn LOCK LEVEL on to prevent mistakes if you are uncertain about a level
    scheme or how a tool works.
    See Also: ACTIVE LEVEL

LOCK SCALE   LOCK SC

Setting: Lock Scale
    Rounds off ACTIVE SCALE entries to the value set in the Active Scale
settings box. For instance, with the scale round-off set to 1.0 and the lock on,
entering AS=1.25 sets the ACTIVE SCALE to 1.0. Selecting the setting toggles
the lock on or off.

Item Selection
    On sidebar menu pick: `Locks`   `Scale`

For window, point to: Settings   Active Scale   Scale Lock
On paper menu see: LOCKS
To show, key in: LOCK
  Locks=SC,
To set, key in: LOCK SCALE

☞ *Tip:* Use the LOCK SCALE setting to automatically round scale entries.
See Also: ACTIVE SCALE

LOCK SELECTION   LOCK SE

Setting: Lock Selection
  Toggles between single-shot and cyclical (repetitive) command execution. For
  example, to execute operations such as COPY, MOVE and DELETE over and over
  without having to reselect the command each time, set LOCK SELECTION ON.
  If you toggle the LOCK SELECTION setting off, each operation executes only once.
  Selecting the setting toggles the lock on or off.
Item Selection
On sidebar menu pick: | Utils | | User | | Pref |
For window, point to: User   Preferences...   Single Click
On paper menu see: n/a
To show, key in: LOCK
  SELECTION SET LOCK : ON
To set, key in: LOCK SELECTION on/off
  Where on/off is ON or OFF.

LOCK SNAP ACS   LOCK SN A

Setting: Lock Snap Auxiliary Coordinate System
  Locks any future input point to the active Auxiliary Coordinate System plane.
  This setting is the same as LOCK SNAP CONSTRUCTION.
Item Selection
On sidebar menu pick: | Locks | | Snap | | ACS |
For window, point to: Settings   Locks   Full   ACS Plane Snap
On paper menu see: LOCKS
To show, key in: LOCK
  ACS Plane Snap: OFF
To set, key in: LOCK SNAP ACS on/off
  Where on/off is either ON or OFF.
See Also: LOCK ASSOCIATION, LOCK SNAP settings

LOCK SNAP CONSTRUCTION

  This setting is the same as LOCK SNAP ACS.
See Also: LOCK SNAP settings

LOCK SNAP INTERSECTION   LOCK SN I

Setting: Snap Lock Intersection
  Snaps an element origination point to an intersection.
Item Selection
On sidebar menu pick: | Locks | | Snap | | Intsec |
For window, point to: Settings   Locks   Full   Mode:
On paper menu see: LOCKS
To show, key in: LOCK
  Locks=PS,SN
To set, key in: LOCK SNAP INTERSECTION on/off
  Where on/off is either ON or OFF.
See Also: LOCK SNAP settings

LOCK SNAP KEYPOINT

LOCK SN K

Setting: Lock Snap Keypoint
   Forces the tentative point to snap to "keypoints" on the nearest element. The
   LOCK SNAP settings work in conjunction with the tentative button on a cursor
   or mouse.
   Element keypoints:
   Arcs: the 0,90,,270 points and center if sweep is more than 90 degrees
   Circles: the center, the 0,90, and 270 degree points along the circle
   Lines: the endpoints
   Line strings: any vertex
   Shapes: any vertex
   Text: the justification point

Item Selection
   On sidebar menu pick: Locks   Snap   Keypt
   For window, point to: Settings   Locks   Full   Mode:
   On paper menu see: LOCKS
   To show, key in: LOCK
      Locks=SN,
   To set, key in: LOCK SNAP KEYPOINT

☞ *Tip:* To have more keypoints on an element, see ACTIVE KEYPNT.
   See Also: ACTIVE KEYPNT, LOCK SNAP PROJECT, REFERENCE SNAP

LOCK SNAP PROJECT

LOCK SN P

Setting: Lock Snap Project
   Forces the tentative point to snap to the closest point on the nearest element.
   The LOCK SNAP settings work in conjunction with the tentative button on a
   cursor or mouse.

Item Selection
   On sidebar menu pick: Locks   Snap   Proj
   For window, point to: Settings   Locks   Full   Mode:
   On paper menu see: LOCKS
   To show, key in: LOCK
      Locks=SN,
   To set, key in: LOCK SNAP PROJECT
   See Also: LOCK SNAP KEYPOINT, REFERENCE SNAP

LOCK TEXTNODE

LOCK T

Setting: Lock Text Node
   Places text at the exact location of text nodes. The text placed at these nodes
   takes on the text width, height, color and font settings of the text node. Selecting
   the setting toggles the lock on or off.

Item Selection
   On sidebar menu pick: Locks   Node
   For window, point to: Settings   Locks   Full   Text Node
   On paper menu see: LOCKS
   To show, key in: LOCK
      Locks=TN,
   To set, key in: LOCK TEXTNODE on/off
      Where on/off is either ON or OFF.

☞ *Tip:* Use LOCK TEXTNODE for more rapid placement of text at text nodes.
   See Also: PLACE NODE tools

LOCK UNIT

LOCK U

Setting: Lock Unit
   Forces all input points to the nearest unit round-off value. Selecting the
   setting toggles the lock on or off.

Item Selection
  On sidebar menu pick: | Locks | | Unit |
  For window, point to: Settings Locks Full Unit Lock
  On paper menu see: LOCKS
  To show, key in: LOCK
    Locks=UN,
  To set, key in: LOCK UNIT on/off
    Where on/off is either ON or OFF.
See Also: ACTIVE UNITROUND

LS=

Alternate Key-in
    Alternate key-in for ACTIVE LINE SPACE. Use it to set the spacing between lines of text.
See Also: ACTIVE LINE SPACE

LT=

Alternate Key-in
    Alternate key-in for ACTIVE TERMINATOR. Use it to set the name of the cell placed at the end of a line.
See Also: ACTIVE TERMINATOR

LV=

Alternate Key-in
    Alternate key-in for ACTIVE LEVEL. Use it to set the active level in the drawing file.
See Also: ACTIVE LEVEL

MARK   MAR

Command: Set Mark
    Places a bookmark in the drawing file for UNDO and REDO operations. Selecting the item places a mark in the drawing file. Keying in UNDO MARK causes MicroStation to undo the element operations up to the mark. The REDO MARK option performs the opposite operation by REDOing all the element operations that where UNDOne. Multiple marks may be placed in the same drawing file.

Item Selection
  On sidebar menu pick: | UnDo | | Mark |
  For window, point to: Edit Set Mark
  On paper menu see: n/a
Command Window Prompts:
  1. Select the item
    Current position MARKed
  To exit item: Exits by itself
See Also: REDO, SET UNDO, UNDO

MATRIX CELL   CM=

Drawing Tool: Place Active Cell Matrix
    Creates an array, or matrix, of several more cells based on an existing cell. For example, the MATRIX CELL tool could be used to lay out an array of building columns on a floor plan. This item requires an input point and four key-ins. The input point identifies the cell to be used in the matrix. The first two key-ins determine the number of rows and columns in the matrix. The last two key-ins control the spacing between the rows and columns.

Item Selection
  On sidebar menu pick: | Cells | | Matrix |

For window, point to: <u>P</u>alettes   <u>M</u>ain   <u>C</u>ells

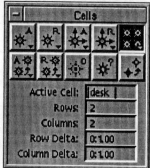

On paper menu see: CELLS
Command Window Prompts:
1. Select the item
```
Matrix Cell (cm=nrow,ncol,drow,dcol)
```
Where nrow is the number of horizontal rows.
Where ncol is the number of vertical columns.
Where drow is the distance between rows.
Where dcol is the distance between columns.

Example(s):
`CM=4,6,3,3`   this syntax produced the example
To exit item: **Exits by itself**
See Also: PLACE CELL tools

MC   MC

Command: Menu Check
  Checks the registration of a paper menu on a digitizing tablet. Key in MC,
and a block outline of the attached menu appears on the screen. Place the cursor
at a known location on the menu. Now verify that the cursor appears on the screen
at the same location on the menu block outline.
  If the menu is shifted, use the AM= key-in to reattach the menu. If the menu
registration is ok, click a RESET to redisplay the drawing file.
Item Selection
  On sidebar menu pick: n/a
  For window, point to: <u>S</u>ettings   Di<u>g</u>itizing   <u>T</u>ablet   <u>C</u>heck Menus
  On paper menu see: n/a
See Also: ATTACH MENU

MDL COMMAND   MD C <command name>

Command: Activate an MDL Application Command
  Starts an MDL application command.
Item Selection
  On sidebar menu pick: n/a
  For window, point to: n/a
  On paper menu see: n/a
See Also: MDL commands

MDL DEBUG   MD D

Command: Debug an MDL Application
  Turns on the debugger for an MDL application already loaded into the
system. This option is for debugging application programs under development.
Item Selection
  On sidebar menu pick: n/a
  For window, point to: n/a

On paper menu see: n/a
See Also: MDL commands

MDL DLOGLOAD

Command: Load MDL Applications
   Same as MDL LOAD.
See Also: MDL LOAD

MDL LOAD   MD L

Command: Load an MDL Application
   Opens the MDL settings box. Click on the MDL application name you want
loaded and then click on Load. A DEBUG or NODEBUG parameter can be added
to the MDL LOAD statement. These options are used as development tools.

Item Selection
   On sidebar menu pick: `Utils`   `User`   `MDL`
   For window, point to: Üser   MDL Applications
   On paper menu see: UTILITIES

✐ Note: If you know you want the calculator application loaded, you could key in MDL
LOAD CALCULAT.
See Also: MDL commands

MDL UNLOAD   MD U

Command: Unload an MDL Application
   Opens the MDL settings box. Click on the MDL application name you want
removed and then click on Unload.

Item Selection
   On sidebar menu pick: `Utils`   `User`   `MDL`
   For window, point to: Üser   MDL Applications
   On paper menu see: n/a

✐ Note: If you know you want the B-spline application unloaded, you could key in MDL
UNLOAD SPLINES.
See Also: MDL commands

MEASURE ANGLE   ME AN

Drawing Tool: Measure Angle Between Lines
   Calculates the angle of two line segments. The lines do not have to be
touching. This item needs two input points. The first point identifies the first leg
of the angle and the second point identifies the other angle leg and starts the
measurement process.

Item Selection
   On sidebar menu pick: `Meas`   `Angle`
   For window, point to: Palettes   Measuring
   On paper menu see: MEASURE
Command Window Prompts:
   1. Select first line
      Pick an angle leg
   2. Select second line
      Pick other angle leg
      Angle = 24.73^
   To exit item: **Make next selection**
See Also: MEASURE RADIUS

MEASURE AREA

Same as MEASURE AREA POINTS.
See Also: MEASURE AREA POINTS

MEASURE AREA ELEMENT   ME AR E

Drawing Tool: Measure Area of Element

Measures the area and perimeter of any closed shape. This item requires two input points. The first point identifies the element to measure and the second point accepts the element and starts the measurement process.

Item Selection

On sidebar menu pick: Meas   Area   Elem

For window, point to: Palettes   Measuring

On paper menu see: MEASURE

Command Window Prompts:
1. Select the item
   Identify Element
2. Pick a closed element
   Accept/Reject (select next input)
3. Give an accept point
   A=173.2989 SQ FT, P=46.6662

To exit item: Make next selection

Note: If the element measured does not highlight and
measure properly, maybe the shape is not a "closed"
shape.

See Also: MEASURE AREA POINTS

MEASURE AREA POINTS   ME AR

Drawing Tool Measure Area by Points

Measures an area and perimeter defined by a series of input points. In order to close the shape, the last point must be placed very close to the first point. Clicking a RESET erases the shape from the screen.

Item Selection

On sidebar menu pick: Meas   Area   Pnts

For window, point to: Palettes   Measuring

On paper menu see: MEASURE

Command Window Prompts:
1. Select the item
   Define are to measure
2. Pick a point
   ⇨   A line string appears. Use this line as a guide to form a
        closed shape.
   A=173.2989 SQ FT, P=46.6662
   Area Closed
   ⇨   These prompts appears after the shape closes.

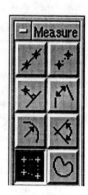

☞ *Tip:* For greater accuracy "snap" to the element you want
measured.

See Also: MEASURE AREA ELEMENT

MEASURE DISTANCE ALONG

ME D A

Drawing Tool: Measure Distance Along Element
Calculates the length of an element. This item requires two input points. The first point identifies the start point of the element and the second point defines the element's end point.

Item Selection
On sidebar menu pick: Meas   Dist   Along
For window, point to: Palettes   Measuring

On paper menu see: MEASURE
Command Window Prompts:

1. Select the item
   Identify Element at first point
2. Pick a point
   Enter end point
3. Pick a second endpoint
   Dist = 16FT 11.702IN
   ⇨ Placing the second point causes the distance to display.
   Measure more points/Reset to reselect
   ⇨ A distance can be accumulated by placing additional points.
To exit item: **Make next selection**

☞ *Tip:* "Snap" to the element you want measured for greater accuracy.

☞ *Tip:* The MEASURE DISTANCE tool only computes the element's distance. Use LABEL LINE in place of MEASURE DISTANCE to measure the length and angle of an element.

See Also: MEASURE DISTANCE tools

MEASURE DISTANCE MINIMUM

ME D MI

Drawing Tool: Measure Minimum Distance Between Elements
Computes the shortest distance between two elements. This item takes three input points. The first two points identify the elements to measure between. The third points initiates the measurement process.

Item Selection
On sidebar menu pick: Meas   Dist   Min
For window, point to: Palettes   Measuring

On paper menu see: MEASURE
Command Window Prompts:

1. Select the item
   Identify first element
2. Pick a point
   Accept, Identify 2nd element/Reject
3. Pick second element
   Accept, Initiate min dist calculation
4. Give an accept point
   Dist = 16FT 11.702IN
   ⇨ Placing the second point causes the distance to display.
To exit item: **Make next selection**

☞ *Tip:* "Snap" to the element you want measured for greater accuracy.

See Also: MEASURE DISTANCE tools

MEASURE DISTANCE PERPENDICULAR

ME D PE

Drawing Tool: Measure Perpendicular Distance From Element
Calculates and shows a temporary perpendicular line between two elements. This item requires an input point to identify the target element and a second point

to define the distance of the perpendicular line. The line erases from the screen by clicking a Reset.

Item Selection

On sidebar menu pick: | Meas | Dist | Perp |

For window, point to: Palettes Measuring

On paper menu see: MEASURE

Command Window Prompts:

1. Select the item
   `Enter first point`
2. Pick a point
   `Enter end point`
3. Pick a second point
4. Dist = 16FT 11.702IN
   ⇨ Placing the second point causes the distance to display.
   `Measure more points/Reset to reselect`
5. Place additional points

To exit item: `Make next selection`

☞ *Tip:* "Snap" to the element you want measured for greater accuracy.

See Also: MEASURE DISTANCE tools

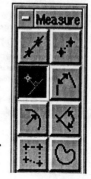

MEASURE DISTANCE POINTS

ME D PO

Drawing Tool: Measure Distance Between Points

Measures the distance between two input points placed on the screen. Place more points to accumulate the distance.

Item Selection

On sidebar menu pick: | Meas | Dist | Pnts |

For window, point to: Palettes Measuring

On paper menu see: MEASURE

Command Window Prompts:

1. Select the item
   `Enter start point`
2. Pick a point
3. Define distance to measure
   `Pick a second endpoint`
4. Dist = 16FT 11.702IN
   ⇨ Placing the second point causes the distance to display.

To exit item: `Make next selection`

☞ *Tip:* "Snap" to the element you want measured for greater accuracy.

☞ *Tip:* The MEASURE DISTANCE POINTS tool only computes the element's distance. Use LABEL LINE instead to measure the length and angle of an element.

See Also: MEASURE DISTANCE tools

MEASURE RADIUS

ME R

Drawing Tool: Measure Radius

Calculates the radius of arcs and circles, or the primary and secondary axes of ellipses. This item needs two input points. The first identifies the element to measure. The second point accepts the element and starts the measurement process.

Item Selection

On sidebar menu pick: | Meas | Rad |

For window, point to: Palettes Measuring
On paper menu see: MEASURE
Command Window Prompts:
  1. Select the item
     Identify element
  2. Pick an element
     Accept/Reject (select next input)
  3. Give an accept point
     Radius = 7:5.126
        ⇨ Placing the second point causes the radius to display.
To exit item: Make next selection
See Also: MEASURE ANGLE

Merged Cross Joint

Drawing Tool: Merged Cross Joint
  Creates an intersecting joint between two multi-line elements. This MDL application tool requires three input points. The first two points identify the multi-line elements you want jointed. The third point processes the two elements and creates the joint.

Item Selection
  On sidebar menu pick: n/a
  For window, point to: Palettes Multi-line Joints

On paper menu see: n/a
Command Window Prompts:
  1. Select the item
     Identify element
  2. Pick first multi-line element
        ⇨ Prompt does not change
  3. Pick second multi-line element
  4. Give an accept point
  To exit item: Make next selection

✍ *Note:* This MDL application is loaded automatically when you start the design session.
See Also: Closed Cross Joint, Open Cross Joint, PLACE MLINE

Merged Tee Joint

Drawing Tool: Merged Tee Joint
  Creates a tee joint between two multi-line elements. This MDL application tool requires three input points. The first two points identify the multi-line elements you want jointed. The third point processes the two elements and creates the tee.
Item Selection
  On sidebar menu pick: n/a

For window, point to: Palettes   Multi-line Joints

On paper menu see: n/a
Command Window Prompts:
1. Select the item
   Identify element
2. Pick first multi-line element
   ⇨ Prompt does not change
3. Pick second multi-line element
4. Give an accept point
To exit item: Make next selection

✍ *Note:* This MDL application is loaded automatically when you start the design session.

*See Also:*Closed Tee Joint, Open Tee Joint, PLACE MLINE

MIRROR COPY HORIZONTAL   MIR C

Drawing Tool: Mirror Element About Horizontal (Copy)

Type: Drawing Tool
   Spins and copies an element about the horizontal axis. This item takes two input points. The first point identifies the element to copy and mirror. The last point accepts the mirror copy operation.
Item Selection
On sidebar menu pick: | Manip | | Mirror | | Copy | | Horz |
For window, point to: Palettes   Main   Mirror Element

On paper menu see: MIRROR ELEMENT
Command Window Prompts:
1. Select the item
   Identify element
2. Pick an element
   Accept/Reject (select next input)
3. Give an accept point or key-in
   ⇨ A mirrored-copy of the element appears.
To exit item: Make next selection
See Also: MIRROR tools

MIRROR COPY LINE   MIR C L

Drawing Tool: Mirror Element About Line (Copy)
   Spins and copies an element about a swing line. This item takes three input points. The first point identifies the element to copy and mirror. The last two points define the swing line.
Item Selection
On sidebar menu pick: | Manip | | Mirror | | Copy | | Line |

For window, point to: Palettes   Main   Mirror Element

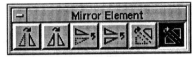

On paper menu see: MIRROR ELEMENT
Command Window Prompts:
  1. Select the item
     Identify element
  2. Pick an element
     Enter 1st pnt on mirror line (or reject)
  3. Pick first swing line point
     Enter 2nd pnt on mirror line
  4. Pick second swing line point
     ⇨   A mirrored-copy of the element appears.
To exit item: Make next selection
See Also: MIRROR tools

MIRROR COPY VERTICAL   MIR C V

Drawing Tool: Mirror Element About Vertical (Copy)
     Spins and copies an element about the vertical axis. This item takes two input points. The first point identifies the element to copy and mirror. The last point accepts the mirror copy operation.

Item Selection
  On sidebar menu pick: | Manip | | Mirror | | Copy | | Vert |
  For window, point to: Palettes   Main   Mirror Element

On paper menu see: MIRROR ELEMENT
Command Window Prompts:
  1. Select the item
     Identify element
  2. Pick an element
     Accept/Reject (select next input)
  3. Give an accept point or key-in
     ⇨   A mirrored-copy of the element appears.
To exit item: Make next selection
See Also: MIRROR tools

MIRROR ORIGINAL HORIZONTAL   MIR

Drawing Tool: Mirror Element About Horizontal (Original)
     Spins or mirrors an element about the horizontal axis. This item takes two input points. The first point identifies the element to mirror. The last point accepts the mirror operation.

Item Selection
  On sidebar menu pick: | Manip | | Mirror | | Orig | | Horz |
  For window, point to: Palettes   Main   Mirror Element

On paper menu see: MIRROR ELEMENT
Command Window Prompts:
  1. Select the item
     Identify element

2. Pick an element
```
Accept/Reject (select next input)
```
3. Give an accept point or key-in
   ⇨ Element is mirrored.

To exit item: Make next selection
See Also: MIRROR tools

MIRROR ORIGINAL LINE
MIR O L

Drawing Tool: Mirror Element About Line (Original)
    Spins or mirrors an element about a swing line. This tool takes three input points. The first point identifies the element to mirror. The last two points define the swing line.

Item Selection
On sidebar menu pick: Manip   Mirror   Orig   Line
For window, point to: Palettes   Main   Mirror Element

On paper menu see: MIRROR ELEMENT
Command Window Prompts:
1. Select the item
```
Identify element
```
2. Pick an element
```
Enter 1st point on mirror line (or reject)
```
3. Pick first swing line point
```
Enter 2nd point on mirror line
```
4. Pick second swing line point
   ⇨ Element is mirrored about the swing line.

To exit item: Make next selection
See Also: MIRROR tools

MIRROR ORIGINAL VERTICAL
MIR O V

Drawing Tool: Mirror Element About Vertical (Original)
    Spins or mirrors an element about the vertical axis. This item takes two input points. The first point identifies the element to mirror. The last point accepts the mirror operation.

Item Selection
On sidebar menu pick: Manip   Mirror   Orig   Vert
For window, point to: Palettes   Main   Mirror Element

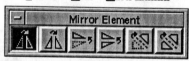

On paper menu see: MIRROR ELEMENT
Command Window Prompts:
1. Select the item
```
Identify element
```
2. Pick an element
```
Accept/Reject (select next input)
```
3. Give an accept point or key-in
   ⇨ Element is mirrored.

To exit item: Make next selection
See Also: MIRROR tools

MODIFY ARC ANGLE

MOD AR AN

Drawing Tool: Modify Arc Angle

Shortens the length of an arc. This item requires one input point to identify the arc and a second point to control the change made to the arc's angle.

Item Selection

On sidebar menu pick: `Modify` `Arc` `Angle`

For window, point to: Palettes Main Arcs

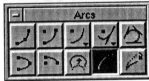

On paper menu see: ARCS

Command Window Prompts:

1. Select the item
   Identify element
2. Pick an arc
   Accept/Reject (select next input)
3. Give an accept point

To exit item: **Make next selection**

See Also: MODIFY ARC tools, MODIFY ELEMENT

MODIFY ARC AXIS

MOD AR AX

Drawing Tool: Modify Arc Axis

Changes the arc's orientation by adjusting the element's axis. This item requires one input point to identify the arc and a second point to control the change made to the arc.

Item Selection

On sidebar menu pick: `Modify` `Arc` `Axis`

For window, point to: Palettes Main Arcs

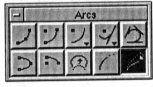

On paper menu see: ARCS

Command Window Prompts:

1. Select the item
   Identify element
2. Pick an arc
   Accept/Reject (select next input)
3. Give an accept point

To exit item: **Make next selection**

See Also: MODIFY ARC tools, MODIFY ELEMENT

MODIFY ARC RADIUS

MOD AR R

Drawing Tool: Modify Arc Radius

Flattens the arc by changing the radius. This item requires one input point to identify the arc and a second point to control the change made to the arc.

Item Selection

On sidebar menu pick: `Modify` `Arc` `Rad`

For window, point to: Palettes   Main   Arcs

On paper menu see: ARCS
Command Window Prompts:
   1. Select the item
      Identify element
   2. Pick an arc
      Accept/Reject (select next input)
   3. Give an accept point
   To exit item: Make next selection
See Also: MODIFY ARC tools, MODIFY ELEMENT

MODIFY ELEMENT   MOD

Drawing Tool: Modify Element
      Modifies the shape of an existing element by moving a single vertex of a line,
line string or shape. It also adjusts the radius of a circle or the primary or secondary
axis of an ellipse. This item requires two input points. The first point identifies
the vertex to move and the second point plots its final resting place.
Item Selection
   On sidebar menu pick: Modify   Elem
   For window, point to: Palettes   Main   Modify Element

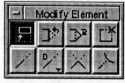

On paper menu see: MODIFY
Command Window Prompts:
   1. Select the item
      Identify element
   2. Pick an element
      Accept/Reject (select next input)
   3. Give an accept point
   To exit item: Make next selection
☞ *Tip:* Use one of the MODIFY ARC tools rather than MODIFY ELEMENT to modify
   arcs.
See Also: MODIFY ARC tools

MODIFY FENCE   MOD F

Drawing Tool: Modify Fence
      Changes the shape of an existing fence. This item expects two input points.
The first point attaches a vertex of the fence to the cursor. The second point
determines the new location of the vertex.
Item Selection
   On sidebar menu pick: Fences   Modify

For window, point to: Palettes   Fence
On paper menu see: PLACE FENCE
Command Window Prompts:
  1. Select the item
     Identify vertex
  2. Pick a fence side
     ⇨   The prompt does not change.
  3. Give an accept point
     ⇨   The fence encloses the new area.
  To exit item: Make next selection
☞ *Tip:* Use tool to stretch a side of the fence, rather than replacing one fence with a larger one.
See Also: MOVE FENCE, PLACE FENCE tools

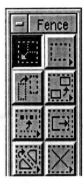

MODIFY TEXT   MOD T

Drawing Tool: Change Text to Active Attributes
   Changes the attributes of existing text in the drawing file. This item sets the size, justification and font of any text in the drawing to the current ACTIVE TXSIZE, ACTIVE TXJ and ACTIVE FONT settings. MODIFY TEXT requires an input point to identify the text and an accept point for the change to take place.

Item Selection
On sidebar menu pick: Text   Attr   Change
For window, point to: Palettes   Main   Text

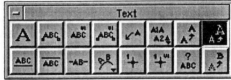

On paper menu see: TEXT
Command Window Prompts:
  1. Select the item
     Identify text element
  2. Pick a text string
     Accept/Reject (select next input)
  3. Give an accept point
  To exit item: Make next selection
See Also: ACTIVE FONT, ACTIVE TXSIZE, ACTIVE TXJ

MOVE ACS   MOV A

Drawing Tool: Move ACS
   Repositions the origin of an Auxiliary Coordinate System. This item expects one input point placed at the new origin.

Item Selection
On sidebar menu pick: Params   ACS   Tools
For window, point to: Palettes   Auxiliary Coordinates

On paper menu see: ACS
See Also: DEFINE ACS settings

MOVE DOWN   MOV D

View Control: Move Down (Scroll)
>   Shifts the view of the drawing down. Pick the item. Now, each time you select
>   a view with an input point, the drawing shifts 25% in that direction.

Item Selection
On sidebar menu pick: | Window | | Move | | Down |
For window, point to: n/a
On paper menu see: VIEW
Command Window Prompts:
1. Select the item
   `Select view`
2. Pick a view
To exit item: `Click a Reset`

Example(s):
MOV D   to move down
See Also: FIT, MOVE, WINDOW and ZOOM view controls

MOVE ELEMENT   MOV

Drawing Tool: Move Element
>   Moves existing elements from one location in the drawing file to another.
>   This tool requires two input points. The first point selects the element you want
>   moved. The second point tells MicroStation where to move the element.

Item Selection
On sidebar menu pick: | Manip | | Move |
For window, point to: Palettes   Main   Copy Element

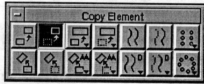

On paper menu see: COPY ELEMENT
Command Window Prompts:
1. Select the item
   `Identify element`
2. Pick an element with an input point
   `Accept/Reject (select next input)`
3. Place the accept point where you want the element
To exit item: `Make next selection`

✍ *Note:* Use the MOVE ELEMENT tool to move an element between levels. First
identify the element. Change the ACTIVE LEVEL with the LV= key-in. Then give
the accept point to move the element to the new level.
See Also: FENCE MOVE, MOVE PARALLEL tools

MOVE FENCE   MOV F

Drawing Tool: Move Fence Block/Shape
>   Relocates an existing fence in the drawing file. This item expects two input points.
>   Place the first point inside the fence. This point is the origin point. The second point tells
>   MicroStation the new fence location.

Item Selection
On sidebar menu pick: | Fences | | MvFnce |

For window, point to: Palettes Fence

On paper menu see: PLACE FENCE

Command Window Prompts:
  1. Select the item
     Define origin
  2. Pick a point within the fence
     Define distance
  3. Pick a destination point with the cursor or key-in.
  To exit item: Make next selection

☞ *Tip:* Use this tool when performing several FENCE operations instead of making a new fence for each operation.

✳ *Error:* No fence defined - If this message appears, you need to use one of the PLACE FENCE tools.

See Also: MODIFY FENCE, PLACE FENCE tools

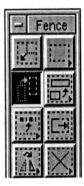

MOVE LEFT

MOV L

View Control: Move Left (Scroll)
  Shifts the drawing to the left. Each time you select the view with an input point, the drawing shifts 25% in that direction.

Item Selection
  On sidebar menu pick: Window Move Left

  For window, point to: n/a
  On paper menu see: VIEW
Command Window Prompts:
  1. Select the item
     Select view
  2. Pick a view
  To exit item: Click a Reset

Example(s):
  MOV L   to move left
See Also: FIT, MOVE, WINDOW and ZOOM view controls

MOVE PARALLEL DISTANCE

MOV P

Drawing Tool: Move Parallel by Distance
  Moves a parallel element from one location to another. This item requires an input point to identify the target element and a second point to define the distance you want the element moved.

Item Selection
  On sidebar menu pick: n/a
  For window, point to: Palettes Main Copy Element

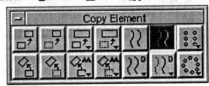

On paper menu see: n/a
Command Window Prompts:
  1. Select the item
     Identify element
  2. Pick an element
     Accept/Reject (select next input)
  3. Give an accept point
  To exit item: Make next selection
See Also: COPY PARALLEL tools, FENCE MOVE, MOVE PARALLEL KEYIN, MOVE ELEMENT

MOVE PARALLEL KEYIN   MOV P V

Drawing Tool: Move Parallel by Key-in

Moves an element parallel from one location to another. The distance the element moves depends on the key-in value. This item requires an input point to identify the target element and a second point (or key-in) to define the distance you want the element moved.

Item Selection

On sidebar menu pick: n/a

For window, point to: Palettes   Main   Copy Element

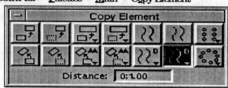

On paper menu see: n/a

Command Window Prompts:

1. Select the item
   Identify element
2. Pick an element
   Accept/Reject (select next input)
3. Give an accept point

To exit item: Make next selection

See Also: COPY PARALLEL tools, FENCE MOVE, MOVE PARALLEL DISTANCE, MOVE ELEMENT

MOVE RIGHT   MOV R

View Control: Move Right (Scroll)

Shifts the drawing to the right. Each time you select the view with an input point, the drawing shifts 25% in that direction.

Item Selection

On sidebar menu pick: Window   Move   Right

For window, point to: n/a

On paper menu see: VIEW

Command Window Prompts:

1. Select the item
   Select view
2. Pick a view

To exit item: Click a Reset

Example(s):

MOV R   to move right

See Also: FIT, MOVE, WINDOW and ZOOM view controls

MOVE UP   MOV U

View Control: Move Up (Scroll)

Shifts the drawing up. Each time you select the view with an input point, the drawing shifts 25% in that direction.

Item Selection

On sidebar menu pick: Window   Move   Up

For window, point to: n/a

On paper menu see: VIEW

Command Window Prompts:

1. Select the item
   Select view
2. Pick a view

To exit item: Click a Reset

Example(s):
MOV U   to move up
See Also: FIT, MOVE, WINDOW and ZOOM view controls

NEWFILE   RD=

Drawing Tool: New File
Allows you to jump from one drawing file to another without ending the design session. Select the new file from the Open Design File dialog box. If you know the name of the file, you may key in NEW [drawing name] or RD=[drawing name]. RD= stands for RETRIEVE DESIGN. This key-in is a holdover from Intergraph's Interactive Graphics Design Software (IGDS).
Item Selection
On sidebar menu pick: n/a
For window, point to: File   Open...
On paper menu see: n/a
See Also: EXCHANGEFILE, FLUSH

NN=

Alternate Key-in
Alternate key-in for ACTIVE NODE. Use it to set the active text node counter.
See Also: ACTIVE NODE

NOECHO   NOE

Setting: No Echo
Stops the message and prompt display in MicroStation's information fields. Use the ECHO setting to turn on the prompts.
Item Selection
On sidebar menu pick: n/a
For window, point to: n/a
On paper menu see: n/a
See Also: ECHO

NULL

Command: Null
Clears MicroStation's command buffer. This insures the last operation is not running.
Item Selection
On sidebar menu pick: n/a
For window, point to: n/a
On paper menu see: n/a

OF=

Type: Alternate Key-in
Alternate key-in for SET LEVELS. Use it to toggle a level's display off.
See Also: SET LEVELS

ON=

Type: Alternate Key-in
Alternate key-in for SET LEVELS. Use it to toggle a level's display on.
See Also: SET LEVELS

Open Cross Joint

Drawing Tool: Open Cross Joint
Creates a clear intersecting joint between two multi-line elements. This MDL application tool requires three input points: two points to identify the multi-line

element you want intersected, and the third to accept the process and create the intersection joint.

Item Selection
On sidebar menu pick: n/a
For window, point to: Palettes Multi-line Joints

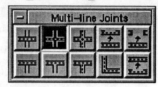

On paper menu see: n/a
Command Window Prompts:
1. Select the item
   Identify element
2. Pick first multi-line element
   ⇨ Prompts do not change
3. Pick second multi-line element
4. Give an accept point
To exit item: Make next selection

📎 *Note:* This MDL application is loaded automatically when you start the design session.
See Also: Closed Cross Joints, Merged Cross Joint, PLACE MLINE

Open Tee Joint

Drawing Tool: Open Tee Joint
Creates an open tee joint (a clear intersecting joint) between two multi-line elements. This MDL application tool requires three input points. The first two points identify the multi-line element you want intersected. The third point accepts the process and creates the intersection joint.

Item Selection
On sidebar menu pick: n/a
For window, point to: Palettes Multi-line Joints

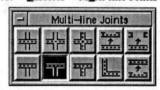

On paper menu see: n/a
Command Window Prompts:
1. Select the item
   Identify element
2. Pick first multi-line element
   ⇨ Prompts do not change
3. Pick second multi-line element
4. Give an accept point
To exit item: Make next selection

📎 *Note:* This MDL application is loaded automatically when you start the design session.
See Also: Closed Tee Joint, Merged Tee Joint, PLACE MLINE

OX=

Type: Alternate Key-in
Alternate key-in for ACTIVE INDEX. Use it to attach a user command index to a drawing file.
See Also: ACTIVE INDEX

PA=

Alternate Key-in
Alternate key-in for ACTIVE PATTERN ANGLE. Use it to set the pattern angle for elements.
See Also: ACTIVE PATTERN ANGLE

PAGE SETUP   PAG

Mac Only
Command: Page Setup
Mac only – The print parameters can be changed by keying in PAGE SETUP. This item opens the Page Setup dialog box.

Item Selection
On sidebar menu pick: n/a
For window, point to: File   Page Setup
On paper menu see: n/a
See Also: PRINT

Pattern

Definition
Patterns help distinguish the sections of a line drawing. A rock pattern could be generated to show a concrete pattern in a cut section of a wall detail. A tuft of grass pattern could show a marsh area. A series of arcs could be strung together in a pattern to show a line of trees on a map. MicroStation provides PATTERN AREA tools to pattern shapes. And PATTERN LINE tools to pattern open chains of elements. Any cell can be used as a pattern cell.
See Also: PATTERN AREA and PATTERN LINEAR tools

PATTERN AREA ELEMENT   PA=

Drawing Tool: Pattern Element Area
Patterns closed shapes that represent areas in the drawing file. The closed shapes can be constructed with lines, line strings, circles, arcs or ellipses.

You can make holes in a pattern by keying in ACTIVE AREA HOLE and placing circles or other shapes. After patterning the area, these holes remain hollow.

The following settings control the appearance of a pattern:

AP=	sets the ACTIVE PATTERN CELL
PS=	sets the ACTIVE PATTERN SCALE
PD=	sets the ACTIVE PATTERN DELTA
PA=	sets the ACTIVE PATTERN ANGLE

Item Selection
On sidebar menu pick: Pattrn
For window, point to: Palettes   Patterning

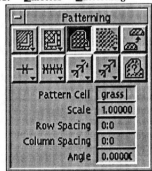

On paper menu see: PATTERNING

Command Window Prompts:
1. Select the item
   `Identify Element, Reset To exit`
2. Pick the element
   ⇨ The shape highlights.
3. Accept/Reject
   `Give an accept point`
   `Patterning in progress...`

To exit item: `Click a Reset`

✍ *Note:* To hide the patterns, key in SET PATTERN off.

☞ *Tip:* To modify part of the pattern, turn the LOCK GGROUP setting off by keying in LOCK GGROUP OFF.

✱ *Error:* `No Patterning Cell Defined` - If this message appears, use the AP= key-in to define the cell name.

See Also: CREATE SHAPE tools, CROSSHATCH, HATCH, PATTERN AREA FENCE

PATTERN AREA FENCE   PA=

Drawing Tool: Pattern Fence Area
    Patterns the inside area defined by a fence shape.

    You can test making holes in a pattern by keying in ACTIVE AREA HOLE and placing circles or other shapes within the fence. After patterning the area, these holes remain hollow.

    The following settings control the appearance of a pattern:

AP=	sets the ACTIVE PATTERN CELL
PS=	sets the ACTIVE PATTERN SCALE
PD=	sets the ACTIVE PATTERN DELTA
PA=	sets the ACTIVE PATTERN ANGLE

Item Selection
On sidebar menu pick: Pattrn
For window, point to: Palettes Patterning

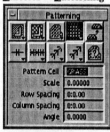

On paper menu see: PATTERNING
Command Window Prompts:
1. Select the item
   `Accept @pattern intersection`
   `point`
2. Give an accept point
   `Loading Patterning Task...`
   `Placing Pattering Elements...`
   `Clipping Pattern elements...`
   `Fence Area Pattering Exited`

To exit item: `Exits by itself`

✍ *Note:* If you want to hide the patterns, key in SET PATTERN off.

☞ *Tip:* Use the Pattern Area Fence tool to test a pattern before patterning the final shape.

☞ *Tip:* To modify part of the pattern, turn the LOCK GGROUP setting off by keying in LOCK GGROUP OFF.

✱ **Error:** No Patterning Cell Defined - **If this message appears, use the AP= key-in to set the pattern cell.**
See Also: CREATE SHAPE tools, CROSSHATCH, HATCH, PATTERN AREA ELEMENT

PATTERN LINE ELEMENT

Same as PATTERN LINEAR TRUNCATED
See Also: PATTERN LINEAR TRUNCATED

PATTERN LINE MULTIPLE

Same as PATTERN LINEAR MULTIPLE
See Also: PATTERN LINEAR MULTIPLE

PATTERN LINE SCALE

Same as PATTERN LINEAR UNTRUNCATED
See Also: PATTERN LINEAR UNTRUNCATED

PATTERN LINE SINGLE

Same as PATTERN LINEAR SINGLE
See Also: PATTERN LINEAR SINGLE

PATTERN LINEAR MULTIPLE

Drawing Tool: Multi-Cycle Segment Linear Pattern
Places multiple pattern cells on an element. The element can be a line, line string, shape, arc, circle, ellipse or curve. This item places as many pattern cells as possible along the element. This tool does not change the scale factor of the pattern cell to fill out an element, so part of the element remains unpatterned.

Item Selection
On sidebar menu pick: `Pattrn`
For window, point to: <u>P</u>alettes <u>P</u>atterning

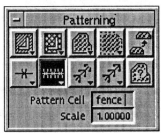

On paper menu see: PATTERNING
Command Window Prompts:
  1. Select the item
     `Identify element`
  2. Pick element to pattern
     `Identify Direction / Reject`
  3. Pick a second point
     ⇨ The second point identifies the flow of the pattern.
  To exit item: `Click a Reset`

📖 *Note:* To hide the patterns, key in SET PATTERN off.

☞ *Tip:* You can set the pattern cell name from the Cells settings box.

✳ *Error:* Element not found - If this message appears, the element identified may not be a linear element. For example, identifying a shape instead of a line would cause this error message to appear.
See Also: PATTERN LINEAR tools, PATTERN AREA tools

PATTERN LINEAR SCALE

Same as PATTERN LINEAR UNTRUNCATED
See Also: PATTERN LINEAR UNTRUNCATED

PATTERN LINEAR SINGLE

Drawing Tool: Single Cycle Segment Linear Pattern
        Places a single pattern cell on an element. The element can be a line, line string, shape, arc, circle, ellipse or curve. This item does not change the scale factor of the pattern cell to fill out an element, so part of the element may go unpatterned.
Item Selection
On sidebar menu pick: Pattrn
For window, point to: Palettes   Patterning

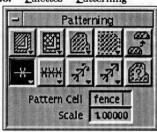

On paper menu see: PATTERNING
Command Window Prompts:
  1. Select the item
     Identify element
  2. Pick element to pattern
     Identify Direction / Reject
  3. Pick a second point
     ⇨   The second point identifies the flow of the pattern.
  To exit item: Click a Reset

✍ *Note:* To hide the pattern, key in SET PATTERN off.

☞ *Tip:* You can set the pattern cell name from the Cells settings box.

✳ *Error:* Element not found - If this message appears, the element identified may not be a linear element. For example, identifying a shape instead of a line would cause this error message to appear.
See Also: PATTERN LINEAR tools, PATTERN AREA tools

PATTERN LINEAR TRUNCATED

Drawing Tool: Truncated Cycle Linear Pattern
        Generates an unscaled linear pattern along an element. The element can be a line, line string, shape, arc, circle, ellipse or curve. MicroStation places as many pattern cells as possible on the element. However, if the pattern does not fit perfectly over an element, part of the cell is truncated.

    AP= sets the ACTIVE PATTERN CELL.
    PS= sets the ACTIVE PATTERN SCALE.
    PD= sets the ACTIVE PATTERN DELTA.
    PA= sets the ACTIVE PATTERN ANGLE.
Item Selection
On sidebar menu pick: Pattrn

For window, point to: Palettes   Patterning

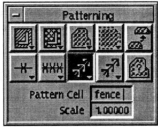

On paper menu see: PATTERNING
Command Window Prompts:
  1. Select the item
     Identify element
  2. Pick element to pattern
     Identify Direction / Reject
  3. Pick a second point
     ⇨ The second point identifies the flow of the pattern.
     Pattering In Progress...
  To exit item: Make next selection

☞ *Tip:* You can set the pattern cell name from the Cells settings box.

✳ *Error:* Element not found - If this message appears, the element identified may
not be a linear element. For example, identifying a shape instead of a line would
cause this error message to appear.
See Also: PATTERN LINEAR tools, PATTERN AREA tools

PATTERN LINEAR UNTRUNCATED

Drawing Tool: Complete Cycle Linear Pattern
     Generates an unscaled linear pattern along an element. The element can be
a line, line string, shape, arc, circle, ellipse or curve. MicroStation places as many
pattern cells as possible on the element. However, if the pattern does not fit
perfectly over an element, MicroStation adjusts the pattern's scale factor until
only complete pattern cells fit on top of the element.

AP= sets the ACTIVE PATTERN CELL.
PS= sets the ACTIVE PATTERN SCALE.
PD= sets the ACTIVE PATTERN DELTA.
PA= sets the ACTIVE PATTERN ANGLE.
Item Selection
  On sidebar menu pick: | Pattrn |
  For window, point to: Palettes   Patterning

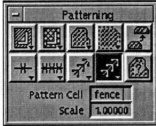

On paper menu see: PATTERNING
Command Window Prompts:
  1. Select the item
     Identify element

2. Pick element to pattern
   `Identify Direction / Reject`
3. Pick a second point
   ⇨ The second point identifies the flow of the pattern.
   `Pattering In Progress...`
To exit item: `Make next selection`

☞ *Tip:* You can set the pattern cell name from the Cells settings box.

✳ *Error:* `Element not found` - If this message appears, the element identified may not be a linear element. For example, identifying a shape instead of a line would cause this error message to appear.
See Also: PATTERN LINEAR tools, PATTERN AREA tools

PAUSE

Command: Pause
    When followed by a number, suspends the processing of a user command for (number) seconds. For instance, PAUSE 5 would suspend a user command for 5 seconds and then resume processing. This item is useful when the user command needs to wait for data, or an event to occur, from an outside source.

PD=

Alternate Key-in
    Alternate key-in for ACTIVE PATTERN DELTA. Use it to set the distance between patterns, hatching and cross hatching.
See Also: ACTIVE PATTERN DELTA

PICTFILE PASTE   PIC P

Mac Only

Command: Paste PICT file data
    Mac only – Imports PICT file data (a picture file) into a drawing file.
Item Selection
    On sidebar menu pick: n/a
    For window, point to: Edit
    On paper menu see: n/a
See Also: PICTFILE commands

PICTFILE SAVE RASTER   PIC S R

Mac Only

Command: Save Raster Data to a PICT file
    Mac only – Exports raster data to an output file in PICT format.
Item Selection
    On sidebar menu pick: n/a
    For window, point to: File   Save As...
    On paper menu see: n/a
See Also: PICTFILE commands

PICTFILE SAVE VECTOR   PIC S V

Mac Only

Command: Save Vector Data to a PICT file
    Mac only – Exports vector data to an output file in PICT format.
Item Selection
    On sidebar menu pick: n/a
    For window, point to: File   Save As...
    On paper menu see: n/a
See Also: PICTFILE commands

Place

Definition

The PLACE drawing tools draw graphics in the design file. For instance, the PLACE ARC EDGE tool generates an arc based on three edge points.

See Also: PLACE tools

PLACE ARC CENTER   PLA A C

Drawing Tool: Place Arc by Center

Constructs an arc about a center point. Requires three input points to construct the arc: a beginning (center) point, a point to define the arc's radius and an endpoint.

Item Selection

On sidebar menu pick: | Place | | Arc | | Center |

For window, point to: Palettes   Main   Arcs

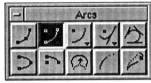

On paper menu see: ARCS

Command Window Prompts:

1. Select the item
   Enter first arc endpoint
2. Pick a point
   Enter point on arc
3. Pick a second point
   Enter second arc endpoint
4. Pick the last point

To exit item: **Make next selection**

See Also: PLACE ARC tools, PLACE ELLIPSE tools

PLACE ARC EDGE   PLA A

Drawing Tool: Place Arc by Edge

Constructs an arc based on an edge-point. It requires three input points to construct the arc: a beginning point (edge), a midpoint and an endpoint.

Item Selection

On sidebar menu pick: | Place | | Arc | | Edge |

For window, point to: Palettes   Main   Arcs

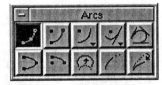

On paper menu see: ARCS

Command Window Prompts:

1. Select the item
   Enter first arc endpoint
2. Pick a point
   Enter point on arc
3. Pick a second point
   Enter second arc endpoint
4. Pick the last point

To exit item: **Make next selection**

See Also: PLACE ARC tools, PLACE ELLIPSE tools

PLACE ARC RADIUS   PLA A R

Drawing Tool: Place Arc by Keyed-in Radius

Constructs an arc based on a keyed in radius. It requires three input points: a beginning point, a point defining the arc's center and an endpoint.

Item Selection

On sidebar menu pick: | Place | | Arc | | Rad |

For window, point to: Palettes Main Arcs

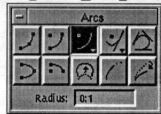

On paper menu see: ARCS

Command Window Prompts:
1. Select the item
   Key in radius
2. Enter a radius
   Enter first arc endpoint
3. Pick a first point
   Enter arc center
4. Pick center point
   Enter second arc endpoint
5. Pick a second point

To exit item: Make next selection

See Also: PLACE ARC tools, PLACE ELLIPSE tools

PLACE ARC TANGENT

Obsolete

An arc tangent to an element. The prompt asks for two key-ins (radius and length). Then enter an input point to define the arc's tangent point. A final input point defines the arc's direction of curvature.

PLACE BLOCK ISOMETRIC   PLA BL I

Drawing Tool: Place Isometric Block

Constructs isometric blocks in 2D or 3D drawing files. This item expects two input points to define the opposite corners of the isometric shape.

Item Selection

On sidebar menu pick: | Place | | Block | | Iso |

For window, point to: Palettes Main Polygons

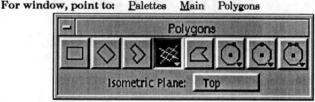

On paper menu see: POLYGONS

Command Window Prompts:
1. Select the item
   Enter first point
2. Pick a point
   Enter opposite corner
3. Pick a second point

To exit item: Make next selection

See Also: PLACE BLOCK tools, PLACE SHAPE tools

PLACE BLOCK ORTHOGONAL · PLA BL O

Drawing Tool: Place Block

Constructs orthogonal blocks in 2D or 3D drawing files. This item expects two input points to define the opposite corners of the shape.

Item Selection

On sidebar menu pick: Place Block Ortho

For window, point to: Palettes Main Polygons

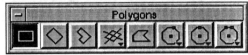

On paper menu see: POLYGONS

Command Window Prompts:

1. Select the item
   Enter first point
2. Pick a point
   Enter opposite corner
3. Pick a second point

To exit item: Make next selection

See Also: PLACE BLOCK tools, PLACE SHAPE tools

PLACE BLOCK ROTATED · PLA BL R

Drawing Tool: Place Rotated Block

Rotates a shape. This tool asks for two points to define the base, and a third to define the shape's rotation.

Item Selection

On sidebar menu pick: Place Block Rotate

For window, point to: Palettes Main Polygons

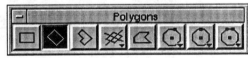

On paper menu see: POLYGONS

Command Window Prompts:

1. Select the item
   Enter first base point
2. Pick a point
   Enter second base point
3. Pick a point
   Enter diagonal point

To exit item: Make next selection

See Also: PLACE BLOCK tools, PLACE SHAPE tools

PLACE BSPLINE CLOSED

Obsolete

The PLACE BSPLINE CLOSED command creates a closed B-spline curve. This curve may contain between three and 15 poles. A second curve may be constructed by placing additional points.

See Also: PLACE BSPLINE tools

PLACE
BSPLINE CURVE LEASTSQUARE

PLA BS C L

Drawing Tool: Place B-spline Curve by Least Squares

This MDL application tool creates a B-spline curve by placing input points along the curve approximates. The newly created curve minimizes the sum of the distances between the points and the curve. The values in the B-splines settings box controls the order, whether its open or closed and the number of poles of the curve.

Item Selection

On sidebar menu pick: | Place | | BSpl | | Curves |

For window, point to: P̲alettes   M̲ain   B-splines

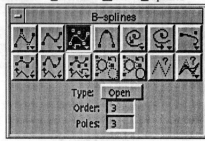

On paper menu see: B-SPLINES 2D 3D

Command Window Prompts:
1. Select the item
   ```
   Enter first point
   Enter point or RESET to complete
   ```
2. Continue to enter points or click Reset

To exit item: **Make next selection**

* *Error:* Invalid character **- If this message appears after your key in, enter MDL LOAD SPLINES or pick the tool from the B-splines palette.**
See Also: PLACE BSPLINE tools, CONSTRUCT BSPLINE tools

PLACE BSPLINE CURVE POINTS   PLA BS C POI

Drawing Tool: Place B-spline Curve by Points

This MDL application tool creates a B-spline curve by placing a series of input points. The curve passes through the points placed. The values in the B-splines settings box controls the order and whether the curve is open or closed.

Item Selection

On sidebar menu pick: | Place | | BSpl | | Curves |

For window, point to: P̲alettes   M̲ain   B-splines

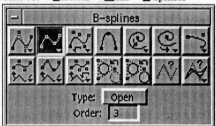

On paper menu see: B-SPLINES 2D 3D

Command Window Prompts:
1. Select the item
   ```
   Enter first point
   Enter point or RESET to complete
   ```
2. Continue to enter points or click Reset

To exit item: **Make next selection**

* *Error:* Invalid character - **If this message appears after your key in, enter MDL LOAD SPLINES or pick the tool from the B-splines palette.**
See Also: PLACE BSPLINE tools, CONSTRUCT BSPLINE tools

PLACE BSPLINE CURVE POLES   PLA BS C POL

Drawing Tool: Place B-spline Curve by Poles
       This MDL application tool creates a B-spline curve by placing input points. The placement of the points defines the poles of the control polygon. The values in the B-splines settings box controls the order and whether the curve is open or closed.

Item Selection
On sidebar menu pick: [Place]   [BSpl]   [Curves]
For window, point to: <u>P</u>alettes   <u>M</u>ain   <u>B</u>-splines

On paper menu see: B-SPLINES 2D 3D
Command Window Prompts:
   1. Select the item
      Enter first point
      Enter point or RESET to complete
   2. Continue to enter points or click Reset
   To exit item: **Make next selection**

* *Error:* Invalid character - **If this message appears after your key in, enter MDL LOAD SPLINES or pick the tool from the B-splines palette.**
See Also: PLACE BSPLINE tools, CONSTRUCT BSPLINE tools

PLACE BSPLINE OPEN

Obsolete
       The PLACE BSPLINE OPEN command creates an open B-spline curve. This curve may contain between three and 15 poles. A second curve may be constructed by placing additional points.
See Also: PLACE BSPLINE tools

PLACE BSPLINE SURFACE LEASTSQUARE   PLA B S L

3D Drawing Tool: Place B-spline Surface by Least Squares
       This MDL application tool creates a B-spline surface by placing input points along the curve approximates. The newly created surface minimizes the sum of the distances between the points and the surface. The values in the B-splines settings box controls the order and whether the surface is open or closed in the u-direction and v-direction.

Item Selection
On sidebar menu pick: [Place]   [BSpl]   [SurFce]
For window, point to: <u>P</u>alettes   3D <u>B</u>-splines   <u>S</u>urfaces
On paper menu see: SURFACES
Command Window Prompts:
   1. Select the item
      Enter first point
   2. Place first point
      Enter first row or RESET to complete
   3. Continue to enter points or click Reset
   To exit item: **Make next selection**

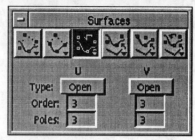

✳ *Error:* Invalid character - **If this message appears after your key in, enter MDL LOAD SPLINES or pick the tool from the Surfaces palette.**

✍ *Note:* The B-spline settings box controls the display of the surface and net.
See Also: PLACE BSPLINE SURFACE tools

PLACE BSPLINE SURFACE POINTS   PLA B S POI

3D Drawing Tool: Place B-spline Surface by Points
      This MDL application tool creates a B-spline surface by placing a series of input points. The surface passes through the input points. The values in the B-splines settings box controls the order and whether the surface is open or closed in the u- and v-directions.

Item Selection
    On sidebar menu pick: Place   BSpl   SurFce
    For window, point to: Palettes   3D B-splines   Surfaces

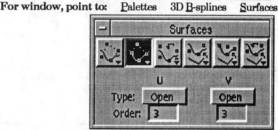

    On paper menu see: SURFACES
Command Window Prompts:
    1. Select the item
        Enter first point
    2. Place first point
        Enter first row or RESET to complete
    3. Continue to enter points or click Reset
    To exit item: Make next selection

✳ *Error:* Invalid character - **If this message appears after your key in, enter MDL LOAD SPLINES or pick the tool from the Surfaces palette.**

✍ *Note:* The B-spline settings box controls the display of the surface and net.
See Also: PLACE BSPLINE SURFACE tools

PLACE BSPLINE SURFACE POLES   PLA B S POL

3D Drawing Tool: Place B-spline Surface by Poles
      This MDL application tool creates a B-spline surface by placing a series of input points. The placement of the points defines the poles of the surface. The values in the B-splines settings box controls the order and whether the curve is open or closed.

Item Selection
    On sidebar menu pick: Place   BSpl   SurFce

For window, point to: Palettes   3D B-splines   Surfaces

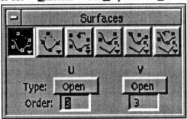

On paper menu see: SURFACES
Command Window Prompts:
1. Select the item
   Enter first point
2. Place first point
   Enter first row or RESET to complete
3. Continue to enter points or click Reset
To exit item: Make next selection

✳ **Error:** Invalid character - **If this message appears after your key in, enter MDL LOAD SPLINES or pick the tool from the Surfaces palette.**

✐ Note: The B-spline settings box controls the display of the surface and net.
See Also: PLACE BSPLINE SURFACE tools

PLACE CELL ABSOLUTE   PLA CE

Drawing Tool: Place Active Cell
Places a standard feature in the drawing file. (Build an active cell before using – see CREATE CELL.) The tool places the elements in the cell on their original levels. For example, create a desk cell with the desk outline on level 10 and the word "DESK" on level 11. When you place the desk cell in the drawing, the desk outline appears on level 10 and the word "DESK" appears on level 11, regardless of your ACTIVE LEVEL setting.

Item Selection
On sidebar menu pick: Cells   Place   Abs
For window, point to: Palettes   Main   Cells

On paper menu see: CELLS
Command Window Prompts:
1. Select the item
   Enter cell origin
   ⇨   A box shape representing the size of the cell attaches itself to the cursor.
2. Pick a point
   ⇨   The cell appears in the drawing.
To exit item: Make next selection

✳ **Error:** No active cell - If this message appears, use the ACTIVE CELL setting to define an active cell.
See Also: PLACE CELL tools, SELECT CELL tools

PLACE CELL ABSOLUTE TMATRX

Drawing Tool: Place Active Cell by Transformation Matrix
   Places a standard cell in the drawing file. Call this item from a user command.
   (Build a cell before using – see CREATE CELL.) The values in the transformation
   matrix control the cell's scale and orientation when placing a cell with this tool.
   This tool places elements on their original levels. For instance, create a desk cell
   with the desk outline on level 10 and the word "DESK" on level 11. Place the desk
   cell in the drawing. The desk outline appears on level 10 and the word "DESK"
   appears on level 11, regardless of the ACTIVE LEVEL setting.

✱   **Error:** No active cell - If this message appears, use the ACTIVE CELL setting
   to define an active cell.
   See Also: PLACE CELL tools, SELECT CELL tools

PLACE CELL INTERACTIVE ABSOLUTE   PLA CE I

Drawing Tool: Place Active Cell (Interactive)
   Changes a cell's scale or rotation while placing the cell in the drawing file.
   (Build a cell before using – see CREATE CELL.) This tool places the elements in
   the cell on their original level. For example, create a desk cell with the desk outline
   on level 10 and the word "DESK" on level 11. Place the desk cell in the drawing.
   The desk outline appears on level 10 and the word "DESK" appears on level 11,
   regardless of the ACTIVE LEVEL setting.

Item Selection
On sidebar menu pick: Cells   IntAct   Abs
For window, point to: Palettes   Main   Cells

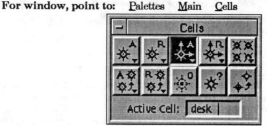

On paper menu see: CELLS
Command Window Prompts:
   1. Select the item
      Enter cell origin
   2. Pick a point
         ⇨   This sets the origin of the cell.
      XSCALE (1.0):
   3. Key in X scale factor or input point.
      YSCALE (1.0):
   4. Key in Y scale factor or input point.
         ⇨   A box shape representing the size of the cell attaches itself to the cursor.
      ROTATION (0):
   5. Key in rotation angle or input point.
         ⇨   The cell appears in the drawing.
   To exit item: Make next selection

✱   **Error:** No active cell - If this message appears, use the ACTIVE CELL setting
   to define an active cell.
   See Also: PLACE CELL tools, SELECT CELL tools

PLACE CELL INTERACTIVE RELATIVE   PLA CE I R

Drawing Tool: Place Active Cell Relative (Interactive)
   Changes a cell's scale or rotation while placing the cell in the drawing file.
   This tool places the elements relative to the ACTIVE LEVEL. For instance, build
   a desk cell with the desk outline on level 10 and the word "DESK" on level 11.
   Place the desk cell in the drawing (with the ACTIVE LEVEL set to 5). The desk
   outline appears on level 5 and the word "DESK" appears on level 6.

Item Selection
On sidebar menu pick: | Cells | | IntAct | | Rel |
For window, point to: Palettes   Main   Cells

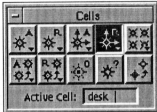

On paper menu see: CELLS
Command Window Prompts:
1. Select the item
   Enter cell origin
2. Pick a point
   ⇨   This sets the cell's origin.
   XSCALE (1.0):
3. Key in X scale factor or input point.
   YSCALE (1.0):
4. Key in Y scale factor or input point.
   ⇨   A box shape representing the size of the cell attaches itself to the cursor.
   ROTATION (0):
5. Key in rotation angle or input point.
   ⇨   The cell appears in the drawing.
To exit item: Make next selection

✷   **Error:** No active cell - If this message appears, use the ACTIVE CELL setting
   to define an active cell.
   See Also: PLACE CELL tools, SELECT CELL tools

PLACE CELL RELATIVE   PLA CE R

Drawing Tool: Place Active Cell Relative
   Places a cell in the drawing file, relative to the ACTIVE LEVEL. (Build a cell
   before using – see CREATE CELL.) For instance, build a desk cell with the desk
   outline on level 10 and the word "DESK" on level 11. Place the desk cell in the drawing
   (with the ACTIVE LEVEL set to 5), the desk outline appears on level 5 and the word
   "DESK" appears on level 6.
Item Selection
On sidebar menu pick: | Cells | | Place | | Rel |
For window, point to: Palettes   Main   Cells

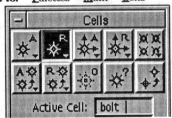

On paper menu see: CELLS
Command Window Prompts:
1. Select the item
   Enter cell origin
   ⇨   A box shape representing the size of the cell attaches itself to the cursor.
2. Pick a point
   ⇨   The cell appears in the drawing.
To exit item: Make next selection

✻ **Error:** No active cell - If this message appears, use the ACTIVE CELL setting to define an active cell.

See Also: PLACE CELL tools, SELECT CELL tools

PLACE CELL RELATIVE TMATRX

Drawing Tool: Place Active Cell Relative by Transformation Matrix

Places a cell in the drawing file, relative to the ACTIVE LEVEL. (Build a cell before using – see CREATE CELL.) The values set in the transformation matrix control the cell's scale and orientation. For example, build a desk cell with the desk outline on level 10 and the word "DESK" on level 11. Place the desk cell in the drawing (with the ACTIVE LEVEL set to 5). The desk outline appears on level 5 and the word "DESK" appears on level 6. Call this tool from a user command.

✻ **Error:** No active cell - If this message appears, use the ACTIVE CELL setting to define an active cell.

See Also: PLACE CELL tools, SELECT CELL tools

PLACE CIRCLE CENTER   PLA CI

Drawing Tool: Place Circle by Center

Constructs a circle based on a center point. Two points are required: the first to determine the circle's center, and the second to define its radius.

Item Selection

On sidebar menu pick: Place   Circle   Center

For window, point to: Palettes   Main   Circles/Ellipses

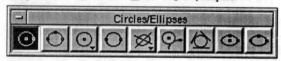

On paper menu see: CIRCLES

Command Window Prompts:

1. Select the item

    Enter circle center

2. Pick a point

    Enter circle edge point

3. Pick a second point

To exit item: Make next selection

See Also: CONSTRUCT TANGENT CIRCLE tools, PLACE CIRCLE tools, PLACE ELLIPSE tools

PLACE CIRCLE DIAMETER   PLA CI D

Drawing Tool: Place Circle by Diameter

Constructs a circle based on its diameter. Two input points are required to define the diameter of the circle.

Item Selection

On sidebar menu pick: Place   Circle   Diam

For window, point to: Palettes   Main   Circles/Ellipses

On paper menu see: CIRCLES

Command Window Prompts:

1. Select the item

    Enter first point on diameter

2. Pick a point

    Enter second point on diameter

3. Pick a second point

To exit item: **Make next selection**
See Also: CONSTRUCT TANGENT CIRCLE tools, PLACE CIRCLE tools, PLACE
ELLIPSE tools

PLACE CIRCLE EDGE   PLA CI E

Drawing Tool: Place Circle by Edge
       Constructs a circle based on three points defining its edge.

Item Selection
  On sidebar menu pick: Place   Circle   Edge
  For window, point to: Palettes   Main   Circles/Ellipses

On paper menu see: CIRCLES
Command Window Prompts:
  1. Select the item
     Enter first circle edge point
  2. Pick a point
     Enter second circle edge point
  3. Pick a second point
     Enter third (final) edge point
  4. Pick a final point
  To exit item: **Make next selection**
See Also: CONSTRUCT TANGENT CIRCLE tools, PLACE CIRCLE tools, PLACE
ELLIPSE tools

PLACE CIRCLE ISOMETRIC   PLA CI I

Drawing Tool: Place Isometric Circle
       Constructs a 2D or 3D circle based on a center point and an edge point.

Item Selection
  On sidebar menu pick: Place   Circle   Iso
  For window, point to: Palettes   Main   Circles/Ellipses

On paper menu see: CIRCLES
Command Window Prompts:
  1. Select the item
     Enter circle center
  2. Pick a point
     Enter circle edge point
  3. Pick a second point
  To exit item: **Make next selection**
See Also: CONSTRUCT TANGENT CIRCLE tools, PLACE CIRCLE tools, PLACE
ELLIPSE tools

PLACE CIRCLE RADIUS   PLA CI R

Drawing Tool: Place Circle by Keyed-in Radius
       Constructs a circle based on an input center point and a radius key-in.

Item Selection
  On sidebar menu pick: Place   Circle   Rad

For window, point to: Palettes   Main   Circles/Ellipses

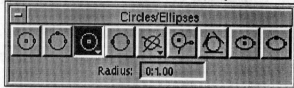

On paper menu see: CIRCLES
Command Window Prompts:
1. Select the item
   Key in radius
2. Key in a radius
   Enter circle center
3. Pick a point
To exit item: Make next selection
See Also: CONSTRUCT TANGENT CIRCLE tools, PLACE CIRCLE tools, PLACE
ELLIPSE tools

PLACE CONE POINTS

Obsolete
      This command assembles a cone based on two input points defining the
centers of the base and top of the cone. Two radius key-ins define the size of the
base and top openings.
See Also: PLACE CONE tools

PLACE CONE RADIUS   PLA CO RA

3D Drawing Tool: Place Right Cone by Keyed-in Radii
      Assembles a right cone based on two input points defining the centers of its
base and top. Two radius key-ins define the size of the base and top openings.
Item Selection
On sidebar menu pick: Place   3-D
For window, point to: Palettes   3D

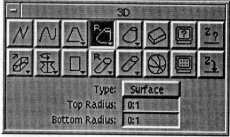

On paper menu see: 3D
Command Window Prompts:
1. Select the item
   Key in radius
2. Key in a base radius
   Key in top radius (or pt for 0 radius)
3. Key in top radius
   Enter center point of base
4. Pick a point or key-in
   Enter center point of top
5. Pick a point or key-in
To exit item: Make next selection

✍ *Note:* To place the cone as a capped surface, turn the ACTIVE CAPMODE setting
ON.
See Also: ACTIVE CAPMODE, PLACE CONE tools

PLACE CONE RIGHT

3D Drawing Tool: Place Right Cone
   Creates a right cone based on four input points. The first two points define the center and radius of the cone's base. The next two points define the center and radius of its top.

Item Selection
   On sidebar menu pick: `Place`   `3-D`
   For window, point to: Palettes   3D

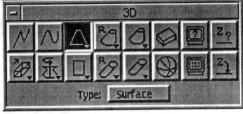

On paper menu see: 3D
Command Window Prompts:
   1. Select the item
      `Enter center point of base`
   2. Pick a point or key-in
      `Enter base radius`
   3. Pick a point or key-in
      `Enter center point of top`
   4. Enter center point or key-in
      `Enter top radius`
   5. Enter radius point or key-in
   To exit item: Make next selection

✍ *Note:* To place the cone as a capped surface, turn the ACTIVE CAPMODE setting ON.
   See Also: ACTIVE CAPMODE, PLACE CONE tools

PLACE CONE SKEWED

3D Drawing Tool: Place Skewed Cone
   Creates a skewed cone based on four input points. The first two points define the center and radius of the cone's base. The next two points define the center and radius of its top.

Item Selection
   On sidebar menu pick: `Place`   `3-D`
   For window, point to: Palettes   3D

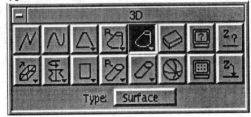

On paper menu see: 3D
Command Window Prompts:
   1. Select the item
      `Enter center point of base`
   2. Pick a point or key-in
      `Enter base radius`
   3. Pick a point or key-in
      `Enter base radius`

4. Enter point or key-in
   Enter center point of top
5. Enter center point or key-in
   Enter top radius
6. Enter top point or key-in
To exit item: Make next selection

✍ *Note:* To place the cone as a capped surface, turn the ACTIVE CAPMODE setting ON.

See Also: ACTIVE CAPMODE, PLACE CONE tools

PLACE CURVE POINT   PLA CU

Drawing Tool: Place Point Curve
   Constructs a single curve segment. Three input points make up each curve segment. A series of curves can be created by placing additional points. Multiple curves placed during one tool operation become a single complex element.

Item Selection
On sidebar menu pick: Place   Curve   Points
For window, point to: Palettes   Main   Line Strings

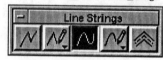

On paper menu see: LINE STRING
Command Window Prompts:
1. Select the item
   Enter first point of curve string
2. Pick a point
   Enter point or RESET to complete
3. Continue to place points
   ⇨ Click a RESET to complete the series of curves.
To exit item: Make next selection
See Also: PLACE ARC tools, PLACE CURVE tools, PLACE ELLIPSE tools

PLACE CURVE SPACE   PLA CU SP

3D Drawing Tool: Place Space Curve
   Constructs a nonplanar 3D curve. This type of curve may be defined by as few as three or as many as 97 input points. A new curve element starts after entering more than 97 points, and the two curves become one complex element.

Item Selection
On sidebar menu pick: Place   3-D
For window, point to: Palettes   3D

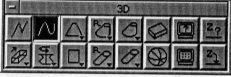

On paper menu see: 3D
Command Window Prompts:
1. Select the item
   Enter first point of curve string
2. Pick a point
   Enter point or RESET to complete
3. Continue to place points
   ⇨ Click a RESET to complete the space curve.
To exit item: Make next selection
See Also: PLACE ARC tools, PLACE ELLIPSE tools

PLACE CURVE STREAM

Drawing Tool: Place Stream Curve

    Places a stream of input points to create curved lines, i.e. contour lines on maps. The stream curve placed during one tool operation is one complex element.

Item Selection

    On sidebar menu pick: `Place` `Curve` `Stream`

    For window, point to: Palettes   Main   Line Strings

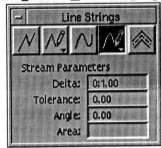

    On paper menu see: LINE STRING

Command Window Prompts:

    1. Select the item

        `Enter first point of curve string`

    2. Pick a point

        `Enter point or RESET to complete`

    3. Continue to move the mouse to place points

        ⇨   Click a RESET to complete the stream curve.

    To exit item: **Make next selection**

✳ **Error:** `Stream delta not set` - If this message appears, the ACTIVE STREAM DELTA (SD=) may be set to zero. Try SD=.1.

    See Also: PLACE ARC tools, PLACE CURVE SPACE, PLACE ELLIPSE tools

PLACE CYLINDER CAPPED

Obsolete

    Turns the ACTIVE CAPMODE setting on and reactivates the last PLACE CYLINDER tool.

    See Also: PLACE CYLINDER tools

PLACE CYLINDER POINTS

Obsolete

    Draws a cylinder from three input points. Two of the points define the centers of the cylinder's base and top. The other input point controls the cylinder skew.

    See Also: PLACE CYLINDER tools

PLACE CYLINDER RADIUS

3D Drawing Tool: Place Right Cylinder by Keyed-in Radius

    Draws a cylinder based on a keyed in radius, and input points for the centers of the base and top.

Item Selection

    On sidebar menu pick: `Place` `3-D`

For window, point to: Palettes 3D

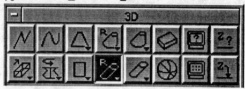

On paper menu see: 3D
Command Window Prompts:
1. Select the item
   Key in radius
2. Key in a base radius
   Enter center point of base
3. Pick base point
   Enter center point of top
4. Pick top point
To exit item: Make next selection

☞ *Tip:* To place the cylinder as a capped surface, turn the ACTIVE CAPMODE setting
ON.

See Also: ACTIVE CAPMODE, PLACE CYLINDER tools

PLACE CYLINDER RIGHT   PLA CY RI

3D Drawing Tool: Place Right Cylinder
    Creates a right cylinder based on three input points. The first two points
determine the base's center point and radius. The third point defines the cylinder's
top surface.
Item Selection
On sidebar menu pick: Place | 3-D
For window, point to: Palettes 3D

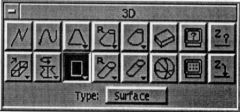

On paper menu see: 3D
Command Window Prompts:
1. Select the item
   Enter center point of base
2. Pick a point
   Enter point on surface
3. Pick a point
   Enter center point of top
4. Pick top point
To exit item: Make next selection

☞ *Tip:* To place the cylinder as a capped surface, turn the ACTIVE CAPMODE setting
ON.

See Also: ACTIVE CAPMODE, PLACE CYLINDER tools

PLACE CYLINDER SKEWED   PLA CY S

3D Drawing Tool: Place Skewed Cylinder
    Creates a slanted right cylinder based on three input points. The first input
point determines the cylinder's base. The second point defines the radius and slant
of the cylinder. The third point defines the cylinder's top surface.

Item Selection
On sidebar menu pick: Place   3-D
For window, point to: Palettes   3D

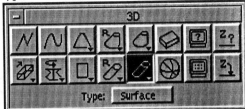

On paper menu see: 3D
Command Window Prompts:
1. Select the item
   `Enter center point of base`
2. Pick a point
   `Enter point on surface`
3. Pick a point
   `Enter center point of top`
4. Pick top point
To exit item: **Make next selection**
See Also: PLACE CYLINDER tools

PLACE CYLINDER UNCAPPED

Obsolete
Turns the ACTIVE CAPMODE setting off and reactivates the last PLACE
CYLINDER tool.
See Also: PLACE CYLINDER tools

PLACE DIALOGTEXT   PLA D

Drawing Tool: Place Text
Places text at a selected point in the drawing file. This item asks you to key
in the text string in the Text Editor window and pick a point to place the text in
the drawing.
Item Selection
On sidebar menu pick: n/a
For window, point to: Palettes   Main   Text

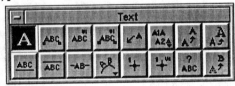

On paper menu see: TEXT
Command Window Prompts:
1. Select the item
   `Enter text`
2. Enter the text string
   `Enter more chars or position text`
3. Pick resting place for text
To exit item: **Make next selection**

✍ *Note:* The ACTIVE LINE SPACE (LS=) setting controls the spacing between the
element and the text.

☞ *Tip:* When you need to change the text size or font style and you've entered the text,
try this: hit Enter and click on the prompt field in the Command Window. Enter
your font change (FT=3). Hit Enter again, and you're ready to place the text string.

☞ *Tip:* If your text line is longer than 60 characters, hit the Enter key and then keep typing. The PLACE TEXT tool places the text string as one line of text.

☞ *Tip:* If you want to intentionally place multiple lines of text, place a line feed character between the lines of text. On a PC, Ctrl+Enter keys generate linefeed characters. On an Intergraph workstation, use the Linefeed key.

☞ *Tip:* When placing text on curved elements like arcs, ellipses or circles, use the PLACE TEXT ALONG tool.

See Also: PLACE DIALOGTEXT tools, PLACE TEXT tools, PLACE NODE

PLACE DIALOGTEXT ABOVE   PLA D AB

Drawing Tool: Place Text Above Element
    Places text, with the aid of the Text Editor window, above a line segment or shape. Placing text above an element requires a key-in for the text string and an input point to determine the text's resting place.

Item Selection
    On sidebar menu pick: n/a
    For window, point to: <u>P</u>alettes   <u>M</u>ain   Text

    On paper menu see: TEXT
Command Window Prompts:
    1. Select the item
       Enter text
    2. Enter the word ABOVE
       Select line segment
    3. Pick a line segment
       Accept/Reject (select next input)
    4. Give an accept point
    To exit item: Make next selection

✐ *Note:* The ACTIVE LINE SPACE (LS=) setting controls the spacing between the element and the text.

☞ *Tip:* When you need to change the text size or font style and you've entered the text, try this: hit Enter and click on the prompt field in the Command Window. Enter your font change (FT=3). Hit Enter again, and you're ready to place the text string.

☞ *Tip:* If your text line is longer than 60 characters, hit the Enter key and then keep typing. The PLACE TEXT tool places the text string as one line of text.

☞ *Tip:* If you want to intentionally place multiple lines of text, place a line feed character between the lines of text. On a PC, Ctrl+Enter keys generate linefeed characters. On an Intergraph workstation, use the Linefeed key.

☞ *Tip:* When placing text on curved elements like arcs, ellipses or circles, use the PLACE TEXT ALONG tool.

See Also: PLACE DIALOGTEXT tools, PLACE TEXT tools, PLACE NODE

PLACE DIALOGTEXT ALONG   PLA D AL

Drawing Tool: Place Text Along Element
    Places text, with the aid of the Text Editor window, along an arc, ellipse or circle. Placing text along an element asks for an optional key-in to change the line spacing of the text. Key in the text string. Next, an input point identifies the target element. A second point determines the text's resting place.

Item Selection
    On sidebar menu pick: n/a

For window, point to: Palettes   Main   Text

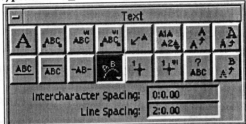

On paper menu see: TEXT
Command Window Prompts:
1. Select the item
   Enter distance between chars (or return)
      ⇨   Hitting return leaves the line spacing at half the text size.
2. Hit return
   Enter text
3. Enter the word ALONG
   Identify element, text location
4. Pick a line segment
   Accept, select text above/below
5. Give an accept point
      ⇨   Accept point will determine text's resting place.
   To exit item: **Make next selection**

🖎 *Note:* The ACTIVE LINE SPACE (LS=) setting controls the spacing between the element and the text.

🖎 *Note:* Characters placed with PLACE TEXT ALONG are placed as individual text characters and not as a string of text characters.

☞ *Tip:* When you need to change the text size or font style and you've entered the text, try this: hit Enter and click on the prompt field in the Command Window. Enter your font change (FT=3). Hit Enter again, and you're ready to place the text string.

☞ *Tip:* If your text line is longer than 60 characters, hit the Enter key and then keep typing. The PLACE TEXT tool places the text string as one line of text.

☞ *Tip:* If you want to intentionally place multiple lines of text, place a line feed character between the lines of text. On a PC, Ctrl+Enter keys generate linefeed characters. On an Intergraph workstation, use the Linefeed key.

See Also: PLACE DIALOGTEXT tools, PLACE TEXT tools, PLACE NODE

PLACE DIALOGTEXT BELOW   PLA D B

Drawing Tool: Place Text Below Element
      Places text, with the aid of the Text Editor window, below a line segment or shape, use the PLACE TEXT BELOW tool. This tool opens the Text Editor window and asks for a key-in for the text string and two input points to determine the text's resting place.

Item Selection
   On sidebar menu pick: n/a
   For window, point to: Palettes   Main   Text

On paper menu see: TEXT
Command Window Prompts:
   1. Select the item
      Enter text

    2. Enter the word BELOW
       Select line segment
    3. Pick a line segment
       Accept/Reject (select next input)
    4. Give an accept point
    To exit item: Make next selection

✍ *Note:* The ACTIVE LINE SPACE (LS=) setting controls the spacing between the element and the text.

✍ *Note:* Characters placed with PLACE TEXT ALONG are placed as individual text characters and not as a string of text characters.

☞ *Tip:* When you need to change the text size or font style and you've entered the text, try this: hit Enter and click on the prompt field in the Command Window. Enter your font change (FT=3). Hit Enter again, and you're ready to place the text string.

☞ *Tip:* If your text line is longer than 60 characters, hit the Enter key and then keep typing. The PLACE TEXT tool places the text string as one line of text.

☞ *Tip:* If you want to intentionally place multiple lines of text, place a line feed character between the lines of text. On a PC, Ctrl+Enter keys generate linefeed characters. On an Intergraph workstation, use the Linefeed key.

☞ *Tip:* When placing text on curved elements like arcs, ellipses or circles, use the PLACE TEXT ALONG tool.

See Also: PLACE DIALOGTEXT tools, PLACE TEXT tools, PLACE NODE

PLACE DIALOGTEXT FITTED   PLA D FI

Drawing Tool: Place Fitted Text
       Places text in a space defined by two input points. Placing fitted text through the Text Editor window requires a key-in for the text string and two input points to determine the text's resting place.

Item Selection
  On sidebar menu pick: n/a
  For window, point to: Palettes Main Text

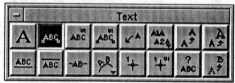

  On paper menu see: TEXT
Command Window Prompts:
    1. Select the item
       Enter text
    2. Enter the word FITTED
       Enter more chars or Position text
    3. Pick text starting point
       Define endpoint of text
    4. Give text endpoint
    To exit item: Make next selection

✍ *Note:* The ACTIVE LINE SPACE (LS=) setting controls the spacing between the element and the text.

✍ *Note:* Characters placed with PLACE TEXT ALONG are placed as individual text characters and not as a string of text characters.

☞ *Tip:* When you need to change the text size or font style and you've entered the text, try this: hit Enter and click on the prompt field in the Command Window. Enter your font change (FT=3). Hit Enter again, and you're ready to place the text string.

☞ *Tip:* If your text line is longer than 60 characters, hit the Enter key and then keep typing. The PLACE TEXT tool places the text string as one line of text.

☞ *Tip:* If you want to intentionally place multiple lines of text, place a line feed character between the lines of text. On a PC, Ctrl+Enter keys generate linefeed characters. On an Intergraph workstation, use the Linefeed key.

☞ *Tip:* When placing text on curved elements like arcs, ellipses or circles, use the PLACE TEXT ALONG tool.

See Also: PLACE DIALOGTEXT tools, PLACE TEXT tools, PLACE NODE

PLACE DIALOGTEXT FVI   PLA D FV

Drawing Tool: Place Fitted View Independent Text
Places fitted text at one view orientation. The text will maintain its original bearing, regardless of subsequent view rotation. Placing fitted text in a view opens the Text Editor window. This item requires a key-in for the text string and two input points to determine the text's resting place.

Item Selection
On sidebar menu pick: n/a
For window, point to: Palettes   Main   Text

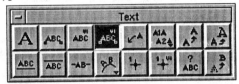

On paper menu see: TEXT
Command Window Prompts:
1. Select the item
   Enter text
2. Enter the words VIEW FITTED
   Enter more chars or Position text
3. Pick text starting point
   Define endpoint of text
4. Pick text endpoint
To exit item: **Make next selection**

✍ *Note:* The ACTIVE LINE SPACE (LS=) setting controls the spacing between the element and the text.

✍ *Note:* Characters placed with PLACE SPACE ALONG are placed as individual text characters and not as a string of text characters.

☞ *Tip:* When you need to change the text size or font style and you've entered the text, try this: hit Enter and click on the prompt field in the Command Window. Enter your font change (FT=3). Hit Enter again, and you're ready to place the text string.

☞ *Tip:* If your text line is longer than 60 characters, hit the Enter key and then keep typing. The PLACE TEXT tool places the text string as one line of text.

☞ *Tip:* If you want to intentionally place multiple lines of text, place a line feed character between the lines of text. On a PC, Ctrl+Enter keys generate linefeed characters. On an Intergraph workstation, use the Linefeed key.

☞ *Tip:* When placing text on curved elements like arcs, ellipses or circles, use the PLACE TEXT ALONG tool.

See Also: PLACE DIALOGTEXT tools, PLACE TEXT tools, PLACE NODE

PLACE DIALOGTEXT ON   PLA D O

Drawing Tool: Place Text On Element
Places text characters on a line segment. Placing text on an element opens the Text Editor window. This item requires a key-in for the text string and two input points to determine the text's resting place.

Item Selection
On sidebar menu pick: n/a

For window, point to: Palettes   Main   Text

On paper menu see: TEXT
Command Window Prompts:
1. Select the item
   Enter text
2. Enter the word ON
   Select line segment
3. Pick a line segment
   Accept/Reject (select next input)
4. Give an accept point
   ⇨   Accept point determines text location on the element.
To exit item: **Make next selection**

✍ *Note:* The ACTIVE LINE SPACE (LS=) setting controls the spacing between the element and the text.

✍ *Note:* Characters placed with PLACE TEXT ALONG are placed as individual text characters and not as a string of text characters.

☞ *Tip:* When you need to change the text size or font style and you've entered the text, try this: hit Enter and click on the prompt field in the Command Window. Enter your font change (FT=3). Hit Enter again, and you're ready to place the text string.

☞ *Tip:* If your text line is longer than 60 characters, hit the Enter key and then keep typing. The PLACE TEXT tool places the text string as one line of text.

☞ *Tip:* If you want to intentionally place multiple lines of text, place a line feed character between the lines of text. On a PC, Ctrl+Enter keys generate linefeed characters. On an Intergraph workstation, use the Linefeed key.

☞ *Tip:* When placing text on curved elements like arcs, ellipses or circles, use the PLACE TEXT ALONG tool.

See Also: PLACE DIALOGTEXT tools, PLACE TEXT tools, PLACE NODE

PLACE DIALOGTEXT TMATRIX

Drawing Tool: Place Text by Transformation Matrix
   Places text according to the values set in the transformation matrix. Keying in this tool name opens the Text Editor window. This tool is usually called from an application program.
See Also: PLACE DIALOGTEXT tools, PLACE TEXT tools, Tmatrix

PLACE DIALOGTEXT VI   **PLA D V**

Drawing Tool: Place View Independent Text
   Places text at one orientation, independent of view rotation. Keying in this tool name opens the Text Editor window. Placing text with a view orientation requires a key-in for the text string and an input point to determine the text's resting place.
Item Selection
On sidebar menu pick: n/a
For window, point to: Palettes   Main   Text
On paper menu see: TEXT
Command Window Prompts:
1. Select the item
   Enter text
2. Enter the word VIEW
   Enter more chars or Position text
3. Pick text location

To exit item: **Make next selection**

✍ *Note:* The ACTIVE LINE SPACE (LS=)setting controls the spacing between the element and the text.

✍ *Note:* Characters placed with PLACE TEXT ALONG are placed as individual text characters and not as a string of text characters.

☞ *Tip:* When you need to change the text size or font style and you've entered the text, try this: hit Enter and click on the prompt field in the Command Window. Enter your font change (FT=3). Hit Enter again, and you're ready to place the text string.

☞ *Tip:* If your text line is longer than 60 characters, hit the Enter key and then keep typing. The PLACE TEXT tool places the text string as one line of text.

☞ *Tip:* If you want to intentionally place multiple lines of text, place a line feed character between the lines of text. On a PC, Ctrl+Enter keys generate linefeed characters. On an Intergraph workstation, use the Linefeed key.

☞ *Tip:* When placing text on curved elements like arcs, ellipses or circles, use the PLACE TEXT ALONG tool.

See Also: PLACE DIALOGTEXT tools, PLACE TEXT tools, PLACE NODE

PLACE ELLIPSE CENTER   PLA E

Drawing Tool: Place Ellipse by Center and Edge
            Constructs an ellipse based on three input points. The first point defines the center of the ellipse, the second defines its axis, and the third determines its size and shape.
Item Selection
On sidebar menu pick: | Place | | Ellips | | Center |

For window, point to: Palettes   Main   Circles/Ellipses
On paper menu see: ELLIPSE
Command Window Prompts:
   1. Select the item
      Enter ellipse center
   2. Enter the center point
      Enter point on axis
      Enter any point on ellipse
         ⇨   The third point may be placed anywhere along the ellipse.
To exit item: **Make next selection**

☞ *Tip:* The three input points needed to form the ellipse may be entered with precision key-ins. (See Precision Key-ins).

☞ *Tip:* You can make an arc by first making an ellipse and then using PARTIAL DELETE or FENCE CLIP to take a bite out of the ellipse.

☞ *Tip:* Create an ellipse by placing a circle, putting a fence around it and stretching it in the X or Y direction.

✳ *Error:* Illegal definition - If this message appears in the error field, it is impossible to draw the ellipse based on the input points. (i.e. three points in a straight line).

See Also: PLACE ARC tools, PLACE CIRCLE tools, PLACE ELLIPSE tools

PLACE ELLIPSE EDGE

PLA E

Drawing Tool: Place Ellipse by Edge Points

Constructs an ellipse based on three input points. The first point is one endpoint of the primary axis. The second point defines the edge of the ellipse. The last point completes the definition of the primary axis.

Item Selection

On sidebar menu pick: Place Ellips Edge

For window, point to: Palettes Main Circles/Ellipses

On paper menu see: ELLIPSE

Command Window Prompts:
1. Select the item
   Enter one end of axis
2. The first point defines one end of the primary axis
   Enter any point on ellipse
3. The second point may be placed anywhere along the ellipse
   Enter other end of axis
4. This endpoint defines the primary axis

To exit item: Make next selection

☞ *Tip:* The three input points needed to form the ellipse may be entered with precision key-ins. (See Precision Key-ins).

☞ *Tip:* You can make an arc by first making an ellipse and then using PARTIAL DELETE or FENCE CLIP to take a bite out of the ellipse.

☞ *Tip:* Create an ellipse by placing a circle, putting a fence around it and stretching it in the X or Y direction.

✳ *Error:* Illegal definition - If this message appears in the error field, it is impossible to draw the ellipse based on the input points. (i.e. three points in a straight line).

See Also: PLACE ARC tools, PLACE CIRCLE tools, PLACE ELLIPSE tools

PLACE ELLIPSE FOURTH

Same as PLACE ELLIPSE QUARTER.
See Also: PLACE ELLIPSE FOURTH

PLACE ELLIPSE HALF

PLA E H

Drawing Tool: Place Half Ellipse

Creates an elliptical arc based on three input points. The first point defines one end of the primary axis. Place the second point anywhere along the elliptical arc. The last point completes the definition of the primary axis.

Item Selection

On sidebar menu pick: Place Ellips Half

For window, point to: Palettes Main Arcs

On paper menu see: ELLIPSE

Command Window Prompts:
1. Select the item
   Enter one edge of axis

2. The first point defines one end of the primary axis
   `Enter any point on ellipse`
3. Place the second point anywhere along the ellipse
   `Enter other end of axis`
4. This endpoint defines the primary axis
   To exit item: Make next selection

☞ *Tip:* The three input points needed to form the ellipse may be entered with precision key-ins. (See Precision Key-ins).

✳ *Error:* `Illegal definition` - If this message appears in the error field, it is impossible to draw the ellipse based on the input points. (i.e. three points in a straight line).
   See Also: PLACE ARC tools, PLACE CIRCLE tools, PLACE ELLIPSE tools

PLACE ELLIPSE QUARTER   PLA EL Q

Drawing Tool: Place Ellipse Quarter
   Creates an elliptical arc based on three input points. The first point is one endpoint of the ellipse. The second point defines its edge. The last point completes the definition of the primary axis.

Item Selection
   On sidebar menu pick: | Place | | Ellips | | Qrtr |
   For window, point to: Palettes   Main   Arcs

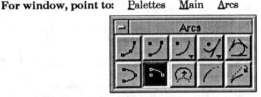

   On paper menu see: ELLIPSE
Command Window Prompts:
   1. Select the item
      `Enter one edge of ellipse quarter`
   2. The first point defines one endpoint of the ellipse
      `Enter center point`
   3. The second point is defines ellipse's center
      `Enter endpoint`
   4. This point defines the other endpoint of the ellipse
      To exit item: Make next selection

☞ *Tip:* The three input points needed to form the ellipse may be entered with precision key-ins. (See Precision Key-ins).

✳ *Error:* `Illegal definition` - If this message appears in the error field, it is impossible to draw the ellipse based on the input points. (i.e. three points in a straight line).
   See Also: PLACE ARC tools, PLACE CIRCLE tools, PLACE ELLIPSE tools

PLACE FENCE BLOCK   PLA F

Drawing Tool: Place Fence Block
   Places a rectangular boundary around a group of elements. Create a fence block by placing two diagonal input points. Edit the fenced elements with one tool. Once you complete a fence operation, use the PLACE FENCE BLOCK or PLACE FENCE SHAPE tool a second time to erase the fence from the screen.

Item Selection
   On sidebar menu pick: | Fences | | Block |

For window, point to: Palettes   Fence   Place Fence

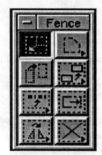

On paper menu see: PLACE FENCE
Command Window Prompts:
1. Select the item
   Enter first point
2. Pick a point
   Enter opposite corner
3. Pick second point
   ⇨ Fence block remains in the highlight color.
To exit item: Exits by itself
See Also: FENCE MOVE, MODIFY FENCE, PLACE FENCE SHAPE

PLACE FENCE SHAPE   PLA F S

Drawing Tool: Place Fence Shape
   Places a multi-sided boundary around a group of elements. Create a fence shape by placing a series of input points; it may consist of as many as 100 sides. Edit the elements in the fence with one tool. Once you complete a fence operation, use the PLACE FENCE BLOCK or PLACE FENCE SHAPE tool a second time to erase the fence from the screen.

Item Selection
On sidebar menu pick: Fences   Shape
For window, point to: Palettes   Fence   Place Fence

On paper menu see: PLACE FENCE
Command Window Prompts:
1. Select the item
   Enter fence points
2. Pick a point
   ⇨ Tool prompt remains the same.
3. Pick several more points
   ⇨ The last point and first point must meet up.
   Fence closed
   ⇨ Fence block remains in the highlighted.
To exit item: Exits by itself
See Also: FENCE MOVE, MODIFY FENCE, PLACE FENCE BLOCK

PLACE HELIX   PLA H

3D Drawing Tool: Place Helix
   Creates a B-spline curve in the form of a helix. This MDL application tool requires four input points. The first two points define the base of the helix center and radius points. The last two input points define the top of the helix center and radius points.

Item Selection
On sidebar menu pick: | Place | | BSpl | | SpcCur |
For window, point to: P̲alettes   3D B̲-splines   Sp̲ace Curves

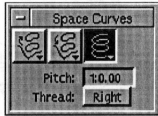

On paper menu see: SPACE CURVE
Command Window Prompts:
1. Select the item
   Enter center point of base
2. Pick base center point
   Enter base radius
3. Pick base radius
   Enter center point of top
4. Pick top center point
   Enter top radius
5. Pick top radius
To exit item: **Make next selection**

* *Error:* Invalid character - **If this message appears after your key in, enter MDL LOAD SPLINES or pick the tool from the Space Curves palette once.**
See Also: PLACE BSPLINE tools

PLACE LINE   PLA LI

Drawing Tool: Place Line
Creates a line based on two input points, defining its end points. You can continue to add more lines by placing additional points. Each line segment is a separate element.

Item Selection
On sidebar menu pick: | Place | | Line | | Pnts |
For window, point to: P̲alettes   M̲ain   L̲ines

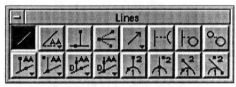

On paper menu see: LINES
Command Window Prompts:
1. Select the item
   Enter first point
2. Enter starting point of the line
   Enter end point
3. Enter a second point
   ⇨ Tool is waiting to draw more lines.
To exit item: **Make next selection**

☞ *Tip:* Use precision key-ins to precisely place the start point and/or end point of the line.
See Also: PLACE LINE ANGLE, PLACE LSTRING tools

PLACE LINE ANGLE

PLA LI A

Drawing Tool: Place Line at Active Angle

Places lines at the ACTIVE ANGLE based on two input points defining its ends. The ACTIVE ANGLE setting controls the angle of the line.

Item Selection

On sidebar menu pick: `Place` `Line`

For window, point to: Palettes Main Lines

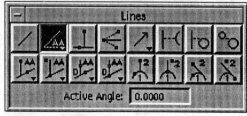

On paper menu see: LINES

Command Window Prompts:

1. Select the item
   Enter first point
2. Enter starting point of the line
   Enter end point
3. Enter a second point
   ⇨ Tool is waiting to draw more lines.

To exit item: Make next selection

☞ *Tip:* Use precision key-ins to precisely place the start point and/or length point of the line.

See Also: PLACE LINE, PLACE LSTRING tools

PLACE LSTRING POINT

PLA LS

Drawing Tool: Place Line String

Forms a line string from a series of associated lines. Each segment of the line string has a beginning and an end point. The maximum number of separate line segments contained in one line string is 100, or 101 points. After entering more than 101 points, MicroStation chains the two line strings together to make a complex chain.

Item Selection

On sidebar menu pick: `Place` `LnStr` `Point`

For window, point to: Palettes Main Line Strings

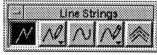

On paper menu see: LINE STRING

Command Window Prompts:

1. Select the item
   Enter first point
2. Enter starting point of first line
   Enter point or RESET to complete
3. Enter a second point
   ⇨ Each input point creates another line segment.

To exit item: Make next selection

☞ *Tip:* Use precision key-ins to precisely place the start point and/or end point of each line segment.

See Also: PLACE LINE tools, PLACE LSTRING tools

PLACE LSTRING SPACE
PLA LS SP

3D Drawing Tool: Place Space Line String

> Creates a line string made of many points occupying more than one plane. The first point starts the line string. Moving the cursor causes the tool to place additional points in the drawing file. ACTIVE STREAM TOLERANCE (SD=) and ACTIVE STREAM DELTA (SD=) settings control the distance between the points. Hitting RESET completes the line string. After placing more than 101 points, MicroStation chains the line strings together to make a complex chain.

Item Selection

On sidebar menu pick: | Place | | 3-D |

For window, point to: Palettes 3D

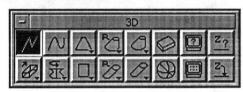

On paper menu see: 3D

Command Window Prompts:

1. Select the item
   `Enter first point`
2. Enter starting point of the line
   `Enter point or RESET to complete`
3. Enter a second point
   ⇨ Moving the cursor causes the tool to place additional points; forming a smooth line.

Examples:

`SD=::3000`	makes the distance between each input point 3000 positional units. (See the SD= key-in)
`ST=::3000`	allows the cursor to wander 3000 positional units before taking a point off the original track of the cursor.

To exit item: **Make next selection**

☞ **Tip:** If you can't create a line string using this tool, check the ACTIVE STREAM DELTA and ACTIVE STREAM TOLERANCE settings. Their values may be too high.

☞ **Tip:** If the line string is not smooth enough, change the SD= and ST= settings to fewer positional units.

See Also: PLACE LINE tools, PLACE LSTRING tools

PLACE LSTRING STREAM
PLA LS ST

Drawing Tool: Place Stream Line String

> Traces existing drawings from a digitizing tablet. The first input point starts the line string. Hold down the DATA button and move the cursor to place additional points in the drawing file. ACTIVE STREAM TOLERANCE (SD=) and ACTIVE STREAM DELTA (SD=) control the spacing between the points. Hitting RESET completes the line string. After placing more than 101 points, MicroStation chains the line strings together to make a complex chain.

Item Selection

On sidebar menu pick: | Place | | LnStr | | Stream |

For window, point to: P̲alettes   M̲ain   Line S̲trings

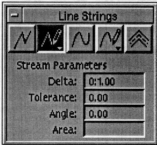

On paper menu see: LINE STRING
Command Window Prompts:
1. Select the item
   Enter first point
2. Enter starting point of the line
   Enter point or RESET to complete
3. Enter a second point
   ⇨   Moving the cursor causes the tool to place additional points; forming a smooth line.

Examples:

SD=::3000   makes the distance between each input point 3000 positional units. (See the SD= key-in)

ST=::3000   allows the cursor to wander 3000 positional units before taking a point off the original track of the cursor.

To exit item: **Make next selection**

☞ *Tip:* If you can't create a line string using this tool, check the ACTIVE STREAM DELTA and ACTIVE STREAM TOLERANCE settings. Their values may be too high.

☞ *Tip:* If the line string is not smooth enough, change the SD= and ST= settings to fewer positional units.

See Also: PLACE LINE tools, PLACE LSTRING tools

PLACE MLINE   PLA M

Drawing Tool: Place Multi-line
   Creates multi-line elements. Each of up to 16 separate lines can have its own color, weight, line style, level and class. The spacing between the multiple lines can vary. You can define the multi-line characteristics from the Multi-line Settings box. This tool requires at least two input points to define a multi-line: a start point and an end point.

Item Selection
On sidebar menu pick: Place   Mline

For window, point to: P̲alettes   M̲ain   Line S̲trings

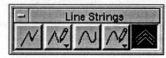

On paper menu see: LINES
Command Window Prompts:
1. Select the item
   Enter first point
2. Pick a starting point
   Enter point or RESET to complete
3. Pick a second point
To exit item: **Make next selection**

✍ *Note:* Modify existing multi-line elements by clicking on P̲alettes, M̲ulti-Lines to use the MDL cutter tools.

✍ *Note:* Change how multi-line elements are placed through the Multi-lines settings box.

☞ *Tip:* Multi-lines can be associated to another element by snapping to that element. Remember to turn your association lock on.

See Also: Cutter Tools, DROP MLINE, FENCE DROP MLINE, FREEZE

■ PLACE NODE   PLA N

Drawing Tool: Place Text Node

Places drop-off points for placing text in the drawing file. Selecting the tool and placing input points create text nodes, which save text attributes such as size, justification, font and line spacing. Later, when a text string is attached to the node, the text takes on the attributes set by the node.

Each text node has a unique number. Application programs use this number to bulk-load text to the text nodes.

Item Selection
On sidebar menu pick: |Text| |Node|
For window, point to: P̲alettes   M̲ain   T̲ext

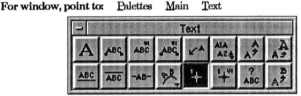

On paper menu see: TEXT
Command Window Prompts:
  1. Select the item
     `Enter text node origin`
  2. Pick a point
     `Define angle or RESET for active angle`
  3. Key in angle, click a RESET or place a point
To exit item: **Make next selection**

✍ *Note:* The LOCK TEXTNODE setting must be enabled to place text at a text node.

✳ *Error:* `Illegal definition` - If this message appears, both input points were probably given at the same location.
See Also: LOCK TEXTNODE, PLACE DIALOGTEXT tools, PLACE NODE tools, SET NODES

■ PLACE NODE TMATRX

Drawing Tool: Place Text Node by Transformation Matrix

Creates drop-off-points for text in the drawing file. The transformation matrix controls the placement of the text nodes. Later when the text string is attached to the node, the text takes on the same characteristics set by the node. Call this tool from a user command.

Each text node has a unique number.

✍ *Note:* The LOCK TEXTNODE setting must be enabled to place text at a text node.

✳ *Error:* `Illegal definition` - If this message appears, both input points were probably given at the same location.
See Also: LOCK TEXTNODE, PLACE NODE tools, SET NODES

PLACE NODE VIEW

Drawing Tool: Place View Independent Text Node

Places drop-off points for text in the drawing file. Rotating the view does not affect the orientation of the text node. This tool places the text node with the active text attributes such as size, justification, font, etc. Later when the text string is attached to the node, the text takes on the attributes set by the node. Create nodes by selecting the tool and placing input points in the drawing file.

Each node has a unique number. Application programs use this number to bulk-load text to the node.

Item Selection

On sidebar menu pick: ⬚Text⬚ ⬚VI Nod⬚
For window, point to: Palettes   Main   Text

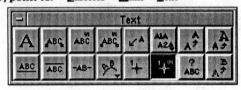

On paper menu see: TEXT
Command Window Prompts:
1. Select the item
   Enter text node origin
2. Pick a point
   Define angle or RESET for active angle
3. Key in angle, click a RESET or pick a point
To exit item: Make next selection

✍ *Note:* The LOCK TEXTNODE setting must be enabled to place text at a text node.

✳ *Error:* Illegal definition - If this message appears, both input points were probably given at the same location.
See Also: LOCK TEXTNODE, PLACE NODE tools, SET NODES

PLACE NOTE

Drawing Tool: Place Note

Places text, a note, a leader line and an arrow. This tool requires a text key-in and two input points. The first point identifies the placement of the arrow symbol. The second point determines the placement and orientation of the text note.

Item Selection

On sidebar menu pick: n/a
For window, point to: Palettes   Dimensioning   Miscellaneous

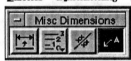

On paper menu see: n/a
Command Window Prompts:
1. Select the item
   Enter text
2. Enter your note
   Enter endpoint
3. Pick an end point
   ⇨   Prompt does not change
4. Pick a second point
To exit item: Make next selection

☞ *Tip:* When you need to change the text size or font style and you've entered the text, try this before placing the text: Enter the text and hit Enter. Now enter your font change (FT=3). Hit Enter again, and you're ready to place the modified text string.
See Also: PLACE NOTE DIALOG, PLACE TEXT tools

PLACE NOTE DIALOG

PLA NOT D

Drawing Tool: Place Dialog Note

Places text, a note, a leader line and an arrow. This tool open the Text Editor window for text entry, and requires two input points to spot the text. The first point identifies the placement of the arrow symbol. The second point determines the placement and orientation of the text note.

Item Selection

On sidebar menu pick: n/a

For window, point to: n/a

On paper menu see: TEXT

Command Window Prompts:

1. Select the item

   `Enter text`
2. Enter your note

   `Enter endpoint`
3. Pick an end point

   ⇨ Prompt does not change
4. Pick a second point

To exit item: `Make next selection`

☞ *Tip:* When you need to change the text size or font style and you've entered the text, try this: hit Enter and click on the prompt field in the Command Window. Enter your font change (FT=3). Hit Enter again, and you're ready to place the text string.

See Also: PLACE NOTE, PLACE TEXT tools

PLACE PARABOLA ENDPOINTS

PLA PAR E

Drawing Tool: Place Parabola by Endpoints

Places a parabolic B-spline curve based on three points. The first and last points define the curve's endpoints. The second point is placed along the curve element.

Item Selection

On sidebar menu pick: Place BSpl Curves

For window, point to: Palettes Main B-splines

On paper menu see: B-SPLINES 2D 3D

Command Window Prompts:

1. Select the item

   `Enter first parabola endpoint`
2. Pick first point

   `Enter point on parabola`
3. Pick second point

   `Enter second parabola endpoint`
4. pick last point

To exit item: `Make next selection`

See Also: FILLET tools, PLACE PARABOLA tools

PLACE
PARABOLA HORIZONTAL MODIFY   PLA PAR H M

Drawing Tool: Horizontal Parabola and Truncate Both

Places a parabolic fillet between two lines, along the horizontal axis of the view. The end points of the fillet are always tangent to the elements. MicroStation trims the elements to tangent points along the fillet. This item requires two input points to identify the target lines, and a third point to accept the truncation operation.

Item Selection

On sidebar menu pick: n/a

For window, point to: <u>P</u>alettes   <u>M</u>ain   <u>F</u>illets

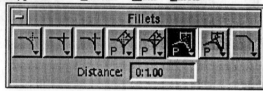

On paper menu see: FILLET PARABOLA

Command Window Prompts:
1. Select the item
   Select first segment
2. Pick an element
   Select second segment
3. Pick a second element
   Accept - initiate truncation
4. Give an accept point

To exit item: Make next selection

See Also: FILLET tools, PLACE PARABOLA tools

PLACE
PARABOLA HORIZONTAL NOMODIFY   PLA PAR H

Drawing Tool: Horizontal Parabola (No Truncation)

Places a parabolic fillet between two lines, along the horizontal axis of the view. The end points of the fillet are always tangent to the line segments. The line segments are not truncated. This item requires two input points to identify the target lines and a third point to draw the parabola.

Item Selection

On sidebar menu pick: n/a

For window, point to: <u>P</u>alettes   <u>M</u>ain   <u>F</u>illets

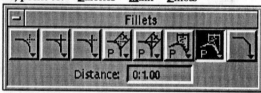

On paper menu see: FILLET PARABOLA

Command Window Prompts:
1. Select the item
   Select first segment
2. Pick an element
   Select second segment
3. Pick a second element
   Accept - initiate construction
4. Give an accept point

To exit item: Make next selection

See Also: FILLET tools, PLACE PARABOLA tools

PLACE PARABOLA MODIFY　　　　　PLA PAR M

Drawing Tool: Symmetric Parabola and Truncate Both
　　Places a parabolic fillet between two lines. The end points of the fillet are always tangent to the line segments. MicroStation truncates elements back to tangent points.

Item Selection
　On sidebar menu pick:　n/a
　For window, point to:　Palettes　Main　Fillets

　On paper menu see:　FILLET PARABOLA
Command Window Prompts:
　1. Select the item
　　 Enter distance
　2. Enter the length of the arc
　　 Select first segment
　3. Pick the first line
　　 Select second segment
　4. Pick the second line
　　 ⇨　Tool highlights the arc.
　5. Give an accept point
To exit item:　Make next selection
See Also: FILLET tools, PLACE PARABOLA tools

PLACE PARABOLA NOMODIFY　　　　PLA PAR

Drawing Tool: Symmetric Parabola (No Truncation)
　　Places a parabolic fillet tangent to two line segments. The tool does not truncate the line segments.

Item Selection
　On sidebar menu pick:　n/a
　For window, point to:　Palettes　Main　Fillets

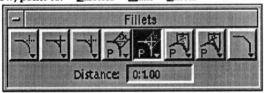

On paper menu see: FILLET PARABOLA
1. Select the item
     Enter distance
2. Enter the length of the arc
     Select first segment
3. Pick the first line
     Select second segment
4. Pick the second line
     ⇨   Tool highlights the arc.
5. Give an accept point
To exit item: **Make next selection**
See Also: FILLET tools, PLACE PARABOLA tools

PLACE POINT   PLA POI

Drawing Tool: Place Active Point
     Creates an active point in the drawing file. An active point can be a line
without length, a single text character or a cell.
Item Selection
     On sidebar menu pick: Place   Point   Single
     For window, point to: Palettes   Main   Points

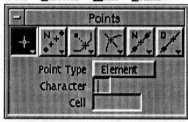

On paper menu see: POINTS
Command Window Prompts:
1. Select the item
     Enter point origin
2. Place a point
To exit item: **Make next selection**
Example(s):

PT=SPOT	to make the active point a cell called SPOT
PT=x	to make the active point the letter x
PT=$	to display the current active point

☞ *Tip:* If you place an active point as an element and have trouble seeing it, boost the
     ACTIVE WEIGHT.
     See Also: ACTIVE POINT, CONSTRUCT POINT tools

PLACE POINT STRING CONTINUOUS

PLA POI S C

Drawing Tool: Place Continuous Point String

Displays a Point String as a series of connected lines. This item requires an input point for each point you want to place. Each point along the point string has an orientation. Application programs use this tool. For example, a walk through application of a floor plan may generate a point string to define all the camera locations.

Item Selection
    On sidebar menu pick: n/a
    For window, point to: n/a
    On paper menu see: n/a
Command Window Prompts:
    1. Select the item
        Enter point origin
    2. Pick a point
    To exit item: Make next selection
See Also: PLACE POINT STRING DISJOINT

PLACE POINT STRING DISJOINT

PLA POI S D

Drawing Tool: Place Disjoint Point String

Displays a Point String as a series of points or dots. This item requires a input point for each point you want to place. Each point along the point string has an orientation. Application programs use this tool. For instance, a walk through application of a floor plan may generate a point string to define all the camera locations.

Item Selection
    On sidebar menu pick: n/a
    For window, point to: n/a
    On paper menu see: n/a
Command Window Prompts:
    1. Select the item
        Enter first point
    2. Place a point
        Enter point or RESET to complete
    3. Pick a second point
    To exit item: Make next selection
☞ *Tip:* To see the dots in the point string better, boost the ACTIVE WEIGHT.
See Also: PLACE POINT STRING CONTINUOUS

PLACE POLYGON CIRCUMSCRIBED

PLA POL C

Drawing Tool: Place Circumscribed Polygon

Creates a regular polygon around a circle. The polygon must have between 3 and 100 sides. The radius of the enclosed circle may be defined by input points or precision key-ins. This tool requires a keyed-in value to determine the number of sides of the polygon, and two input points to determine its center and outside edge.

Item Selection
    On sidebar menu pick: Place   Poly   Circm
    For window, point to: Palettes   Main   Polygons

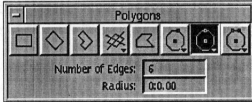

On paper menu see: POLYGONS

Command Window Prompts:
1. Select the item
   `Enter number of polygon Edges`
2. Enter a number between 3 and 100
   `Enter circle point`
3. Place an input point to define the center
   `Enter radius or pnt on circle`
4. Place a second point to define the radius

To exit item: `Click Reset, Reset`

See Also: PLACE BLOCK tools, PLACE POLYGON tools, PLACE SHAPE tools

PLACE POLYGON EDGE   PLA POL E

Drawing Tool: Place Polygon by Edge

Creates a polygon that has between 3 and 100 sides. This tool requires a keyed-in value to determine the number of sides of the polygon and two input points to determine its center and outside edge.

Item Selection

On sidebar menu pick: | Place | | Poly | | Edge |

For window, point to: <u>P</u>alettes <u>M</u>ain Polygons

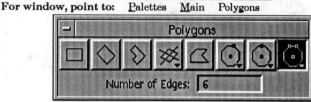

On paper menu see: POLYGONS

Command Window Prompts:
1. Select the item
   `Enter number of polygon edges`
2. Enter 5
   `Enter first edge point`
3. Place an input point
   `Enter next (CCW) edge point`
4. Place a second point

To exit item: `Click Reset, Reset`

See Also: PLACE BLOCK tools, PLACE POLYGON tools, PLACE SHAPE tools

PLACE POLYGON INSCRIBED   PLA POL I

Drawing Tool: Place Inscribed Polygon

Creates a regular polygon inside a circle. The polygon must have between 3 and 100 sides. This tool requires a keyed-in value to determine the number of sides of the polygon and two input points to determine its center and outside edge.

Item Selection

On sidebar menu pick: | Place | | Poly | | Inscr |

For window, point to: <u>P</u>alettes <u>M</u>ain Polygons

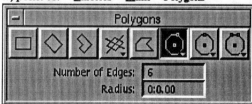

On paper menu see: POLYGONS

Command Window Prompts:
1. Select the item
   `Enter number of polygon Edges`
2. Enter 7

Enter center point
3. Place an input point
   RADIUS:
4. Enter a radius
To exit item: Click Reset, Reset
   See PLACE POLYGON examples.
See Also: PLACE BLOCK tools, PLACE POLYGON tools, PLACE SHAPE tools

PLACE SHAPE   PLA S

Drawing Tool: Place Shape
   Creates a closed shape based on a series of input points. The last point placed
   must coincide with the first input point in order to close the shape. The shape may
   contain as many as 100 sides.
Item Selection
   On sidebar menu pick: Place   Shape   Points
   For window, point to: Palettes   Main   Polygons

On paper menu see: POLYGONS
Command Window Prompts:
   1. Select the item
      Enter shape vertex
   2. Place the first input point
      ⇨ The prompt message remains the same as each leg of the shape appears.
   3. Place several more points.
      ⇨ Place the last input point by the first input point.
      Element closed
         ⇨ This message only flashes by.
   To exit item: Make next selection
✍ ***Note:*** The shape remains highlighted until closed.
   See Also: PLACE BLOCK tools, PLACE SHAPE ORTHOGONAL

PLACE SHAPE ORTHOGONAL   PLA S O

Drawing Tool: Place Orthogonal Shape
   Draws a closed shape in which each side is either parallel or perpendicular
   to all the other sides. The first side can be placed at any angle. The last point must
   touch the first point to close the shape. This shape can have up to 100 sides.
Item Selection
   On sidebar menu pick: Place   Shape   Orth
   For window, point to: Palettes   Main   Polygons

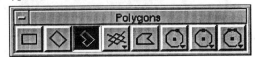

On paper menu see: POLYGONS
Command Window Prompts:
   1. Select the item
      Enter shape vertex
   2. Place the first input point
      ⇨ The prompt message remains the same as each leg of the shape appears.
   3. Place several more points.
      ⇨ Place the last input point by the first input point.
      Element closed
         ⇨ This message only flashes by.
   To exit item: Make next selection

✍ *Note:* The shape remains highlighted until closed.
See Also: PLACE BLOCK tools, PLACE SHAPE

PLACE SLAB   PLA SL

3D Drawing Tool: Place Slab

Creates a 3D rectangular volume of projection. This item requires three input points. The first two points define the opposite corners of the slab. The third point defines the slab's depth, or thickness.

Item Selection

On sidebar menu pick: Place   3-D

For window, point to: Palettes   3D

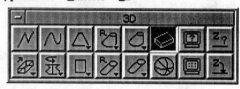

On paper menu see: 3D
Command Window Prompts:
1. Select the item
   Enter first point
   Enter opposite corner
2. Pick the opposite corner
   Define slab depth
3. Pick the slab thickness
To exit item: Make next selection
See Also: SURFACE PROJECTION

PLACE SPHERE   PLA SP

3D Drawing Tool: Place Sphere

Creates a 3D volume of rotation. This MDL application tool requires two input points. The first point defines the sphere's center. The second point defines its diameter.

Item Selection

On sidebar menu pick: Place   3-D

For window, point to: Palettes   3D

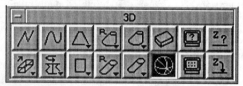

On paper menu see: 3D
Command Window Prompts:
1. Select the item
   Enter center point
2. Pick the sphere's center point
   Enter edge point
3. Pick the edge point
To exit item: Make next selection

✳ *Error:* Invalid character - **If this message appears after your key in, enter MDL LOAD SPLINES or pick the tool from the Space Curves palette once.**
See Also: SURFACE REVOLUTION

PLACE SPIRAL ANGLE

PLA SPI A

3D Drawing Tool: Place Spiral by Sweep Angle

Creates a B-spline element in the shape of a spiral. This MDL application tool requires keyed-in radii: an angle and two input points. The first input point defines the spiral's origin. The last point determines its orientation. Set the starting and ending radius, the length, and the angle in the B-splines sub-palette.

Item Selection

On sidebar menu pick: Place | BSpl | Curves

For window, point to: Palettes   Main   B-splines

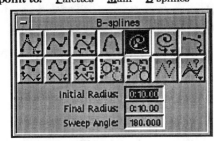

On paper menu see: B-SPLINES 2D 3D

Command Window Prompts:

1. Select the item
   Enter first spiral endpoint
2. Pick first endpoint
   Define spiral orientation
3. Pick orientation

To exit item: Make next selection

* **Error:** Invalid character **- If this message appears after your key in, enter MDL LOAD SPLINES.**

See Also: PLACE B-SPLINE tools, PLACE SPIRAL tools

PLACE SPIRAL ENDPOINTS

PLA SPI E

3D Drawing Tool: Place Spiral by Endpoints

Creates a B-spline element in the shape of a spiral. This MDL application tool requires two keyed-in radii and three input points. The first input point defines the spiral's first endpoint and the second point determines the orientation of the spiral. The last point defines the spiral's last endpoint. Set the starting and ending radius, and the degree of the angle in the B-splines sub-palette.

Item Selection

On sidebar menu pick: Place | BSpl | Curves

For window, point to: Palettes   Main   B-splines

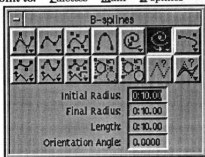

On paper menu see: B-SPLINES 2D 3D

Command Window Prompts:

1. Select the item

```
       Enter first spiral endpoint
```
2. Pick first endpoint
```
       Define spiral orientation
```
3. Pick orientation
```
       Enter second spiral endpoint
```
4. Pick second endpoint
To exit item: Make next selection

✳ *Error:* `Invalid character` - **If this message appears after your key in, enter MDL LOAD SPLINES.**
See Also: PLACE B-SPLINE tools, PLACE SPIRAL tools

PLACE SPIRAL LENGTH   PLA SPI L

3D Drawing Tool: Place Spiral by Length

    Creates a B-spline element in the shape of a spiral. This MDL application tool requires keyed-in radii, length, angle and two input points. The first input point defines the spiral's origin, and the last point sets its orientation. Set the starting and ending radius, the length, and the angle in the B-splines sub-palette.

Item Selection

On sidebar menu pick: [Place] [BSpl] [Curves]

For window, point to: <u>P</u>alettes   <u>M</u>ain   <u>B</u>-splines

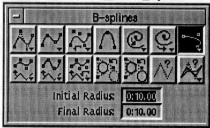

On paper menu see: B-SPLINES 2D 3D
Command Window Prompts:
1. Select the item
```
       Enter first spiral endpoint
```
2. Pick first endpoint
```
       Define spiral orientation
```
3. Pick orientation
To exit item: Make next selection

✳ *Error:* `Invalid character` - **If this message appears after your key in, enter MDL LOAD SPLINES or pick the tool from the Space Curves palette once.**
See Also: PLACE B-SPLINE tools, PLACE SPIRAL tools

PLACE TERMINATOR   PLA TER

Drawing Tool: Place Active Line Terminator

    Places a cell at the end of a line, line string or arc. This tool requires two input points. The first point identifies the target element and the second point serves as an accept point. The ACTIVE LINE TERMINATOR (LT=) setting determines the name of the cell. The ACTIVE TSCALE setting controls the terminator's scale factor.

Item Selection

On sidebar menu pick: [Place] [Term] [Place]

For window, point to: Palettes Main Lines

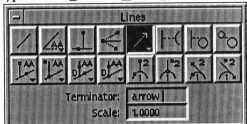

On paper menu see: LINES
Command Window Prompts:
  1. Select the item
     Identify element
  2. Select the line segment
     Accept/Reject
  3. Give an accept point
  To exit item: Make next selection

✳ Error: No active cell - If this message appears, use the LT= key-in to declare an active terminator cell.
See Also: ACTIVE TERMINATOR, ACTIVE TSCALE

PLACE TEXT   PLA TEX

Drawing Tool: Place Text
   Places text in the drawing file. This tool asks you to key in the text string and pick a point to place the text in the drawing.
Item Selection
  On sidebar menu pick: [Text] [Place]
  For window, point to: n/a
  On paper menu see: n/a
Command Window Prompts:
  1. Select the item
     Enter text
  2. Enter the text string
     Enter more chars or position text
  3. Pick resting place for text
  To exit item: Make next selection

✍ Note: The ACTIVE LINE SPACE (LS=) setting controls the spacing between the element and the text.

☞ Tip: When you need to change the text size or font style and you've entered the text, try this before placing the text: Enter the text and hit Enter. Now enter your font change (FT=3). Hit Enter again, and you're ready to place the modified text string.

☞ Tip: If your text line is longer than 60 characters, hit the Enter key and then keep typing. The PLACE TEXT tool places the text string as one line of text.

☞ Tip: If you want to intentionally place multiple lines of text, place a line feed character between the lines of text. On a PC, Ctrl+Enter keys generate linefeed characters. On an Intergraph workstation, use the Linefeed key.

☞ Tip: When placing text on curved elements like arcs, ellipses or circles, use the PLACE TEXT ALONG tool.
See Also: PLACE DIALOG TEXT tools, PLACE TEXT tools, PLACE NODE

PLACE TEXT ABOVE   PLA TEX AB

Drawing Tool: Place Text Above Element
   Places text above a line segment or shape. This tool requires a key-in for the text string and an input point to determine the text's resting place.

Item Selection

On sidebar menu pick: | Text | | Above |

For window, point to: n/a

On paper menu see: n/a

Command Window Prompts:

1. Select the item
   Enter text
2. Enter the word ABOVE
   Select line segment
3. Pick a line segment
   Accept/Reject (select next input)
4. Give an accept point

To exit item: Make next selection

✍ *Note:* The ACTIVE LINE SPACE (LS=) setting controls the spacing between the element and the text.

☞ *Tip:* When you need to change the text size or font style and you've entered the text, try this before placing the text: Enter the text and hit Enter. Now enter your font change (FT=3). Hit Enter again, and you're ready to place the modified text string.

☞ *Tip:* If your text line is longer than 60 characters, hit the Enter key and then keep typing. The PLACE TEXT tool places the text string as one line of text.

☞ *Tip:* If you want to intentionally place multiple lines of text, place a line feed character between the lines of text. On a PC, Ctrl+Enter keys generate linefeed characters. On an Intergraph workstation, use the Linefeed key.

☞ *Tip:* When placing text on curved elements like arcs, ellipses or circles, use the PLACE TEXT ALONG tool.

See Also: PLACE DIALOG TEXT tools, PLACE TEXT tools, PLACE NODE

PLACE TEXT ALONG   PLA TEX AL

Drawing Tool: Place Text Along Element

Places text along an arc, ellipse or circle. Placing text along an element asks for an optional key-in to change the line spacing of the text. Key in the text string. Next, an input point identifies the target element, and a second point determines the text's resting place.

Item Selection

On sidebar menu pick: | Text | | Along |

For window, point to: n/a

On paper menu see: n/a

Command Window Prompts:

1. Select the item
   Enter distance between chars (or return)
   ➪ Hitting return leaves the line spacing at half the text size.
2. Hit return
   Enter text
3. Enter the word ALONG
   Identify element, text location
4. Pick a line segment
   Accept, select text above/below
5. Give an accept point
   ➪ Accept point will determine text's resting place.

To exit item: Make next selection

✍ *Note:* The ACTIVE LINE SPACE (LS=) setting controls the spacing between the element and the text.

✍ *Note:* Characters placed with PLACE TEXT ALONG are placed as individual text characters and not as a string of text characters.

☞ *Tip:* When you need to change the text size or font style and you've entered the text, try this before placing the text: Enter the text and hit Enter. Now enter your font change (FT=3). Hit Enter again, and you're ready to place the modified text string.

☞ *Tip:* If your text line is longer than 60 characters, hit the Enter key and then keep typing. The PLACE TEXT tool places the text string as one line of text.

☞ *Tip:* If you want to intentionally place multiple lines of text, place a line feed character between the lines of text. On a PC, Ctrl+Enter keys generate linefeed characters. On an Intergraph workstation, use the Linefeed key.

See Also: PLACE DIALOG TEXT tools, PLACE TEXT tools, PLACE NODE

PLACE TEXT BELOW   PLA TEX B

Drawing Tool: Place Text Below Element
Places text below a line segment or shape. This tool requires a key-in for the text string and two input points to determine the text's resting place.

Item Selection
On sidebar menu pick: | Text | | Below |
For window, point to: n/a
On paper menu see: n/a
Command Window Prompts:
1. Select the item
   `Enter text`
2. Enter the word BELOW
   `Select line segment`
3. Pick a line segment
   `Accept/Reject (select next input)`
4. Give an accept point
To exit item: Make next selection

✍ *Note:* The ACTIVE LINE SPACE (LS=) setting controls the spacing between the element and the text.

✍ *Note:* Characters placed with PLACE TEXT ALONG are placed as individual text characters and not as a string of text characters.

☞ *Tip:* When you need to change the text size or font style and you've entered the text, try this before placing the text: Enter the text and hit Enter. Now enter your font change (FT=3). Hit Enter again, and you're ready to place the modified text string.

☞ *Tip:* If your text line is longer than 60 characters, hit the Enter key and then keep typing. The PLACE TEXT tool places the text string as one line of text.

☞ *Tip:* If you want to intentionally place multiple lines of text, place a line feed character between the lines of text. On a PC, Ctrl+Enter keys generate linefeed characters. On an Intergraph workstation, use the Linefeed key.

☞ *Tip:* When placing text on curved elements like arcs, ellipses or circles, use the PLACE TEXT ALONG tool.

See Also: PLACE DIALOG TEXT tools, PLACE TEXT tools, PLACE NODE

PLACE TEXT FITTED   PLA TEX FI

Drawing Tool: Place Fitted Text
Places text in a space defined by two input points. This tool requires a key-in for the text string and two input points to determine the text's resting place.

Item Selection
On sidebar menu pick: | Text | | Fitted |
For window, point to: n/a
On paper menu see: n/a
Command Window Prompts:
1. Select the item
   `Enter text`
2. Enter the word FITTED
   `Enter more chars or Position text`
3. Pick text starting point
   `Define endpoint of text`
4. Give text endpoint
To exit item: Make next selection

✍ *Note:* The ACTIVE LINE SPACE (LS=) setting controls the spacing between the element and the text.

✍ *Note:* Characters placed with PLACE TEXT ALONG are placed as individual text characters and not as a string of text characters.

☞ *Tip:* When you need to change the text size or font style and you've entered the text, try this before placing the text: Enter the text and hit Enter. Now enter your font change (FT=3). Hit Enter again, and you're ready to place the modified text string.

☞ *Tip:* If your text line is longer than 60 characters, hit the Enter key and then keep typing. The PLACE TEXT tool places the text string as one line of text.

☞ *Tip:* If you want to intentionally place multiple lines of text, place a line feed character between the lines of text. On a PC, Ctrl+Enter keys generate linefeed characters. On an Intergraph workstation, use the Linefeed key.

☞ *Tip:* When placing text on curved elements like arcs, ellipses or circles, use the PLACE TEXT ALONG tool.

See Also: PLACE DIALOG TEXT tools, PLACE TEXT tools, PLACE NODE

PLACE TEXT FVI   PLA TEX FV

Drawing Tool: Place Fitted View Independent Text
    Places fitted text at one view orientation, independent of view rotation. Placing fitted text in a view requires a key-in for the text string and two input points to determine the text's resting place.

Item Selection
    On sidebar menu pick: | Text | | VI Fit |
    For window, point to: n/a
    On paper menu see: n/a
Command Window Prompts:
  1. Select the item
    Enter text
  2. Enter the words VIEW FITTED
    Enter more chars or Position text
  3. Pick text starting point
    Define endpoint of text
  4. Pick text endpoint
    To exit item: Make next selection

✍ *Note:* The ACTIVE LINE SPACE (LS=) setting controls the spacing between the element and the text.

✍ *Note:* Characters placed with PLACE SPACE ALONG are placed as individual text characters and not as a string of text characters.

☞ *Tip:* When you need to change the text size or font style and you've entered the text, try this before placing the text: Enter the text and hit Enter. Now enter your font change (FT=3). Hit Enter again, and you're ready to place the modified text string.

☞ *Tip:* If your text line is longer than 60 characters, hit the Enter key and then keep typing. The PLACE TEXT tool places the text string as one line of text.

☞ *Tip:* If you want to intentionally place multiple lines of text, place a line feed character between the lines of text. On a PC, Ctrl+Enter keys generate linefeed characters. On an Intergraph workstation, use the Linefeed key.

☞ *Tip:* When placing text on curved elements like arcs, ellipses or circles, use the PLACE TEXT ALONG tool.

See Also: PLACE DIALOG TEXT tools, PLACE TEXT tools, PLACE NODE

PLACE TEXT ON   PLA TEX O

Drawing Tool: Place Text On Element
    Places text characters on a line segment. This tool requires a key-in for the text string and two input points to determine the text's resting place.

Item Selection
    On sidebar menu pick: | Text | | On |
    For window, point to: n/a
    On paper menu see: n/a

Command Window Prompts:
1. Select the item
   Enter text
2. Enter the word ON
   Select line segment
3. Pick a line segment
   Accept/Reject (select next input)
4. Give an accept point
   ⇨ Accept point determines text location on the element.
To exit item: Make next selection

✐ *Note:* The ACTIVE LINE SPACE (LS=) setting controls the spacing between the element and the text.

✐ *Note:* Characters placed with PLACE TEXT ALONG are placed as individual text characters and not as a string of text characters.

☞ *Tip:* When you need to change the text size or font style and you've entered the text, try this before placing the text: Enter the text and hit Enter. Now enter your font change (FT=3). Hit Enter again, and you're ready to place the modified text string.

☞ *Tip:* If your text line is longer than 60 characters, hit the Enter key and then keep typing. The PLACE TEXT tool places the text string as one line of text.

☞ *Tip:* If you want to intentionally place multiple lines of text, place a line feed character between the lines of text. On a PC, Ctrl+Enter keys generate linefeed characters. On an Intergraph workstation, use the Linefeed key.

☞ *Tip:* When placing text on curved elements like arcs, ellipses or circles, use the PLACE TEXT ALONG tool.
See Also: PLACE DIALOG TEXT tools, PLACE TEXT tools, PLACE NODE

PLACE TEXT TMATRIX

Drawing Tool: Place Text by Transformation Matrix
Places text by the values set in the transformation matrix. This item is usually called from an application program.
See Also: PLACE DIALOG TEXT tools, PLACE TEXT tools, Tmatrix

PLACE TEXT VI   PLA TEX V

Drawing Tool: Place View Independent Text
Places text at one orientation, independent of view rotation. Placing text with a view orientation requires a key-in for the text string and an input point to determine the text's resting place.
Item Selection
On sidebar menu pick: `Text`   `VI Txt`
For window, point to: n/a
On paper menu see: n/a
Command Window Prompts:
1. Select the item
   Enter text
2. Enter the word VIEW
   Enter more chars or Position text
3. Pick text location
To exit item: Make next selection

✐ *Note:* The ACTIVE LINE SPACE (LS=) setting controls the spacing between the element and the text.

✐ *Note:* Characters placed with PLACE TEXT ALONG are placed as individual text characters and not as a string of text characters.

☞ *Tip:* When you need to change the text size or font style and you've entered the text, try this before placing the text: Enter the text and hit Enter. Now enter your font change (FT=3). Hit Enter again, and you're ready to place the modified text string.

☞ *Tip:* If your text line is longer than 60 characters, hit the Enter key and then keep typing. The PLACE TEXT tool places the text string as one line of text.

☞ *Tip:* If your text line is longer than 60 characters, hit the Enter key and then keep typing. The PLACE TEXT tool places the text string as one line of text.

☞ *Tip:* If you want to intentionally place multiple lines of text, place a line feed character between the lines of text. On a PC, Ctrl+Enter keys generate linefeed characters. On an Intergraph workstation, use the Linefeed key.

☞ *Tip:* When placing text on curved elements like arcs, ellipses or circles, use the PLACE TEXT ALONG tool.

See Also: PLACE DIALOG TEXT tools, PLACE TEXT tools, PLACE NODE

PLOT

Obsolete

Opens the Preview Plot dialog box. You can create or adjust your plot files from this window.

See Also: PLOT @

PLOT @   Ctrl+T

Command: Plot

Creates a plot file while bypassing the Preview Plot dialog box. By default, the plot file extension is .000. For example, the drawing file CGSI.DGN produces a plot file with the name CGSI.000. If you wish to change the name of this filename, type a new filename along with the PLOT tool (Example: PLOT NEWNAME.PLT). After creating the plot file, use the MicroStation Plotting Facility menu to send the file to the plotter. The directory used for DOS plots is \USTATION\SCR. It's /mstation/scr for Unix.

Item Selection

On sidebar menu pick: `Utils` `Plot`

For window, point to: F̲ile P̲lot...

On paper menu see: TOP BORDER

Command Window Prompts:

1. Select the item
   Select view
2. Place an input point in the view to plot
   ⇨ Preview Plot dialog box appears.
   <CREATING PLOT FILE>
   Display complete

To exit item: **Exits by itself**

✱ *Error:* Cannot open plot config. file - If this messages appears, make sure your plotter is configured. Use \USTATION\USCONFIG to configure a plotter.

See Also: SET PLOTTER, SHOW PLOTTER

POINT ABSOLUTE   XY=

Key-in: Enter input point - absolute coordinates

Places an input point at a specified XY location in a two dimensional file or a specific XYZ location in a three dimensional drawing. Track the cursor's position in the Precision Input settings box. The value keyed in can be in working units, decimals or fractions. When keying in fractions, a space must separate the whole number from the fraction. (Example: 23 3/4).

A mapping application would use this tool to precisely place monument points. Also, place the beginning and end points of lines to represent street center lines or property lines with this tool.

If the X, Y or Z value is absent, it defaults to zero. For instance, XY=,6 places a point at the coordinates (0,6). XY=2 places a point at (2,0)

Item Selection

On sidebar menu pick: n/a

For window, point to: S̲ettings Precision I̲nput

On paper menu see: n/a

☞ *Tip:* If you place points with an XY= key-in and can't seem to see the results, check your location in the drawing file. This can be done by placing a tentative point in the view to see the location of the cursor in XY coordinates. (For example, the readout may say 12:8.000, 8:8.000) This means you placed the tentative point near the coordinates X=12, Y=8).

☞ *Tip:* Use negative numbers to go in the opposite direction of the last point placed instead of using distance and direction key-ins. For instance, DI=5,180 draws a line 5 units long at an angle of 180 degrees. DI=-5 delivers the same result.

See Also: POINT key-ins

POINT ACSABSOLUTE   AZ=

Key-in: Enter input point - absolute auxiliary coordinates

Places absolute coordinate points for an Auxiliary Coordinate System based on key-ins. Track the cursor's position in the Precision Input settings box.

Item Selection
  On sidebar menu pick: n/a
  For window, point to: Settings   Precision Input
  On paper menu see: n/a
See Also: POINT key-ins

POINT ACSDELTA   AD=

Key-in: Enter input point - delta auxiliary coordinates

Places an input point at coordinates relative to the previous tentative or data point placed in the current Auxiliary Coordinate System. Track the cursor's position in the Precision Input settings box. This tool requires a key-in.

Item Selection
  On sidebar menu pick: n/a
  For window, point to: Settings   Precision Input
  On paper menu see: n/a
See Also: POINT key-ins

POINT DELTA   DL=

Key-in: Enter input point - delta coordinates

Places an input point at some known distance and direction in a two or three dimensional file. The tool format is DL=[delta-x,delta-y,delta-z]. This tool places points relative to the drawing – not the view. Track the cursor's position in the Precision Input settings box.

The value keyed in with the DL= tool may be in working units, decimals or fractions. When keying in fractions, a space must separate the whole number from the fraction.

As an example, use the POINT DELTA tool to construct a quick 5 inch by 4 inch box. The procedure would go something like this: Pick a PLACE BLOCK tool. Place an input point somewhere in the view. Now put the cursor down and key in DL=5,4. If you can't see the entire box use FIT ACTIVE. If one of the [distance] values is absent, the value defaults to zero.

Item Selection
  On sidebar menu pick: n/a
  For window, point to: Settings   Precision Input
  On paper menu see: n/a

✍ *Note:* The coordinate value of the points placed does not change when the view rotates.
See Also: POINT key-ins

POINT DISTANCE   DI=

Key-in: Enter input point - distance, direction

Places an input point at some known distance and direction in a two or three dimensional file. DI= is an alternate key-in for the POINT DISTANCE tool. Track the cursor's position in the Precision Input settings box. DI=[distance],[direction] is the format for the POINT DISTANCE key-in. The tool calculates distance from the last input point placed. It measures the direction in degrees counterclockwise from the horizontal direction of the last point.

As an example, use the POINT DISTANCE tool in constructing a "B" size drawing border with the dimensions of 22 inches by 17 inches. The procedure would go something like this: Pick the PLACE LINE tool. Place an input point somewhere in the view. Now put the cursor down and key in the remaining four DI= tools one at a time: DI=22 DI=17,90 DI=22,180 and DI=17,270. If you can't see the border, do a FIT ACTIVE.

The value keyed in with the DI= tool may be in working units, decimals or fractions. When keying in fractions, a space must separate the whole number from the fraction. If the distance value is absent, the value defaults to zero. Leaving the direction out, defaults the value to the current angle of the last point.

Item Selection
On sidebar menu pick: n/a
For window, point to: Settings   Precision Input
On paper menu see: n/a
See Also: POINT key-ins

POINT VDELTA   DX=

Key-in: Enter input point - delta view coordinates
Places an input point at some known distance and direction in a two or three dimensional file. The tool format is DX=[vdelta-x,vdelta-y,vdelta-z]. This tool places points relative to the view instead of the drawing. Track the cursor's position in the Precision Input settings box. The value keyed in with the DX= tool may be in working units, decimals or fractions. When keying in fractions, a space must separate the whole number from the fraction.

As an example, use the POINT DELTA tool to construct a quick 5 inch by 4 inch box. The tool syntax would go something like this: Pick the PLACE BLOCK tool. Place an input point somewhere in the view. Now put the cursor down and key in DX=5,4. If you can't see the entire box, use FIT ACTIVE. If one of the [distance] values is absent, the value defaults to zero.

✍ *Note:* The coordinate value of the points placed changes based on whether or not the view rotates.

Item Selection
On sidebar menu pick: n/a
For window, point to: Settings   Precision Input
On paper menu see: n/a
See Also: POINT key-ins

Precision key-ins

Definition
There are several precision key-in tools within MicroStation. A precision key-in places an input point at some known location in the drawing file instead of using the cursor or mouse to graphically place the input point. They are POINT ABSOLUTE (XY=), POINT ACSABSOLUTE (AX=), POINT ACSDELTA (AD=), POINT DISTANCE (DI=), POINT DELTA (DL=) and POINT VDELTA (DX=).
See Also: POINT tools

PRINT

Mac Only
Command: Print
Mac only – opens the Print Dialog box. This tool is similar to the print screen command on a PC.
See Also: PAGE SETUP

PS=

Alternate Key-in
Alternate key-in for ACTIVE PATTERN SCALE. Use it to control the scale factor for patterns.
See Also: ACTIVE PATTERN SCALE

PT=

Alternate Key-in
Alternate key-in for ACTIVE POINT. Use it to define the active point.
See Also: ACTIVE POINT

PX=

Alternate Key-in
Alternate key-in for DELETE ACS. Use it to delete a previously stored Auxiliary Coordinate System.
See Also: DELETE ACS

QUIT   Ctrl+Q

Command: Quit
Quits and ends the design session. MicroStation writes to disk the last few elements input to the drawing file. Then it closes the file and returns the control back to the computer's operating system.

Item Selection
On sidebar menu pick: Exit

For window, point to: File   Exit
On paper menu see: TOP BORDER
See Also: BACKUP, COMPRESS Tools, EXIT Tools, QUIT tools

QUIT NOCLEAR

Command: Quit NoClear
Ends the design session, but does not clear the screen. MicroStation writes to disk the last few elements to the drawing file. This option is run from a user command, and works on MicroStation PC only.
See Also: BACKUP, COMPRESS Tools, EXIT tools, QUIT tool

QUIT NOUC

Command: Quit NoUnclear
Ends the design session, but does not run the EXITUC step of a user command upon exiting. MicroStation writes to disk the last few elements input to the drawing file. This option is run from a user command.
See Also: BACKUP, COMPRESS Tools, EXIT tools, QUIT tools

RA=

Alternate Key-in
Alternate key-in for ACTIVE REVIEW. Use it to control the name of an SQL SELECT statement used during the review process.
See Also: ACTIVE REVIEW

RC=

Alternate Key-in
Alternate key-in for ATTACH LIBRARY. Use it to attach a cell library to the drawing file.
See Also: ATTACH LIBRARY

RD=

Alternate Key-in
Alternate key-in for NEWFILE. Use it to jump from one drawing file to another without ending the current design session. The letters RD stand for Retrieve Design. This term is a holdover from Intergraph's Interactive Graphics Design Software (IGDS).
See Also: NEWFILE

RECORD

REC

Setting: Record Input on/off

Toggles the recording of all input points and keyboard entries. Key in RECORD ON name to direct the recording session to an output file. This item is great for recording demonstrations or drawing techniques. Use LISTEN to play back your recording.

Item Selection
On sidebar menu pick: n/a
For window, point to: n/a
On paper menu see: UTILITIES

✍ *Note:* This recording process creates about 10 KB of disk space per minute.

✍ *Note:* Make a copy of the design file you are recording to insure the windows and palettes remain in the same locations for play back later.

✍ *Note:* Before you can play back a recording session you must turn record off by keying in RECORD OFF.

See Also: LISTEN

REDO

Alt+Ins

Command: Redo

REDOes what was previously UNDOne – during the current design session.

Item Selection
On sidebar menu pick: Undo Redo
For window, point to: Edit Redo
On paper menu see: TOP BORDER
Command Window Prompts:
1. Select the item
   <COMMAND NAME> Redone
To exit item: Make next selection

✳ *Error:* Nothing has been UNDONE - If this message appears, execute UNDO first.

See Also: MARK, UNDO

REFERENCE ATTACH

RF=

Setting: Attach Reference File

Attaches another design file behind the active drawing file. Up to thirty-two reference files can be attached to the current drawing.

Selecting the tool opens the Attach Reference File dialog box. You may either pick through the directory and file columns, or key the drawing name in the Name box. Then click on OK, and a second dialog box opens on the screen. In this window, you can select a logical name, description and coincident point for the reference file attachment. Again, make your selections and click on OK.

Reference files can be attached to the drawing file two ways: Files attached coincidentally have the same coordinates. If a reference file is attached non-coincidentally (offset, rotated or scaled from the main file), a named view must be saved in the reference file.

Reference files can be plotted as though they exist within the active drawing file. You can copy the elements from a reference file into the active file. You CANNOT copy, delete, move around or change the elements in an attached reference file.

Item Selection
On sidebar menu pick: Refrnc Attach

For window, point to: Palettes   Reference Files
On paper menu see: REFERENCE FILES
To show, key in: `Click on File Reference`
   ⇨   See SHOW REFERENCE tool.
To set, key in: `RF=filename`
   Where filename is the name of the file.
Command Window Prompts:
   1. Select the item
      `Enter logical name (RESET = Coincident)`
   2. Enter a logical (simple) name
      ⇨   The logical name can be twenty characters long.
      `Enter description (RESET = Coincident)`
   3. Enter a short name or click Reset
      ⇨   Clicking a RESET attaches the reference file at the same orientation as the drawing file.
      ⇨   The description can be forty characters long.
   4. Enter the description
      `Enter view name (RESET = Coincident)`
   5. Enter saved view name
      ⇨   Attach the reference file with an offset based on a saved view within the reference file.
      `Reference file attached`
To exit item: `Exits by itself`
Example(s):
   `RF=CGSI.REF`

✍ *Note:* The MS_RFDIR environment variable controls the default directory for reference files.

☞ *Tip:* In a networked environment you can actively watch another person work by attaching his or her active file as a reference file to your active drawing file. Then periodically use UPDATE ALL.

☞ *Tip:* Let several people attach to the same base drawing and speed the project along.

☞ *Tip:* Attach the drawing border to all the files as a reference file instead of copying the border information in each file. This technique saves disk space.

☞ *Tip:* Attaching the drawings from different disciplines strengthens communication.

☞ *Tip:* Attach the active drawing to itself when detailing. This saves having to update the detail files that support a model.

✳ *Error:* `Incompatible dimensions` - If this message appears, it is because 2D and 3D files cannot be referenced together.

✳ *Error:* `Logical name must be unique` - If this message appears, it means the logical name keyed in already exists.

✳ *Error:* `Named view UUUUUU not found` - If this message appears, you did not save the named view (UUUUUU) for the reference file or check your typing.

See Also: ATTACH REFERENCE, REFERENCE DETACH, SHOW REFERENCE

REFERENCE CLIP BACK   REF CLI BA

3D Setting: Define Reference File Back Clipping Plane
   Simulates taking a slice out of an object and displaying it with the active drawing file. Selecting this item prompts you to identify the back clipping plane by entering the name of a reference drawing or by identify some element in the drawing.

Item Selection
On sidebar menu pick: `Refrnc` `ClipBk`
For window, point to: Palettes Reference Files

On paper menu see: REFERENCE FILES
Command Window Prompts:
1. Select the item
   `Identify (or key in) ref file`
2. Enter filename, logical or pick some element in the drawing file
   `Display complete`
To exit item: Exits by itself

✍ *Note:* A FENCE shape must be placed in the drawing file before selecting REFER-ENCE CLIP.
See Also: REFERENCE CLIP settings, REFERENCE LEVELS

REFERENCE CLIP BOUNDARY   REF C

3D Setting: Define Reference File Clipping Boundary
"Clips out" a certain area of a reference file for display with the active drawing file. Selecting this item prompts you to enter the name of the reference drawing or to identify some element in the drawing.

The clipped area may be a shape containing as many as sixty sides. However, when MicroStation displays or plots this clipped area, it adjusts the multi-sided shape to a rectangle. Therefore, no current advantage exists to placing multi-sided clip boundary shapes.

Item Selection
On sidebar menu pick: `Refrnc` `ClipBd`
For window, point to: Palettes Reference Files

On paper menu see: REFERENCE FILES
Command Window Prompts:
1. Select the item
   `Identify (or key in) ref file`
2. Enter filename
   `Display complete`
To exit item: Exits by itself

✍ *Note:* A FENCE shape must be placed in the drawing file before selecting REFER-ENCE CLIP.
See Also: REFERENCE CLIP settings, REFERENCE LEVELS, SET REFBOUND

REFERENCE CLIP FRONT   REF CLI FR

3D Setting: Define Reference File Front Clipping Plane
Simulates taking a slice out of an object and displaying it with the active drawing file. Selecting this item prompts you to identify the front clipping plane by entering the name of a reference drawing or by identifying some element in the drawing.

Item Selection
  On sidebar menu pick: Refrnc ClipFr
  For window, point to: Palettes Reference Files

  On paper menu see: REFERENCE FILES
Command Window Prompts:
  1. Select the item
     Identify (or key in) ref file
  2. Enter filename
     Display complete
  To exit item: **Exits by itself**

✍ *Note:* A FENCE shape must be placed in the drawing file before selecting REFER-
ENCE CLIP.
See Also: REFERENCE CLIP settings, REFERENCE LEVELS

REFERENCE CLIP MASK   REF C M

Setting: Define Reference File Clipping Mask
     Hides a portion of a reference file inside the clipping boundary. The mask
area is defined by placing a fence. More than one mask can be defined for a single
view. Only the reference file elements inside the clipping boundary and outside
the masked areas display. This item requires a key-in or an input point to identify
the target reference file for masking.
Item Selection
  On sidebar menu pick: n/a
  For window, point to: Palettes Reference Files

  On paper menu see: n/a
Command Window Prompts:
  1. Select the item
     Identify (or keyin) ref file
  2. Pick ref file of key in logical name
  To exit item: **Exist by itself**
See Also: PLACE FENCE, REFERENCE CLIP BOUNDARY

REFERENCE DETACH   REF DE

Setting: Detach Reference File
     Detaches a reference file from the active drawing file. Selecting this item
prompts you to enter the logical name of the reference drawing, or to identify some
element in the drawing.
Item Selection
  On sidebar menu pick: Refrnc Detach

For window, point to: Palettes   Reference Files

On paper menu see: REFERENCE FILES
Command Window Prompts:
1. Select the item
   ```
   Identify (or key in) ref file
   ```
2. Select the file to detach
   ```
   Reference file detached
   ```
To exit item: Exits by itself
See Also: ATTACH REFERENCE, REFERENCE ATTACH

REFERENCE DISPLAY   REF DI

Setting: Toggle Reference Display On/Off
      Toggles display of the reference file on or off. Selecting this item prompts you
to enter the name of the reference drawing or identify some element in the drawing.
Item Selection
   On sidebar menu pick: Refrnc   Disply
   For window, point to: File   Reference   Display
   On paper menu see: REFERENCE FILES
Command Window Prompts:
1. Select the item
   ```
   Identify (or key in) ref file
   ```
2. Select the file to display
   ```
   Display complete
   ```
To exit item: Exits by itself
See Also: REFERENCE LEVELS, REFERENCE RELOAD

REFERENCE FIT

View Control: Fit Reference
      Same as FIT REFERENCE.
See Also: FIT REFERENCE

REFERENCE LEVELS   REF LE

Setting: Reference File Levels On
      Turns off display of selected levels. Selecting this item prompts you to enter
the name of the reference drawing or identify some element in the drawing. You
then enter the level numbers and pick a view. You see the level changes.
      The reference file levels remain unchanged. If several people have attached
the same reference file, each person can decide what levels they want to see.
Item Selection
   On sidebar menu pick: Refrnc   LvOn
   For window, point to: File   Reference   Settings   Levels...
   On paper menu see: REFERENCE FILES
Command Window Prompts:
1. Select the item
   ```
   Identify (or key in) ref file
   ```
2. Key in filename
   ```
   Enter levels
   ```
3. Enter level numbers
   ```
   Select view
   ```
4. Pick the view with an input point
To exit item: Click a Reset
See Also: REFERENCE CLIP settings, REFERENCE DISPLAY

REFERENCE LOCATE   REF LO

Setting: Toggle Reference Locate On/Off
Copies reference file elements into the active drawing file with FENCE COPY tools. Selecting this item prompts you to enter the name of the reference drawing, or to identify some element in the drawing.

Elements can also be identified in the reference file and used as a base to create new elements in the active file.

Item Selection
On sidebar menu pick: Refrnc Locate
For window, point to: File Reference Files   Locate
On paper menu see: REFERENCE FILES
Command Window Prompts:
  1. Select the item
     Identify (or key in) ref file
  2. Key in filename
To exit item: Exits by itself
See Also: REFERENCE CLIP settings

REFERENCE MIRROR HORIZONTAL   REF MI H

Drawing Tool: Mirror Reference File About Horizontal
Mirror the contents of a reference file about the horizontal axis. This item requires one input point or a key-in to identify the target reference file for mirroring.

Item Selection
On sidebar menu pick: n/a

For window, point to: File   Reference

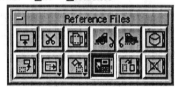

On paper menu see: REFERENCE FILES
Command Window Prompts:
  1. Select the item
     Identify (or keyin) ref file
  2. Pick reference file or key in logical name
To exit item: Make next selection
See Also: REFERENCE MIRROR VERTICAL

REFERENCE MIRROR VERTICAL   REF MI V

Drawing Tool: Mirror Reference File About Vertical
Use REFERENCE MIRROR VERTICAL to mirror the contents of a reference file about the vertical axis. This item requires one input point or a key-in to identify the target reference file for mirroring.

Item Selection
On sidebar menu pick: n/a
For window, point to: File   Reference

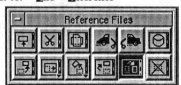

On paper menu see: REFERENCE FILES

Command Window Prompts:
1. Select the item
   Identify (or key in) ref file
2. Pick reference file or key in logical name
To exit item: Make next selection
See Also: REFERENCE MIRROR HORIZONTAL

REFERENCE MOVE   REF M

Setting: Move Reference File
    Shifts the reference file into position behind the active drawing file. This setting prompts you to enter the name of the reference drawing or identify some element in the drawing. Then enter two input points. The first point identifies some point in the reference file. The second point identifies the coincident point in the active drawing file.

Item Selection
  On sidebar menu pick: | Refrnc | | Move |
  For window, point to: Palettes   Reference Files

  On paper menu see: REFERENCE FILES
Command Window Prompts:
1. Select the item
   Identify (or key in) ref file
2. Enter filename
   Enter point to move from
3. Pick a point in the reference file
   Enter point to move to
4. Pick the common point in the active file
   Display complete
To exit item: Exits by itself
See Also: REFERENCE ROTATE

REFERENCE RELOAD   REF RE

Setting: Reload Reference File
    Reloads a reference file attached to a drawing file. MicroStation attempts to cache the drawing file and attached reference files in memory. This speeds up the display screen functions. However, if the reference file is accessed for modification by a second user, the copy cached in memory becomes out-dated. Reloading the reference file keeps the cached copy as up to date as possible. Selecting this item prompts you to enter the name of the reference drawing or identify some element in the drawing.

Item Selection
  On sidebar menu pick: n/a
  For window, point to: Palettes   Reference Files

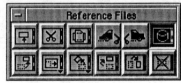

  On paper menu see: n/a
Command Window Prompts:
1. Select the item
   Identify (or key in) ref file

2. Key in filename or pick a point
To exit item: **Exits by itself**
See Also: REFERENCE DISPLAY

REFERENCE ROTATE   REF R

Setting: Rotate Reference File
Rotates the display of the reference file. This item expects a logical name, a rotation angle key-in, and an input point.
Item Selection
On sidebar menu pick: Refrnc   Rotate
For window, point to: Palettes   Reference Files

On paper menu see: REFERENCE FILES
Command Window Prompts:
1. Select the item
   Identify (or key in) ref file
2. Enter filename
   Enter rotation angle(s)
3. Pick the pivot point in a view
   ⇨   Enter point to rotate ref file about.
   Display complete
To exit item: **Exits by itself**
See Also: REFERENCE MOVE

REFERENCE SCALE   REF SC

Setting: Scale Reference File
Scales the display of the reference file. This item expects a logical name, a scale factor for the reference file, and an input point.
Item Selection
On sidebar menu pick: Refrnc   Scale
For window, point to: Palettes   Reference Files

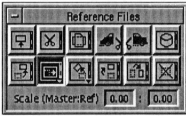

On paper menu see: REFERENCE FILES
Command Window Prompts:
1. Select the item
   Identify (or key in) ref file
2. Enter filename
   Enter master:ref. file scale factor
3. Key in scale factors
   ⇨   Enter the scale ratio between the two files.
   Enter point to scale ref file about
4. Pick the point in a view
   Display complete
To exit item: **Exits by itself**

Example(s):
```
RF Scale (Mast:Ref) >1:2
```
                    makes the reference file elements appear half
                    their normal size.
```
RF Scale (Mast:Ref) >1:.5
```
                    makes the reference file elements appear
                    twice their normal size.

✳ *Error:* Specify des file MU:ref file MU - If this messages appears, check
the scale factor key-in.
See Also: ACTIVE SCALE

REFERENCE SNAP   REF SN

Setting: Toggle Reference Snap On/Off
      Elements in a reference file can be "snapped to" by turning on REFERENCE
SNAP. Selecting this item prompts you to enter the name of the reference drawing
or identify some element in the drawing.
      This item is useful when using the geometry within a reference file to build
new elements in the active drawing file. Keying in REFERENCE SNAP toggles
the setting on or off.

Item Selection
  On sidebar menu pick: Refrnc   Snap
  For window, point to: File   Reference Files   Snap
  On paper menu see: REFERENCE FILES
  To set, key in: REF SN on/off
      Where on/off is ON or OFF.
Command Window Prompts:
  1. Select the item
     Identify (or key in) ref file
  2. Enter filename
  To exit item: Exits by itself
See Also: LOCK SNAP settings

RENAME CELL   CR=

Command: Rename Cell name
      Changes the name of an existing cell in a cell library. The cell name can be
changed by opening the Cell dialog box and then clicking Edit. Change the Name
field and select Modify.

Item Selection
  On sidebar menu pick: Cells   Rename
  For window, point to: Settings   Cells
  On paper menu see: n/a
  To set, key in: CR=uuu,UUU
      Where uuu is the old cell name and UUU is the new cell name.
See Also: DELETE CELL, REPLACE CELL

RENDER ALL CONSTANT   REND A C

Command: Constant Shading (All Views)
      Displays a filled hidden line rendering of the elements contained in all active
views. Curved surfaces are broken down into polygons. Each polygon is shaded
with a single constant color. Selecting this item starts the rendering process. The
rendering remains on the screen until the next update view command takes place.

Item Selection
  On sidebar menu pick: n/a
  For window, point to: n/a
  On paper menu see: n/a
Command Window Prompts:
  1. Select the item
     Update in progress
     Display complete

To exit item: **Exits by itself**

🖐 *Note:* This item is dimmed if you are not in a 3D drawing file.

☞ *Tip:* Tap a Reset when you want to terminate the rendering process early.
   See Also: RENDER FENCE commands, RENDER VIEW commands

■ RENDER ALL FILLED   REND A F

Command: Filled Hidden Line Removal (All Views)
        Displays a filled hidden line rendering of the elements contained in all active
views. The objects appear cartoon-like. Curved surfaces are broken down into
polygons. Each polygon is shaded with a single color. Selecting this item starts the
rendering process. The rendering remains on the screen until the next update view
command takes place.

Item Selection
On sidebar menu pick: n/a
For window, point to: n/a
On paper menu see: n/a
Command Window Prompts:
   1. Select the item
      Update in progress
      Update complete
To exit item: Exits by itself

🖐 *Note:* This item is dimmed if you are not in a 3D drawing file.

☞ *Tip:* Tap a Reset when you want to terminate the rendering process early.
   See Also: RENDER FENCE commands, RENDER VIEW commands

■ RENDER ALL HIDDEN   REND A H

Command: Hidden Line Removal (All Views)
        Displays a hidden line rendering of the elements contained in all active views.
Only the elements lines which are not hidden by other elements appear. Selecting
this item starts the rendering process. The rendering remains on the screen until
the next update view command takes place.

Item Selection
On sidebar menu pick: n/a
For window, point to: n/a
On paper menu see: n/a
Command Window Prompts:
   1. Select the item
      Update in progress
      Update complete
To exit item: Exits by itself

🖐 *Note:* This item is dimmed if you are not in a 3D drawing file.

☞ *Tip:* Tap a Reset when you want to terminate the rendering process early.
   See Also: RENDER FENCE commands, RENDER VIEW commands

■ RENDER ALL PHONG   REND A P

Command: Phong Shading (All Views)
        Displays the most realistic rendering of the elements contained in all active
views. This option takes much longer to process. The objects appear transparent,
and do not obscure other elements on the display screen. Selecting this item starts
the rendering process. The rendering remains on the screen until the next update
view command takes place.

Item Selection
On sidebar menu pick: n/a
For window, point to: n/a
On paper menu see: n/a
Command Window Prompts:
   1. Select the item
      Update in progress

```
Update complete
```
To exit item: **Exits by itself**

🖎 *Note:* This item is dimmed if you are not in a 3D drawing file.

☞ *Tip:* Tap a Reset when you want to terminate the rendering process early.
See Also: RENDER FENCE commands, RENDER VIEW commands

RENDER ALL SECTION   REND A SE

Command: Cross Section Display (All Views)
     Displays a cross-section rendering of the elements contained in all active
views. The cross-section cut is made at the Active Depth setting. These objects
appear transparent, and do not obscure other elements on the display screen.
Selecting this item starts the rendering process. The rendering remains on the
screen until the next update view command takes place.

Item Selection
  On sidebar menu pick: n/a
  For window, point to: n/a
  On paper menu see: n/a
Command Window Prompts:
  1. Select the item
     ```
     Update in progress
     Update complete
     ```
To exit item: **Exits by itself**

🖎 *Note:* This item is dimmed if you are not in a 3D drawing file.

☞ *Tip:* Tap a Reset when you want to terminate the rendering process early.
See Also: RENDER FENCE commands, RENDER VIEW commands

RENDER ALL SMOOTH   REND A SM

Command: Smooth Shading (All Views)
     Displays a smooth (Gouraud) rendering of the elements contained in all active
views. A smooth shaded model is more realistic than a constant shaded model.
These objects appear transparent, and do not obscure other elements on the
display screen. Selecting this item starts the rendering process. The rendering
remains on the screen until the next update view command takes place.

Item Selection
  On sidebar menu pick: n/a
  For window, point to: n/a
  On paper menu see: n/a
Command Window Prompts:
  1. Select the item
     ```
     Update in progress
     Update complete
     ```
To exit item: **Exits by itself**

🖎 *Note:* This item is dimmed if you are not in a 3D drawing file.

☞ *Tip:* Tap a Reset when you want to terminate the rendering process early.
See Also: RENDER FENCE commands, RENDER VIEW commands

RENDER ALL STEREO   REND A ST

Command: Stereo Image (All Views)
     Displays a stereo (3D glasses required) rendering of the elements contained
in all active views. These objects appear transparent, and do not obscure other
elements on the display screen. Selecting this item starts the rendering process.
The rendering remains on the screen until the next update view command takes
place.

Item Selection
  On sidebar menu pick: n/a
  For window, point to: n/a
  On paper menu see: n/a

Command Window Prompts:
1. Select the item
   Update in progress
   Update complete
To exit item: Exits by itself

✍ *Note:* This item is dimmed if you are not in a 3D drawing file.

☞ *Tip:* Tap a Reset when you want to terminate the rendering process early.
See Also: RENDER FENCE commands, RENDER VIEW commands

RENDER ALL WIREMESH   REND A W

Command: Wiremesh Display (All Views)
   Displays a wireframe rendering of the elements contained in all active views.
These objects appear transparent, and do not obscure other elements on the
display screen. Selecting this item starts the rendering process. The rendering
remains on the screen until the next update view command takes place.
Item Selection
On sidebar menu pick: n/a
For window, point to: n/a
On paper menu see: n/a
Command Window Prompts:
1. Select the item
   Update in progress
      @EXERCOMP = Update complete
To exit item: Exits by itself

✍ *Note:* This item is dimmed if you are not in a 3D drawing file.

☞ *Tip:* Tap a Reset when you want to terminate the rendering process early.
See Also: RENDER FENCE commands, RENDER VIEW commands

RENDER FENCE CONSTANT   REND F C

Command: Constant Shading (Fence)
   Displays a filled hidden line rendering of the elements contained within a
fence. Curved surfaces are broken down into polygons. Each polygon is shaded
with a single constant color. Selecting this item starts the rendering process. The
rendering remains on the screen until the next update view command takes place.
Item Selection
On sidebar menu pick: n/a
For window, point to: n/a
On paper menu see: n/a
Command Window Prompts:
1. Select the item
   Update in progress
   Display complete
To exit item: Exits by itself

✍ *Note:* This item is dimmed if you are not in a 3D drawing file.

☞ *Tip:* Tap a Reset when you want to terminate the rendering process early.

✳ *Error:* No fence defined - **If this message appears, place a fence before
reselecting this item.**
See Also: RENDER ALL commands, RENDER VIEW commands

RENDER FENCE FILLED   REND F F

Command: Filled Hidden Line Removal (Fence)
   Displays a filled hidden line rendering of the elements contained within a
fence. The objects appear cartoon-like. Curved surfaces are broken down into
polygons. Each polygon is shaded with a single color. Selecting this item starts the
rendering process. The rendering remains on the screen until the next update view
command takes place.

Item Selection
On sidebar menu pick: n/a
For window, point to: n/a
On paper menu see: n/a
Command Window Prompts:
1. Select the item
   Update in progress
   Update complete
To exit item: Exits by itself

✍ *Note:* This item is dimmed if you are not in a 3D drawing file.

☞ *Tip:* Tap a Reset when you want to terminate the rendering process early.

✳ *Error:* No fence defined - **If this message appears, place a fence before reselecting this item.**
See Also: RENDER ALL commands, RENDER VIEW commands

RENDER FENCE HIDDEN   REND F H

Command: Hidden Line Removal (Fence)
Displays a hidden line rendering of the elements contained within a fence. Only the elements lines which are not hidden by other elements appear. Selecting this item starts the rendering process. The rendering remains on the screen until the next update view command takes place.
Item Selection
On sidebar menu pick: n/a
For window, point to: n/a
On paper menu see: n/a
Command Window Prompts:
1. Select the item
   Update in progress
   Update complete
To exit item: Exits by itself

✍ *Note:* This item is dimmed if you are not in a 3D drawing file.

☞ *Tip:* Tap a Reset when you want to terminate the rendering process early.

✳ *Error:* No fence defined - **If this message appears, place a fence before reselecting this item.**
See Also: RENDER ALL commands, RENDER VIEW commands

RENDER FENCE PHONG   REND F P

Command: Phong Shading (Fence)
Displays the most realistic rendering of the elements contained within a fence. This option takes much longer to process. These objects appear transparent, and do not obscure other elements on the display screen. Selecting this item starts the rendering process. The rendering remains on the screen until the next update view command takes place.
Item Selection
On sidebar menu pick: n/a
For window, point to: n/a
On paper menu see: n/a
Command Window Prompts:
1. Select the item
   Update in progress
   Update complete
To exit item: Exits by itself

✍ *Note:* This item is dimmed if you are not in a 3D drawing file.

☞ *Tip:* Tap a Reset when you want to terminate the rendering process early.

✳ *Error:* No fence defined - **If this message appears, place a fence before reselecting this item.**
See Also: RENDER ALL commands, RENDER VIEW commands

RENDER FENCE SECTION

REND F SE

Command: Cross Section Display (Fence)

Displays a cross-section rendering of the elements contained within a fence. The cross-section cut is made at the Active Depth setting. These objects appear transparent, and do not obscure other elements on the display screen. Selecting this item starts the rendering process. The rendering remains on the screen until the next update view command takes place.

Item Selection
On sidebar menu pick: n/a
For window, point to: n/a
On paper menu see: n/a
Command Window Prompts:
1. Select the item
   Update in progress
   Update complete
To exit item: Exits by itself

✍ *Note:* This item is dimmed if you are not in a 3D drawing file.

☞ *Tip:* Tap a Reset when you want to terminate the rendering process early.

✴ *Error:* No fence defined - **If this message appears, place a fence before reselecting this item.**
See Also: RENDER ALL commands, RENDER VIEW commands

RENDER FENCE SMOOTH

REND F SM

Command: Smooth Shading (Fence)

Displays a smooth (Gouraud) rendering of the elements contained within a fence. A smooth shaded model is more realistic than a constant shaded model. These objects appear transparent, and do not obscure other elements on the display screen. Selecting this item starts the rendering process. The rendering remains on the screen until the next update view command takes place.

Item Selection
On sidebar menu pick: n/a
For window, point to: n/a
On paper menu see: n/a
Command Window Prompts:
1. Select the item
   Update in progress
   Update complete
To exit item: Exits by itself

✍ *Note:* This item is dimmed if you are not in a 3D drawing file.

☞ *Tip:* Tap a Reset when you want to terminate the rendering process early.

✴ *Error:* No fence defined - **If this message appears, place a fence before reselecting this item.**
See Also: RENDER ALL commands, RENDER VIEW commands

RENDER FENCE STEREO

REND F ST

Command: Stereo Image (Fence)

Displays a stereo (3D glasses required) rendering of the elements contained within a fence. These objects appear transparent, and do not obscure other elements on the display screen. Selecting this item starts the rendering process. The rendering remains on the screen until the next update view command takes place.

Item Selection
On sidebar menu pick: n/a
For window, point to: n/a
On paper menu see: n/a
Command Window Prompts:
1. Select the item
   Update in progress

        Update complete
    To exit item: Exits by itself

✍ *Note:* This item is dimmed if you are not in a 3D drawing file.

☞ *Tip:* Tap a Reset when you want to terminate the rendering process early.

✳ *Error:* No fence defined **- If you receive this message, place a fence before reselecting this item.**
See Also: RENDER ALL commands, RENDER VIEW commands

RENDER FENCE WIREMESH   REND F W

Command: Wiremesh Display (Fence)
        Displays a wireframe rendering of the elements contained within a fence. These objects appear transparent, and do not obscure other elements on the display screen. Selecting this item starts the rendering process. The rendering remains on the screen until the next update view command takes place.

Item Selection
    On sidebar menu pick: n/a
    For window, point to: n/a
    On paper menu see: n/a
Command Window Prompts:
    1. Select the item
        Update in progress
            @EXERCOMP = Update complete
    To exit item: Exits by itself

✍ *Note:* This item is dimmed if you are not in a 3D drawing file.

☞ *Tip:* Tap a Reset when you want to terminate the rendering process early.

✳ *Error:* No fence defined **- If you receive this message, place a fence before reselecting this item.**
See Also: RENDER ALL commands, RENDER VIEW commands

RENDER VIEW CONSTANT   REND V C

Command: Constant Shading (View)
        Displays a filled hidden line rendering of the elements contained within a view. This item requires an input point in each view where the rendering will take place. Curved surfaces are broken down into polygons. Each polygon is shaded with a single constant color. Selecting this item starts the rendering process. The rendering remains on the screen until the next update view command takes place.

Item Selection
    On sidebar menu pick: | Window | | Render | | Const |
    For window, point to: Views   Render   Constant Shading
    On paper menu see: VIEW
Command Window Prompts:
    1. Select the item
        Select view
    2. Pick a view
    To exit item: Make next selection

✍ *Note:* This item is dimmed if you are not in a 3D drawing file.

☞ *Tip:* Tap a Reset when you want to terminate the rendering process early.
See Also: RENDER ALL commands, RENDER FENCE commands

RENDER VIEW FILLED   REND V F

Command: Filled Hidden Line Removal (View)
        Displays a filled hidden line rendering of the elements contained within a view. This item requires an input point in each view where the rendering will take place. The objects appear cartoon-like. Curved surfaces are broken down into polygons. Each polygon is shaded with a single color. Selecting this item starts the rendering process. The rendering remains on the screen until the next update view command takes place.

Item Selection
   On sidebar menu pick: Window   Render   Filled
   For window, point to: Views   Render   Filled Hidden Line
   On paper menu see: VIEW
Command Window Prompts:
   1. Select the item
      Select view
   2. Pick a view
   To exit item: Make next selection

✍ *Note:* This selection is dimmed if you are not in a 3D drawing file.

☞ *Tip:* Tap a Reset when you want to terminate the rendering process early.
   See Also: RENDER ALL commands, RENDER FENCE commands

RENDER VIEW HIDDEN   REND V H

Command: Hidden Line Removal (View)
   Displays a hidden line rendering of the elements contained within a view.
This item requires an input point in each view where the rendering will take place.
Only the elements lines which are not hidden by other elements appear. Selecting
this item starts the rendering process. The rendering remains on the screen until
the next update view command takes place.

Item Selection
   On sidebar menu pick: Window   Render   Hidden
   For window, point to: Views   Render   Hidden Line
   On paper menu see: VIEW
Command Window Prompts:
   1. Select the item
      Select view
   2. Pick a view
   To exit item: Make next selection

✍ *Note:* This item is dimmed if you are not in a 3D drawing file.

☞ *Tip:* Tap a Reset when you want to terminate the rendering process early.
   See Also: RENDER ALL commands, RENDER FENCE commands

RENDER VIEW PHONG   REND V P

Command: Phong Shading (View)
   Displays the most realistic rendering of the elements contained within a view.
This item requires an input point in each view where the rendering will take place.
This option takes much longer to process. These objects appear transparent, and
do not obscure other elements on the display screen. Selecting this item starts the
rendering process. The rendering remains on the screen until the next update view
command takes place.

Item Selection
   On sidebar menu pick: Window   Render   Phong
   For window, point to: Views   Render   Phong Shading
   On paper menu see: VIEW
Command Window Prompts:
   1. Select the item
      Select view
   2. Pick a view
   To exit item: Make next selection

✍ *Note:* This item is dimmed if you are not in a 3D drawing file.

☞ *Tip:* Tap a Reset when you want to terminate the rendering process early.
   See Also: RENDER ALL commands, RENDER FENCE commands

RENDER VIEW SECTION

REND V SE

Command: Cross Section Display (View)

Displays a cross-section rendering of the elements contained within a view. The cross-section cut is made at the Active Depth setting. This item requires an input point in each view where the rendering will take place. These objects appear transparent, and do not obscure other elements on the display screen. Selecting this item starts the rendering process. The rendering remains on the screen until the next update view command takes place.

Item Selection

On sidebar menu pick: | Window | | Render | | Sect |

For window, point to: Views   Render   Cross-section

On paper menu see: VIEW

Command Window Prompts:

1. Select the item
   Select view
2. Pick a view
To exit item: Make next selection

✍ *Note:* This item is dimmed if you are not in a 3D drawing file.

☞ *Tip:* Tap a Reset when you want to terminate the rendering process early.

See Also: RENDER ALL commands, RENDER FENCE commands

RENDER VIEW SMOOTH

REND V SM

Command: Smooth Shading (View)

Displays a smooth (Gouraud) rendering of the elements contained within a view. A smooth shaded model is more realistic than a constant shaded model. This item requires an input point in each view where the rendering will take place. These objects appear transparent, and do not obscure other elements on the display screen. Selecting this item starts the rendering process. The rendering remains on the screen until the next update view command takes place.

Item Selection

On sidebar menu pick: | Window | | Render | | Smooth |

For window, point to: Views   Render   Smooth Shading

On paper menu see: VIEW

Command Window Prompts:

1. Select the item
   Select view
2. Pick a view
To exit item: Make next selection

✍ *Note:* This item is dimmed if you are not in a 3D drawing file.

☞ *Tip:* Tap a Reset when you want to terminate the rendering process early.

See Also: RENDER ALL commands, RENDER FENCE commands

RENDER VIEW STEREO

REND V ST

Command: Stereo Image (View)

Displays a stereo (3D glasses required) rendering of the elements contained within a view. This item requires an input point in each view where the rendering will take place. These objects appear transparent, and do not obscure other elements on the display screen. Selecting this item starts the rendering process. The rendering remains on the screen until the next update view command takes place.

Item Selection

On sidebar menu pick: | Window | | Render | | Stereo |

For window, point to: Views   Render   Stereo

On paper menu see: VIEW

Command Window Prompts:

1. Select the item
   Select view
2. Pick a view

To exit item: **Make next selection**

🖎 *Note:* This item is dimmed if you are not in a 3D drawing file.

☞ *Tip:* Tap a Reset when you want to terminate the rendering process early.
See Also: RENDER ALL commands, RENDER FENCE commands

RENDER VIEW WIREMESH   REND V W

Command: Wiremesh Display (View)
Displays a wireframe rendering of the elements contained within a view. This item requires an input point in each view where the rendering will take place. These objects appear transparent, and do not obscure other elements on the display screen. Selecting this item starts the rendering process. The rendering remains on the screen until the next update view command takes place.

Item Selection
On sidebar menu pick: Window   Render   WMesh
For window, point to: Views   Render   Wiremesh
On paper menu see: VIEW
Command Window Prompts:
  1. Select the item
     Select view
  2. Pick a view
  To exit item: **Make next selection**

🖎 *Note:* This item is dimmed if you are not in a 3D drawing file.

☞ *Tip:* Tap a Reset when you want to terminate the rendering process early.
See Also: RENDER ALL commands, RENDER FENCE commands

REPLACE CELL   REP

Drawing Tool: Replace Cell
This tool replaces a cell in the drawing file with another cell. The new cell from the library must have the same name as the one it is replacing. Selecting this item prompts you to identify the cell to replace and then accept that replacement with a second input point.

Item Selection
On sidebar menu pick: Cells   Replac
For window, point to: Palettes   Main   Cells

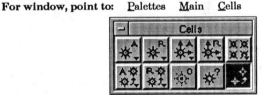

On paper menu see: CELLS
Command Window Prompts:
  1. Select the item
     Identify element
  2. Select the cell to replace
     Accept/Reject
  3. Click an input point
  To exit item: **Make next selection**
See Also: ACTIVE RCELL, DELETE CELL, RENAME CELL

RESET

Command: Reset
Performs the same function as clicking a RESET on the cursor or mouse. Use this command in a user command.

REVIEW   REV

Drawing Tool: Review Database Attributes of Element

REVIEW identifies an element and displays the database information concerning that particular element. A database must be attached to the drawing file. Selecting this item prompts you to identify and accept the element with two input points. The SET DATABASE setting controls the screen information that appears.

Item Selection

On sidebar menu pick: dBase   Tools

For window, point to: Palettes   Database

On paper menu see: DATABASE

Command Window Prompts:
1. Select the item
   Identify element
2. Select the element in question
   Accept/Reject
3. Confirm the choice

To exit item: Follow the REVIEW window prompts

See Also: SET DATABASE

RF=

Alternate Key-in

Alternate key-in for REFERENCE ATTACH. Use it to attach a reference file to the current drawing file. Thirty-two reference files may be attached to a drawing file at any one time.

See Also: REFERENCE ATTACH

ROTATE 3PTS

Obsolete

The ROTATE 3PTS view control rotates a view based on three input points. Place the first two points along the X-axis. The third and subsequent points identify the view or views you want rotated.

See Also: ROTATE view controls

ROTATE ACS ABSOLUTE   ROT ACS AB

Drawing Tool: Rotate ACS Absolute

Rotates the active Auxiliary Coordinate System (ACS) from its unrotated top view. This rotation does not affect the origin point of the ACS. In a two dimensional file, the ACS is rotated in the top or plan view. In three dimensional files the ACS is rotated for the X, Y and Z axes.

Item Selection

On sidebar menu pick: Params   ACS   Tools

For window, point to: Palettes   Auxiliary Coordinates

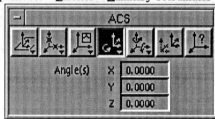

On paper menu see: ACS
Command Window Prompts:
1. Select the item
   `Select Auxiliary System (@origin)`
2. Pick first point
   `Enter second point on X-axis`
3. Pick second point
   `Select view(s) for rotations`
4. Pick view(s) to rotate
To exit item: Click a Reset

Example(s):
ROT ACS AB 15   rotates the top view of a 2D file fifteen degrees
                                 from its unrotated view.

ROT ACS AB 15,15,15
                                 rotates the X,Y,Z axes of a 3D file fifteen degrees

See Also: ROTATE ACS RELATIVE

ROTATE ACS RELATIVE   ROT ACS RE

Drawing Tool: Rotate ACS Relative
       Rotates the active Auxiliary Coordinate System (ACS) from its current
orientation. This rotation does not affect the origin point of the ACS. In a two
dimensional file, the ACS is rotated in the top or plan view. In three dimensional
files the ACS is rotated for the X, Y and Z axes.

Item Selection
On sidebar menu pick: | Params | | ACS | | Tools |
For window, point to: Palettes   Auxiliary Coordinates

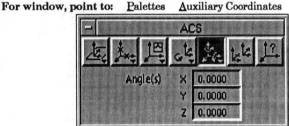

On paper menu see: ACS
Example(s):
ROT ACS RE 15   rotates the orientation of a 2D file fifteen degrees

ROT ACS RE 15,15,15
                                 rotates the X,Y,Z axes of a 3D file fifteen degrees

See Also: ROTATE ACS ABSOLUTE

ROTATE COPY   RO C

Drawing Tool: Rotate Element by Active Angle (Copy)
       ROTATE COPY copies an existing element and rotates the copy. Change the
ACTIVE ANGLE (AA=) setting to some non-zero value before attempting to rotate
and copy the element. Selecting this item prompts you to identify and accept the
target element to rotate with two input points.

Item Selection
On sidebar menu pick: | Manip | | Rotate | | Copy |

For window, point to: Palettes   Main   Copy Element

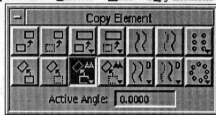

On paper menu see: COPY ELEMENT
Command Window Prompts:
1. Select the item
   Reset the ACTIVE ANGLE
2. AA=nnn
   Where nnn in a number between -360 and +360 (but not zero).
   Identify element
1. Pick the element to rotate
   Accept/Reject
2. Click an accept point
To exit item: **Make next selection**
See Also: ROTATE ORIGINAL, SPIN tools, FENCE ROTATE tools

ROTATE ORIGINAL   RO

Drawing Tool: Rotate Element by Active Angle (Original)
       ROTATE ORIGINAL rotates an element. Change the ACTIVE ANGLE
(AA=) setting to some non-zero value before attempting to rotate the element.
Selecting this item prompts you to identify and accept the target element to rotate
with two input points.
Item Selection
On sidebar menu pick: Manip   Rotate   Orig
For window, point to: Palettes   Main   Copy Element

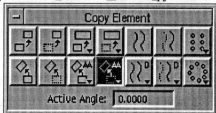

On paper menu see: COPY ELEMENT
Command Window Prompts:
1. Select the item
   Reset the ACTIVE ANGLE
2. AA=nnn
   Where nnn is a number between -360 and +360.
   Identify element
1. Pick the element to rotate
   Accept/Reject
2. Click an accept point
To exit item: **Make next selection**
See Also: ROTATE COPY, SPIN tools, FENCE ROTATE tools

ROTATE VIEW ABSOLUTE   ROT VI A

View Control: Rotate View (Absolute)
       The ROTATE VIEW ABSOLUTE option rotates a view from zero. Either set
the degree of rotate in the View Rotation settings box or key-in the rotation value

after the view control. An input point selects the target view for rotation. RV= is the alternate key-in.

Item Selection
On sidebar menu pick: n/a
For window, point to: Yiew   Rotation   Step
On paper menu see: n/a
Command Window Prompts:
1. Select the item
   Select view(s) for rotation
2. Place an input point in the view
To exit item: Click a Reset

☞ *Tip:* To get a 2D rotated
    view back to it's zero rotation, key in VI=TOP.

See Also: ROTATE VIEW view controls, VIEW view controls

ROTATE VIEW ELEMENT   ROT VI E

View Control: Rotate View (Aligned with Element)
    ROTATE VIEW ELEMENT orients a view relative to an element's position. This item requires one input point to identify the target element and a second point to identify the view for rotation.

Item Selection
On sidebar menu pick: n/a
For window, point to: n/a
On paper menu see: n/a
Command Window Prompts:
1. Select the item
   Identify element
2. Pick an element
3. Select view(s) for rotation
   Pick views
To exit item: Make next selection

☞ *Tip:* To get a 2D rotated view back to it's zero rotation, key in VI=TOP.
    See Also: ROTATE VIEW view controls, VIEW view controls

ROTATE VIEW POINTS   RO 3

View Control: Rotate View by Points
    ROTATE VIEW POINTS spins a view based on a series of input points. This item requires an RV= degree key-in to calculate the rotation and three input points. The first two points define the X-axis and the third point defines the view for rotation.

Item Selection
On sidebar menu pick: Window   View   Rotate   3 Pt
For window, point to: n/a
On paper menu see: n/a
Command Window Prompts:
1. Select the item
   Enter first point on X-axis
2. Place the first input point
   Enter second point on X-axis
3. Place the second input point
   Select view(s) for rotation
4. Place an input point in the view
To exit item: Click a Reset

☞ *Tip:* To get a 2D rotated
    view back to it's zero rotation, key in VI=TOP.

✳ *Error:* No rotation specified - If this message appears, use the RV= key-in, or the input points placed
    do not define a new viewing angle.

See Also: VIEW view controls

ROTATE VIEW RELATIVE   RO 3

View Control: Rotate View (Relative)

ROTATE VIEW RELATIVE rotates a view counterclockwise around a point in the center of the active view. This item requires a degree key-in to calculate the rotation. An input point selects the target view for rotation. RV= is an alternate key-in for the ROTATE VIEW RELATIVE control.

Item Selection

On sidebar menu pick: | Window | | View | | Rotate | | Key |

For window, point to: n/a

On paper menu see: n/a

Command Window Prompts:

1. Select the item
   `Enter first point on X-axis`
2. Place the first input point
   `Enter second point on X-axis`
3. Place the second input point
   `Select view(s) for rotation`
4. Place an input point in the view

To exit item: `Click a Reset`

☞ *Tip:* To get a 2D rotated view back to it's zero rotation, key in VI=TOP.

✳ *Error:* `No rotation specified` - If this message appears, use the RV= key-in, or the input points placed
do not define a new viewing angle.

See Also: ROTATE VIEW view controls, VIEW view controls

ROTATE VMATRX

View Control: Rotate View by Transformation Matrix

ROTATE VMATRX uses the current definition in the transformation matrix to rotate a view. Execute this item from within a user command.

See Also: ROTATE VIEW view controls

RS=

Alternate Key-in

Alternate key-in for ACTIVE REPORT. Use it to specify the name of the report table created during a fence operation.

See Also: ACTIVE REPORT

RV=

Alternate Key-in

Alternate key-in for ROTATE VIEW. Use it to rotate the top view in a 2D file, or any of the standard views defined in a 3D drawing file.

See Also: ROTATE VIEW

RX=

Alternate Key-in

Alternate key-in for ATTACH ACS. Use it to attach an Auxiliary Coordinate System.

See Also: ATTACH ACS

SAVE   SAVE

Mac Only

Command: Save

This command saves the active drawing file settings.

Item Selection

On sidebar menu pick: n/a

For window, point to: File

On paper menu see: n/a
See Also: FILEDESIGN

SAVE ACS
SX=[acsname],[description]

Setting: Save Auxiliary Coordinate System
SAVE ACS saves the current Auxiliary Coordinate System for future refer-
ence. The saved ACS name can be six characters. You can tag on a thirty character
description.
Item Selection
On sidebar menu pick: Params   ACS   Set
For window, point to: Settings   Auxiliary Coordinates   Save
On paper menu see: ACS
See Also: ATTACH ACS, DEFINE ACS ELEMENT, DELETE ACS, SHOW ACS

SAVE FUNCTION_KEY
SA F

Setting: Save Function Keys
Drawing tools can be attached to function keys (F#) on the keyboard.
MicroStation comes with default function key assignments, but
these key commands can be changed any time through the Function
Keys settings box. You will lose new key assignments when you exit
the design session unless you use SAVE FUNCTION_KEY. The
default function key menu is \USTATION\DATA\FUNCKEY.MNU
Item Selection
On sidebar menu pick: Params   FKey   Save
For window, point to: User   Function Keys
On paper menu see: n/a
Command Window Prompts:
1. Select the item
   Function Keys settings box appears
To exit item: Close the settings box

✍ *Note:* If you decide to change the name of the function key menu, you need to attach
the new menu by using the AM= key-in.
See Also: SET FUNCTION

SAVE IMAGE
SA I

Command: Save Image
Saves an image of a shaded view to disk. This image can be stored in either
Intergraph RGB, PICT, Targa or TIFF. The image can be saved as mono on stereo
(3D appearance). Keying in SAVE IMAGE opens the Save Image dialog box. Select
the View, Format, Shading, Resolution and Stereo; then click Save. This invokes
the Save Image As... dialog box. Pick the Name and click OK.
Item Selection
On sidebar menu pick: n/a
For window, point to: File   Save Image As...
On paper menu see: CAMERA
Command Window Prompts:
1. Select the item
   Save Image dialog box appears
To exit item: Click Save or Cancel

✍ *Note:* To display a saved image click on File, Display, Image... to open the Display
File Image dialog box.
See Also: VIEW CLEAR, VIEW IMAGE

SAVE VIEW
SV=UUU,uuu

View Control: Save Named View
SAVE VIEW names and saves views for later recall. To save a view, just key
in SAVE VIEW, or click View Saved. The Saved Views settings box appears. The
name of the saved view can be six characters long. A description of 27 characters can

accompany the saved view to describe the picture. Pick the Name and Description. Then click Save.

Take snapshots of the drawing during construction or rotate the model to just the right perspective and save the view. Later use the VI= key-in to redisplay these saved views. Use SAVE VIEW to call up common areas in the drawing sheet faster. Save a view of the title block information and name it TITLE. Save another view of the design notes and call this saved view NOTES. Now use the VI= key-in to jump to these areas quicker.

Item Selection

On sidebar menu pick: | Window | View | Name | Save |

For window, point to: <u>V</u>iew Sa<u>v</u>ed Save
On paper menu see: n/a

Where UUU is six character user defined view name and uuu is 27 character user defined description.

Command Window Prompts:
1. Select the item
   Saved Views settings box appears
To exit item: Close the window

☞ *Tip:* Window in on the title block and save a view called TITLE. Zoom out until you see the whole drawing sheet and name another view BORDER. Window in on the area where you keep notes and name this saved view NOTES. Now whenever you need to see those areas, key in VI=TITLE, VI=BORDER or VI=NOTES.

See Also: DELETE VIEW, SHOW VIEWS, VIEW

SCALE COPY   SC C

Drawing Tool: Scale Element (Copy)
The SCALE COPY tool scales and copies an element in one operation. This tool requires two input points. The first point identifies the element to copy. The second point starts the process of scaling and copying the new element. Remember to set the ACTIVE SCALE to the scale factor you want before selecting this tool.

Item Selection

On sidebar menu pick: | Manip | Scale | Copy |

For window, point to: <u>P</u>alettes <u>M</u>ain C<u>o</u>py Element

On paper menu see: COPY ELEMENT
Command Window Prompts:
1. Select the item
   Identify element
2. Pick element to scale
   Accept/Reject
3. Give an accept point
To exit item: Make next selection

✍ *Note:* The element's size will not change if the ACTIVE SCALE setting at 1.0.

✍ *Note:* Cells and graphic groups can be scaled and copied with this tool.
See Also: LOCK SCALE, SCALE ORIGINAL

SCALE ORIGINAL   SC

Drawing Tool: Scale Element (Original)
SCALE ORIGINAL changes the size of an element in a two-step process. The first step is to set the ACTIVE SCALE to the scale factor you want. After setting ACTIVE SCALE, select the SCALE ORIGINAL tool. This tool requires two input points. The first point identifies the target element and the second point starts the scaling process.

For example, an ACTIVE SCALE of .5 would scale the element down to half its original size. Choosing an ACTIVE SCALE of 2 would scale the object up to twice its original size.

Item Selection

On sidebar menu pick: Manip Scale Orig

For window, point to: Palettes Main Copy Element

On paper menu see: COPY ELEMENT

Command Window Prompts:
1. Select the item
   Identify element
2. Pick element to scale
   Accept/Reject
3. Give an accept point

To exit item: Make next selection

✐ *Note:* The element's size will not change if the ACTIVE SCALE setting is 1.0. The element does not change its size.

✐ *Note:* Cells and graphic groups can be scaled with this tool.

☞ *Tip:* You can avoid rescaling some elements, like cells, by setting ACTIVE SCALE before placing them.

See Also: LOCK SCALE, SCALE COPY

SD=

Alternate Key-in

Alternate key-in for ACTIVE STREAM DELTA. Use it to set the distance between sampled input points in stream mode.

See Also: ACTIVE STREAM DELTA

SELECT CELL ABSOLUTE   SELE C

Drawing Tool: Select and Place Cell

This tool allows you to graphically define the active cell. For instance, instead of keying in AC=SPOT to activate the SPOT cell, you could execute the SELECT CELL ABSOLUTE tool and identify an existing SPOT cell in the drawing file. SELECT CELL ABSOLUTE then invokes the PLACE CELL ABSOLUTE tool. The existing cell's library needs to be attached to the drawing file for the SELECT CELL ABSOLUTE tool to work.

Item Selection

On sidebar menu pick: Cells Select Abs

For window, point to: Palettes Main Cells

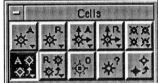

On paper menu see: CELLS

Command Window Prompts:
1. Select the item
   Identify element
2. Pick a cell
   Accept/Reject

3. This next input point places the cell
    ⇨ Choose the spot for the cell before you give the second point.
To exit item: Make next selection

✍ *Note:* A cell library can be attached to the drawing file through the Cells settings box or key in RC=LIBRARYNAME.

✳ *Error:* No cell library attached - If this message appears, there is no cell library attached to the drawing file. Use the RC=$ key-in to display the attached cell library.
See Also: PLACE CELL tools, SELECT CELL tools

SELECT CELL ABSOLUTE TMATRX

Drawing Tool: Select Cell and Place Cell by Transformation Matrix
    This tool allows you to graphically define the active cell from a user command.
SELECT CELL ABSOLUTE TMATRX invokes the PLACE CELL ABSOLUTE
TMATRX tool. The values in the transformation matrix control the cell's scale and
orientation. The existing cell's library needs to be attached to the drawing file for
the SELECT CELL ABSOLUTE TMATRX tool to work.

✍ *Note:* A cell library can be attached to the drawing file through the Cells settings box or key in RC=LIBRARYNAME.

✳ *Error:* No cell library attached - If this message appears, there is no cell library attached to the drawing file. Use the RC=$ key-in to display the attached cell library.
See Also: PLACE CELL tools, SELECT CELL tools

SELECT CELL INTERACTIVE ABSOLUTE

Obsolete
    Selects an existing cell from the drawing file, sets the ACTIVE CELL name
to that cell, and then places the cell interactively. Interactively means you can
change the cell's scale and rotation while placing the cell. The interactive cell is
placed absolute.
See Also: PLACE CELL tools, SELECT CELL tools

SELECT CELL INTERACTIVE RELATIVE

Obsolete
    Allows you to select an existing cell from the drawing file, set the ACTIVE
CELL name to that cell and then place the cell interactively. Interactively means
you can change the cell's scale and rotation while placing the cell. The interactive
cell is place relative.
See Also: PLACE CELL tools, SELECT CELL tools

SELECT CELL RELATIVE   SELE C R

Drawing Tool: Select and Place Cell (Relative)
    This tool allows you to graphically define the active cell. For example, instead
of keying in AC=SPOT to activate the SPOT cell, you could execute the SELECT
CELL RELATIVE tool and identify an existing SPOT cell in the drawing file.
SELECT CELL RELATIVE then invokes the PLACE CELL RELATIVE tool. The
existing cell's library needs to be attached to the drawing file for the SELECT
CELL RELATIVE tool to work.
Item Selection
On sidebar menu pick: | Cells | | Select | | Rel |

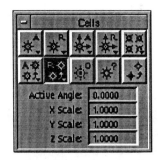

For window, point to: Palettes   Main   Cells
On paper menu see: CELLS
Command Window Prompts:
1. Select the item
   Identify element
2. Pick a cell
   Accept/Reject
3. This next input point places the cell
   ⇨ Choose the spot for the cell before you give the second point.
To exit item: Make next selection

🖉 *Note:* A cell library can be attached to the drawing file through the Cells settings box or by keying in RC=LIBRARYNAME.

✳ *Error:* No cell library attached - If this message appears, there is no cell library attached to the drawing file. Use the RC=$ key-in to display the attached cell library.
See Also: PLACE CELL tools, SELECT CELL tools

SELECT CELL RELATIVE TMATRX

Drawing Tool: Select Cell and Place Cell Relative by Transformation Matrix
This tool allows you to graphically define the active cell from a user command. SELECT CELL RELATIVE TMATRX then invokes the PLACE CELL RELATIVE TMATRX tool. The values in the transformation matrix control the cell's scale and orientation. The existing cell's library needs to be attached to the drawing file for the SELECT CELL RELATIVE TMATRX tool to work.

🖉 *Note:* A cell library can be attached to the drawing file through the Cells settings box or by keying in RC=LIBRARYNAME.

✳ *Error:* No cell library attached - If this message appears, there is no cell library attached to the drawing file. Use the RC=$ key-in to display the attached cell library.
See Also: PLACE CELL tools, SELECT CELL tools

SELVIEW

Command: Select View
SELVIEW is equivalent to placing an input point in a view. For instance, after selecting the SET GRID tool, it asks you to select the view to toggle the display grid on or off. You can either place a point in that view or key in SELVIEW 1.
Item Selection
On sidebar menu pick: n/a
For window, point to: n/a
On paper menu see: n/a
For short key-in use: **SELV**

☞ *Tip:* Attach SELVIEW to a function key. Now you can identify the active view with either your mouse or by clicking a function key.

SET ACSDISPLAY   SET ACS

Setting: Toggle ACS Display On/Off
      SET ACSDISPLAY is an Auxiliary Coordinate System setting that controls
the display of the coordinate triads in each view.

Item Selection
    On sidebar menu pick: n/a
    For window, point to: View   Attributes   ACS Triad
    On paper menu see: VIEW
    To set, key in: SET ASC on/off
        Where on/off is either ON or OFF.
See Also: DEFINE ACS VIEW

SET AUTOPAN   SET AUT

Unix Only

      Allows you to use the digitizing pad as a panning control.

Item Selection
    For window, point to: n/a
    On sidebar menu pick: n/a

    On paper menu see: n/a
See Also: DIGITIZING settings and commands

SET AUXINPUT   SET AUX

Setting: Set Auxiliary Input
      Some Third-Party software companies have developed programs that need
to feed MicroStation input from an auxiliary device. For instance, a stereo plotter
generating maps would require this kind of input. Turning SET AUXINPUT on
allows communication between the stereo plotter's device driver and MicroStation.

Item Selection
    On sidebar menu pick: n/a
    For window, point to: n/a
    On paper menu see: n/a
    To show, key in: SET AUXINPUT
      Auxiliary Input: OFF
    To set, key in: SET AUXINPUT on/off
        Where on/off is either ON or OFF.
See Also: DIGITIZER tools and settings

SET BACKGROUND   SET BA

DOS and Unix Only
Setting: Toggle Background Display On/Off
      MicroStation can display an image behind any active view. The SET BACK-
GROUND settings toggles the image display on or off. Select the setting and place
an input point in the view you want toggled on or off.

Item Selection
    On sidebar menu pick: Params   View   Backgr
    For window, point to: View   Attributes   Background
    On paper menu see: VIEW
    To show, key in: SET BACKGROUND
      Background: 1
    To set, key in: SET BACKGROUND on/off
        Where on/off is either On or OFF.
See Also: ACTIVE BACKGROUND

SET BORDER

Obsolete
   Places a thin border line around the edge of the screen. This border separates the view area from the information fields and sidebar menu.

SET BUTTON   SE BU

Setting: Remap Digitizer Buttons
   The SET BUTTON setting remaps the buttons on the cursor or mouse. This setting will ask you what buttons you want assigned as the Data, Reset and Tentative positions on the cursor or mouse.

Item Selection
On sidebar menu pick: `Params` `Set`   `Cursor`
For window, point to: <u>S</u>ettings <u>D</u>igitizing <u>T</u>ablet <u>B</u>utton Assignment
On paper menu see: n/a
To set, key in: SET BUTTON
Command Window Prompts:
   1. Select the item
      Press Data button
   2. Pick the button you want to be the Data button
      Press RESET button
   3. Pick the button you want to be the Reset button
      Press Tentative button
   4. Pick the button you want to be the Tentative button
      When all buttons assigned, hit return
      Digitizer buttons remapped
   5. Hit the Enter key
To exit item: Exits by itself

✍ *Note:* These button positions cannot be saved using the FILEDESIGN tool.

✳ *Error:* Button already assigned! - If this message appears, you have attempted to assign two options to one button.
   See Also: DIGITIZER BUTTONS

SET CAMERA   SET CA

Setting: Set Camera on/off
   Turns a view camera on or off. This item requires an input point in each view you want affected.

Item Selection
On sidebar menu pick: `Params` `Camera` `On`
For window, point to: <u>V</u>iew <u>A</u>ttributes Camera
On paper menu see: CAMERA

✍ *Note:* This menu item is dimmed if the file is not 3D.
   See Also: SET CAMERA settings

SET CAMERA DEFINITION   SET CA DE

Setting: Set Camera Definition
   Define your camera with this setting. This item requires input points to designate the view, the camera's position, and the front and back clipping planes.

Item Selection
On sidebar menu pick: `Params` `Camera` `Def`
For window, point to: <u>V</u>iew Ca<u>m</u>era <u>S</u>etup
On paper menu see: CAMERA
Command Window Prompts:
   1. Select the item
      Select view
   2. Pick view
      Define Camera Target Point
   3. Pick subject

```
             Define Camera Position
       4. Pick camera location
             Define front clipping plane
       5. Pick front clipping plane
             Define back clipping plane
       6. Pick back clipping plane
```
          ⇨ the active depth is set to the display depth
    To exit item: Exits by itself

✍ *Note:* This item is dimmed if you are not in a 3D file.
    See Also: SET CAMERA settings

SET CAMERA DISTANCE   SET CA DI

3D Setting: Show Camera Distance
        The camera's distance from the object is displayed in the MicroStation
Command Window. This setting expects a single input point placed in the target
view.
Item Selection
    On sidebar menu pick: | Params | | Camera | | Dist |
    For window, point to: n/a
    On paper menu see: CAMERA
    To show, key in: **SET CAMERA DISTANCE**
       Camera Distance : 425:9.34

✍ *Note:* This menu item is dimmed if the file is not 3D.
    See Also: SET CAMERA settings, SHOW CAMERA settings

SET CAMERA LENGTH   SET CA LEN

Setting: Set Camera Focal Length
        Sets a custom view camera focal length. This item requires a length key-in,
and an input point in each view you want affected.
Item Selection
    On sidebar menu pick: | Params | | Camera | | Lens | | Keyin |
    For window, point to: View Camera Lens Focal Length
    On paper menu see: CAMERA

✍ *Note:* This menu item is dimmed if the file is not 3D.
    See Also: SET CAMERA settings

SET CAMERA LENS   SET CA L

3D Setting: Set View Camera Lens
        Selects the type of camera lens used for a view. This setting can be changed
or viewed through the Camera Lens settings box or by a key-in. The lens angle
and focal length can also be adjusted through this settings box.
Choices:
        Lens Type, Lens Angle, Focal Length(mm)
        Fisheye, 93.3°, 20
        Extra Wide, 74.3°, 28
        Wide, 62.4°, 35
        Normal, 46.0°, 50
        Portrait, 28.0°, 85
        Telephoto, 12.1°, 200
        Telescopic, 2.4°, 1000
Item Selection
    On sidebar menu pick: | Params | | Camera | | Lens |
    For window, point to: View Camera Lens Standard
    On paper menu see: CAMERA
    To show, open: **Camera Lens settings box**
    To set, key in: **SET CAMERA LENS Choice**
        Where Choice is a valid camera lens.

✍ *Note:* This menu item is dimmed if the file is not 3D.
· *See Also:* SHOW CAMERA LENS

SET CAMERA LENS ANGLE   SET CA LEN AN

Setting: Set Camera Lens Angle
   Sets a custom view camera lens angle. This item requires an angle key-in
between 0 and 180 degrees, and an input point in each view you want affected.
Item Selection
   On sidebar menu pick: n/a
   For window, point to: View   Camera   Lens   Angle
   On paper menu see: CAMERA

✍ *Note:* This menu item is dimmed if the file is not 3D.
See Also: SET CAMERA settings

SET CAMERA POSITION   SET CA P

3D Setting: Set View Camera Position
   Changes the camera position relative to the object being viewed. This item
requires three input points. First pick the view, then pick the front and back
clipping planes.
Item Selection
   On sidebar menu pick: | Params | | Camera | | Positn |
   For window, point to: View   Camera   Position
   On paper menu see: CAMERA
Command Window Prompts:
   1. Select the item
      Select view
      Define Camera Position
   2. Pick new vantage point
      Define front clipping plane
   3. Pick front plane
      Define back clipping plane
   4. Pick back plane
   To exit item: Make next selection

✍ *Note:* This menu item is dimmed if the file is not 3D.
See Also: SHOW CAMERA POSITION

SET CAMERA TARGET   SET CA T

3D Setting: Set View Camera Target
   Lets you select the target for a view camera. For example, think about holding
a camera and moving it from side to side to view different subjects. This item
requires four input points. This first two points define the view and the target.
The last two define the front and back clipping planes.
Item Selection
   On sidebar menu pick: | Params | | Camera | | Target |
   For window, point to: View   Camera   Target
   On paper menu see: CAMERA
Command Window Prompts:
   1. Select the item
      Select view
   2. Pick a view
      Define Camera Target Point
   3. Pick the subject
      Define front clipping plane
   4. Pick front plane
      Define back clipping plane
   5. Pick back plane
   To exit item: Make next selection

✍ *Note:* This menu item is dimmed if the file is not 3D.

See Also: SHOW CAMERA TARGET

SET COLOR

Obsolete
> Sets any color shade can be assigned to a display color number. MicroStation has the ability of displaying 256 colors. Most graphics boards display fewer than 256 colors. Display color one is normally mapped to element color one.

See Also: ACTIVE COLOR

SET COMPATIBLE   SET COM

Setting: Set Compatibility On/Off
> Turns MicroStation compatibility mode on or off. Turning compatibility on changes the Compatibility setting to 3.X in the Preferences dialog box. The default compatibility setting is 4.0 (or off).

Item Selection
On sidebar menu pick: `Utils` `User` `Pref`
For window, point to: U̲ser P̲references...   Compatibility:
On paper menu see: LINEAR DIMENSION
To show, open: **Preferences dialog box**
To set, key in: **SET COMPATIBLE on/off**
> Where on/off is ON or OFF.

See Also: SET COMPATIBLE settings

SET COMPATIBLE DIMENSION   SET COM D

Setting: Toggle Compatibility Dimension On/Off
> In order to be compatible with Intergraph's Interactive Graphic Design Software (IGDS), toggle SET COMPATIBLE DIMENSION on when placing dimension data. With compatibility on, these element types are placed as graphic groups. Key in SET COMPATIBLE off if compatibility to IGDS is not needed.

Item Selection
On sidebar menu pick: n/a
For window, point to: n/a
On paper menu see: n/a
To show, key in: **SET COM D**
  Compatible Dimensions : ON
To set, key in: **SET COMPATIBLE DIMENSION on/off**
> Where on/off is ON or OFF.

See Also: SET COMPATIBLE settings

SET COMPATIBLE MLINE   SET COM M

Setting: Toggle Compatibility Multi-line On/Off
> In order to be compatible with Intergraph's Interactive Graphic Design Software (IGDS), toggle SET COMPATIBLE MLINE on when placing Multi-line elements. With compatibility on, these element types are placed as graphic groups. Key in SET COMPATIBLE off if compatibility to IGDS is not needed.

Item Selection
On sidebar menu pick: n/a
For window, point to: n/a
On paper menu see: n/a
To show, key in: **SET COM M**
  Compatible Multi-lines : ON
To set, key in: **SET COMPATIBLE MLINE on/off**
> Where on/off is ON or OFF.

See Also: SET COMPATIBLE settings

SET CONSTRUCT

Setting: Toggle Construction Display On/Off
  The SET CONSTRUCT setting controls the display of construction elements.
Keying in SET CONSTRUCT toggles the display between on and off. This setting
requires one input point in each view to toggle the construction display.

Item Selection
  On sidebar menu pick: `Params` `Set` `Const`
  For window, point to: Y̲iew A̲ttributes Constructions
  On paper menu see: n/a
  To show, key in: **SET CONSTRUCT**
    Const. Elements : 1,2,3,4,5,6,7,8
  To set, key in: **SET CONSTRUCT on/off**
    Where on/off is ON or OFF.
Command Window Prompts:
  1. Select the item
    Select view
  2. Pick a view
    ⇨ If the view number does not appear, the display is off.
  To exit item: **Click a Reset**
See Also: ACTIVE CLASS CONSTRUCTION

SET CONTROL

DOS Only
Obsolete
  Links a drawing file or group of drawing files to a database through a
database control file. The control file provides information relating these files and
keeps track of the auxiliary database files.
See Also: ACTIVE DATABASE, SET DATABASE

SET COORDINATES

Obsolete
  Opens or closes the Precision Input settings box. Keying in SET COORDI-
NATES ON allows a continuous display of the screen cursor's XY location on the
design plane. SET COORDINATES OFF closes the settings box.
  The SET TPMODE settings and the Coordinates Readout settings box in
Design Options determine how the coordinates display.
See Also: SET TPMODE settings

SET CURSOR FULL

DOS and Unix Only
Setting: Set Full View Pointer
  Displays a cross-hair cursor the entire size of the view. This setting can be
viewed or changed through the Preferences dialog box or a key-in.

Item Selection
  On sidebar menu pick: `Params` `Set` `Cursor`
  For window, point to: U̲ser P̲references... Pointer Size:
  On paper menu see: n/a
  To show, open: **Preference setting box**
  To set, key in: **SET CURSOR F**

☞ *Tip:* Use the full screen cursor when placing tight fitting fences or when lining up
  elements for placement.
  See Also: SET CURSOR settings, SET TPMODE settings

SET CURSOR ISOMETRIC

Setting: Set Cursor Isometric
  The screen cursor can have an isometric or orthogonal orientation. Key in
SET CURSOR ISOMETRIC for an isometric orientation.

Item Selection
On sidebar menu pick: | Params | Set | | Cursor |
For window, point to: Uɪser Preferences... Pointer Type
On paper menu see: n/a
To show, key in: SET CURSOR
   Cursor : SMALL, ORTHOGONAL
To set, key in: SET CURSOR ISOMETRIC.

☞ *Tip:* Use the ISO cursor when creating isometric perspectives.
See Also: SET CURSOR settings, SET TPMODE settings

SET CURSOR ORTHOGONAL   SET CURS O

Setting: Set Cursor Orthogonal
   The screen cursor can have an isometric or orthogonal orientation. Key in
SET CURSOR ORTHOGONAL for an orthogonal orientation.

Item Selection
On sidebar menu pick: | Params | Set | | Cursor |
For window, point to: Uɪser Preferences... Pointer Type
On paper menu see: n/a
To show, key in: SET CURSOR
   Cursor : SMALL, ORTHOGONAL
To set, key in: SET CURSOR ORTHOGONAL
See Also: SET CURSOR settings, SET TPMODE settings

SET CURSOR SMALL   SET CURS S

DOS and Unix Only
Setting: Set Normal Pointer
   Displays the normal cross-hair cursor. This setting can be viewed or changed
through the Preferences dialog box or a key-in.

Item Selection
On sidebar menu pick: | Params | Set | | Cursor |
For window, point to: Uɪser Preferences... Pointer Size:
On paper menu see: n/a
To show, key in: SET CURSOR
   Cursor : SMALL, ORTHOGONAL
To set, key in: SET CURSOR NORMAL

☞ *Tip:* Use the full screen cursor when placing tight fitting fences or when lining up
   elements for placement.
See Also: SET CURSOR settings, SET TPMODE settings
      settings

SET CURVES   SET CURV

Setting: Toggle Curve Display Fast/Slow
   Curve lines appear either smooth or rough depending on how many points
define the curve lines. MicroStation displays the same line two ways: fast or slow.
With the fast display turned on, curves appear rough, showing each vertex of the
curved line. Turning on the slow display smoothes the curved line as if the curves
contained many more vertices.
   The slow and fast display can be independently set in each view. If the slow
display is on in view one, the fast display must be off in that view. This setting
expects an input point in each view selected for slow or fast display.

Item Selection
On sidebar menu pick: | Params | Disp | Slow | Curves |
For window, point to: Uɪser Preferences... Fast Curves
On paper menu see: DISPLAY
To show, key in: SET CURVES
   Fast curves : 1,3
To set, key in: SET CURVES opt
   Where opt is SLOW or FAST.

Command Window Prompts:
1. Select the item
   Select view
2. Pick the view you want changed
   Fast curves : 1,3
   ⇨ If the view number shows up, fast is on.
To exit item: Make next selection

☞ *Tip:* Use this display technique to smooth out rough curves.
See Also: PLACE CURVE tools

SET DATABASE   SET DA

DOS Only
Setting: Set Database
SET DATABASE specifies a database file associated with the drawing file. After keying in SET DATABASE, a window opens. Each database file connected to the drawing file appears on a separate screen. Several fields of information must be filled in before a database can work with a drawing file.

Item Selection
On sidebar menu pick: | dBase | | Setup | | Enviro |
For window, point to: n/a
On paper menu see: n/a
To show, key in: SET DATABASE
To set, key in: SET DATABASE
See Also: ACTIVE DATABASE, SET CONTROL

SET DDEPTH ABSOLUTE   DP=

3D Setting: Set Display Depth (Absolute)
Control how much depth you see on the screen. This setting requires two key-in values to define the front and back clipping planes. Then place an input point in each view to change the display.

Item Selection
On sidebar menu pick: | Params | | Disp | | Depth |
For window, point to: n/a
On paper menu see: DEPTH
Command Window Prompts:
1. Select the item
   Select view
2. Pick a view
To exit item: Exits by itself

☞ *Tip:* You can also change the absolute display depth with the DP= key-in. For example, DP=-10,10 sets the display depth to 10 master units in front of and behind the object.

✳ *Error:* Illegal depth definition - If this message appears, MicroStation may not be able to calculate the depth of the display area you have defined.
See Also: DEPTH ACTIVE, DEPTH DISPLAY, SET DDEPTH RELATIVE

SET DDEPTH RELATIVE   DD=

3D Setting: Set Display Depth (Relative)
Control how much depth you see on the screen. This setting requires three input points. The first point tells MicroStation which view's depth you want to adjust. The next two points define the clipping plane in that view.

Item Selection
On sidebar menu pick: | Params | | Disp | | Depth |
For window, point to: n/a
On paper menu see: n/a
Command Window Prompts:
1. Select the item
   Select view for display depth
2. Pick a view

```
    Define front clipping plane.
3. Pick your first point
    Define back clipping plane
4. Pick your second point
```
To exit item: `Exits by itself`

☞ *Tip:* You can also change the relative display depth with the DD= key-in. For example, DD=10,-10 sets the display depth to 10 master units in front of and behind the object.

✳ *Error:* `Illegal depth definition` - If this message appears, MicroStation may not be able to calculate the depth of the display area you have defined.
See Also: DEPTH ACTIVE, DEPTH DISPLAY, SET DDEPTH ABSOLUTE

SET DEBUG   SET DEB

Setting: Set Debug
An application development setting for debugging programs. SET DEBUG ON turns the setting on. SET DEBUG OFF turns debug off.

Item Selection
On sidebar menu pick: n/a
For window, point to: n/a
On paper menu see: n/a

SET DELETE   SET DE

Setting: Toggle Delete On/Off
Flip the SET DELETE setting on if you want database records deleted when you delete their corresponding graphics. When SET DELETE is off, MicroStation keeps the database records. Keying in SET DELETE toggles the setting between on and off.

Item Selection
On sidebar menu pick: dBase   Settng
For window, point to: Settings   Database   Delete Linked Database Rows
On paper menu see: n/a
To show, key in: `SET DELETE`
   `Delete Entity: OFF`
To set, key in: `SET DELETE on/off`
   Where on/off is either ON or OFF.
See Also: DELETE ELEMENT

SET DEPTHCUE   SET DEP

Setting: Toggle Depth Cueing On/Off
Dims distant objects for visual effect in a shared view. This setting can be changed or viewed with a key-in or through the View Attributes settings box.

Item Selection
On sidebar menu pick: n/a
For window, point to: View   Attributes   Depth Cueing
On paper menu see: VIEW
To show, open: `View Attributes settings box`
To set, key in: `SET DEPTHCUE on/off`
   Where on/off is either ON or OFF.
See Also: RENDER commands

SET DIMENSION   SET DI

Setting: Toggle Dimension Display On/Off
Display or hide dimension data with the SET DIMENSION setting. The dimension data display can be set on or off in each view. This setting requires an input point in each view to set the dimension display. Keying in SET DIMENSION toggles the display between on and off.

Item Selection
On sidebar menu pick: Params   Disp   Dimen

For window, point to: Views   Attributes   Dimensions
On paper menu see: DISPLAY
To show, key in: SET DIMENSION
   Dimension data : 1,2,3,4,5,6,7,8
To set, key in: SET DIMENSION on/off
   Where on/off is either ON or OFF.
Command Window Prompts:
   1. Select the item
      Select view
   2. Pick the views with input points
      Dimension data : 1,2,3,4,5,6,7,8
        ⇨   If the view number does not appear, that view's dimension display is off.
To exit item: Click a Reset
See Also: DIMENSION LEVEL

SET DYNAMIC   SET DI

Setting: Toggle Dynamic Update On/Off
   SET DYNAMIC allows you to see elements dynamically as you construct
them on the screen. Keying in SET DYNAMIC toggles the display between on and
off. This setting requires one input point in each view to set the dynamic display
on or off.
Item Selection
   On sidebar menu pick: | Params | | View | | Dynamc |
   For window, point to: Views   Attributes   Dynamics
   On paper menu see: DISPLAY
Command Window Prompts:
   1. Select the item
   To show, key in: SET DYNAMIC FAST
      Fast Dynamic update : 1,2,3,4,5,6,7,8
   To show, key in: SET DYNAMIC SLOW
      Slow Dynamic update : 1,2,3,4,5,6,7,8
   To set, key in: SET DYNAMIC FAST on/off
      Where on/off is either ON or OFF.
   To exit item: Exits by itself
See Also: VIEW view controls

SET DYNOSIZE

Obsolete
   Controls the buffer size (in Kbytes) of the buffer used to dynamically display
elements. If you have memory to spare, set the buffer size above the limit of the
largest element manipulation. The default size of the dynamic buffer is 32 Kbytes.

SET ED   SET ED

Setting: Toggle Enter Data Field Display On/Off
   SET ED is short for SET ENTER DATA FIELD. This setting controls the
display of enter data field characters. The enter data field display can be set on or
off in each view separately. This setting requires one input point in each view to
set the enter data field display. Keying in SET ED toggles the display between on
and off.
Item Selection
   On sidebar menu pick: | Params | | Disp | | ED Fld |
   For window, point to: Views   Attributes   Data Fields
   On paper menu see: DISPLAY
   To show, key in: SET ED
      Enter data on : 1,2,3,4,5,6,7,8
   To set, key in: SET ED on/off
      Where on/off is ON or OFF.
Command Window Prompts:
   1. Select the item
      Select view

2. Pick the view
```
Enter data on : 1,2,3,4,5,6,7,8
```
    ⇨ If the view number appears, that view's enter data field character display is off.

To exit item: Click a Reset
See Also: EDIT tools, PLACE TEXT tools

SET EDCHAR   SET EDC

Setting: Set Enter Data Field Character
The SET EDCHAR setting defines the enter data field character. The default character is the under bar ( ). Only one character at a time can represent an enter data field.

Item Selection
On sidebar menu pick: | Utils | | User | | Pref |
For window, point to: U̲ser P̲references... ED Character:
On paper menu see: n/a
To show, key in: SET EDCHAR
```
Enter data field character='_'
```
To set, key in: SET EDCHAR u
    Where u is the user defined character.

✍ *Note:* The EDCHAR setting cannot be saved with the FILEDESIGN operation.
See Also: Enter Data Fields

SET FILL   SET FI

Setting: Toggle Fill Display On/Off
SET FILL on causes closed shapes to fill with solid color. You can SET FILL on or off separately in each view. This setting requires one input point in each view to set the fill display. Keying in SET FILL toggles the display between on and off.

Item Selection
On sidebar menu pick: | Params | | View | | Fill |
For window, point to: V̲iews A̲ttributes Area Fill
On paper menu see: DISPLAY
To show, key in: SET FILL
```
Fill : 1
```
To set, key in: SET FILL on/off
    Where on/off is either ON or OFF.
Command Window Prompts:
1. Select the item
```
Select view
```
2. Pick the view
```
Enter data on : 1,2,3,4,5,6,7,8
```
    ⇨ If the view number does not appear, the display is off.

To exit item: Click a Reset
See Also: ACTIVE FILL, CHANGE FILL

SET FONT   SET FO

Setting: Toggle Font Fast/Slow
Text can be placed in the drawing with different lettering styles. SET FONT tells MicroStation whether to display the various lettering styles on the screen (SLOW) or to use its default display font (FAST). This setting requires one input point in each view to set the font display. Keying in SET FONT toggles the display between slow and fast.

Item Selection
On sidebar menu pick: | Text | | Displ | | On/Off |
For window, point to: V̲iews A̲ttributes Fast Font
On paper menu see: DISPLAY
To show, key in: SET FONT
```
Slow font : 1,2,3,4,5,6,7,8
```
To set, key in: SET FONT opt

Where opt is FAST or SLOW.
Command Window Prompts:
1. Select the item
   ```
   Select view
   ```
2. Pick the view
   ```
   Slow font : 1,2,3,4,5,6,7,8
   ```
   ⇨   If the view number does not appear, that view's display is fast.
To exit item: Click a Reset
See Also: PLACE TEXT Tools, SET TEXT

SET FUNCTION

Obsolete
Drawing functions can be attached to function keys (F#) on the keyboard. MicroStation comes with default function key assignments, but these key commands or settings can be changed at any time by using SET FUNCTION.
See Also: SAVE FUNCTION_KEY

SET GRID   SET GR

Setting: Toggle Grid Display On/Off
When making flow diagrams, charts or forms, use SET GRID to turn on MicroStation's display grid. This is similar to placing a clear piece of grid paper on the display screen to help you draw straight lines and right angles. The display grid can be set on or off in each view. Keying in SET GRID and identifying the view(s) toggles the display on and off. To force all the input points on to the display grid, the LOCK GRID setting must be activated.

Item Selection
On sidebar menu pick: Params   Disp   Grid
For window, point to: Views   Attributes   Grid
On paper menu see: DISPLAY
To show, key in: SET GRID
   ```
   Grids on : 1,2,3,4,5,6,7,8
   ```
To set, key in: SET GRID on/off
   Where on/off is ON or OFF.
Command Window Prompts:
1. Select the item
   ```
   Select view
   ```
2. Pick the view
   ```
   Grids on : 1,2,3,4,5,6,7,8
   ```
   ⇨   If the view number does not appear, that view's grid display is off.
To exit item: Click a Reset
See Also: ACTIVE GRIDMODE, ACTIVE GRIDRATIO, ACTIVE GRIDREF, ACTIVE GRIDUNIT, LOCK GRID

SET HELP   SET HE

Mac and Unix Only
Setting: Toggle Help On/Off
Summons the help routine. Key in SET HELP on, and the help window appears in the upper right hand corner of the display screen. Each time you request help for a specific command, the information is shown in this window. The help utility may be deactivated by typing in SET HELP off.

Item Selection
On sidebar menu pick: n/a
For window, point to: n/a
On paper menu see: n/a
See Also: HELP

SET HILITE   SET H

Setting: Set Hilite Color
When you select elements using one of MicroStation's tools, the element's color changes to the highlight color. The hilite color can be changed to any color

with the SET HILITE setting. The default highlight color is white. On monochrome monitors elements flash instead of changing color.

Item Selection
On sidebar menu pick: | Utils | | User | | Pref |
For window, point to: User Preferences... Highlight:
On paper menu see: n/a
To show, key in: **SET HILITE**
   Hilite color = 14
To set, key in: **SET HILITE nnn**
   Where nnn is a valid color number.
See Also: ACTIVE COLOR, SET XOR

SET ISOPLANE ALL   SET I A

Lets you place objects in the left, right or top isometric view. The PLACE BLOCK ISOMETRIC and PLACE CIRCLE ISOMETRIC tools default to the top isometric view only.

Item Selection
On sidebar menu pick: n/a
For window, point to: Settings Locks Full Isometric Plane:
On paper menu see: DISPLAY
To show, open: **Locks setting box**
To set, key in: **SET ISOPLANE LEFT**
See Also: ACTIVE GRIDMODE, SET ISOPLANE settings

SET ISOPLANE LEFT   SET I L

Setting: Set Isometric Drawing Plane (Left)
Locks elements into a left isometric plane. This setting can be view or changed by a key-in or through the Locks settings box.

Item Selection
On sidebar menu pick: n/a
For window, point to: Settings Locks Full Isometric Plane:
On paper menu see: DISPLAY
To show, open: **Locks settings box**
To set, key in: **SET ISOPLANE LEFT**
See Also: ACTIVE GRIDMODE, SET ISOPLANE settings

SET ISOPLANE RIGHT   SET I R

Setting: Set Isometric Drawing Plane (Right)
Locks elements into a right isometric plane. This setting can be view or changed by a key-in or through the Locks settings box.

Item Selection
On sidebar menu pick: n/a
For window, point to: Settings Locks Full Isometric Plane:
On paper menu see: DISPLAY
To show, open: **Locks settings box**
To set, key in: **SET ISOPLANE RIGHT**
See Also: ACTIVE GRIDMODE, SET ISOPLANE settings

SET ISOPLANE TOP   SET I T

Setting: Set Isometric Drawing Plane (Top)
Locks elements into the top isometric plane. This setting can be view or changed by a key-in or through the Locks settings box.

Item Selection
On sidebar menu pick: n/a
For window, point to: Settings Locks Full Isometric Plane:
On paper menu see: DISPLAY
To show, open: **Locks settings box**

To set, key in: **SET ISOPLANE TOP**
See Also: ACTIVE GRIDMODE, SET ISOPLANE settings

SET LEVELS   Ctrl+E

Setting: Toggle Level Display On/Off
Elements can be displayed on the screen or hidden from view. The SET
LEVELS setting controls whether the display of each of the 63 levels in the
drawing file are on or off. Any combination of levels can be turned on or off in each
view. Click an input point in each view to change the level settings. ON= and OF=
are short key-ins for SET LEVELS.

Item Selection

vels

Levels   Isometric Plane:

ιs
nnn
ber.

< appears
.dow
hey appear without the need of a screen update.
level numbers to turn off several levels at one

umbers to turn off a range of levels at one time.
'ashes can be combined).

Key in 1-63 to turn all the levels but the active level on or off at one time.
Example: ON=1-63
See Also: ACTIVE LEVEL

SET LINEFILL

Setting: Toggle Fill Display On/Off
Causes lines drawn with SET LINEWIDTH ON to be filled, else they would
appear hollow. SET LINEFILL can be turned on or off in each view. Keying in
SET LINEFILL and identifying the view(s) toggles the display on and off.
Item Selection
On sidebar menu pick: | Params | | View | | Fill |
For window, point to: View   Attributes   Area Fill
On paper menu see: n/a
Short key-in: **SET LINEF**
To show, key in: **SET LINEFILL**
   Fill : 1,2
To set, key in: **SET LINEFILL on/off**
   Where on/off is ON or OFF.
Command Window Prompts:
   1. Select the item
   Select view
   2. Pick the view
   Fill : 1,2,3,4
      ⇨ If the view number does not appear, that view's linefill display is off.
To exit item: **Click a Reset**
See Also: SET LINEWIDTH

SET LINEWIDTH   SET LINEW

Setting: Toggle Line Width On/Off
The SET LINEWIDTH setting allows you to draw elements with lines that
have width. When SET LINEWIDTH is on, lines appear tubular. With this setting
off, the elements are drawn with normal solid lines.

Item Selection
   On sidebar menu pick: n/a
   For window, point to: n/a
   On paper menu see: n/a
   To show, key in: SET LINEWIDTH
      Lines with width : ON
   To set, key in: SET LINEWIDTH on/off
      Where on/off is ON or OFF.
See Also: ACTIVE LINEWIDTH, ACTIVE WEIGHT, SET LINEFILL

SET LOCATE   SET LO

Setting: Set Locate Tolerance
   The SET LOCATE setting controls the size of the element selection range.
   The element selection range is the area around the cursor which MicroStation
   searches when looking to snap to an element. The default size is ten units. The
   maximum size is 100 units. A unit is an arbitrary size. It is not a working unit.
Item Selection
   On sidebar menu pick: | Params | | Set | | Locate |
   For window, point to: User   Preferences...   Locate Tolerance
   On paper menu see: n/a
   To show, key in: SET LOCATE
      Locate Tolerance=10
   To set, key in: SET LOCATE nnn
      Where nnn is a whole number between 1 and 100.
See Also: LOCK SNAP settings

SET LVLSYMB   SET LV

Setting: Toggle Level Symbology On/Off
   The SET LVLSYMB setting controls whether the level symbology display is
   on or off. With SET LVLSYMB on, the level symbology table controls the element's
   display in each view. SET LVLSYMB off turns symbology display off. Keying in
   SET LVLSYMB toggles the display between on and off. This setting requires one
   input point for each view.
Item Selection
   On sidebar menu pick: | Params | | View | | LvlSym |
   For window, point to: View   Attributes   Level Symb.
   On paper menu see: DISPLAY
   To show, key in: SET LVLSYMB
      Level Symbology on : 1,2
   To set, key in: SET LVLSYMB on/off
      Where on/off is ON or OFF.
Command Window Prompts:
   1. Select the item
      Select view
   2. Pick the view
      Level Symbology on : 1
      ⇨   If the view number does not appear, that view's symbology display is off.
   To exit item: Click a Reset
✍ *Note:* See the Review/Change Level Symbology portion in Design Options settings
   box.
See Also: CHANGE SYMBOLOGY

SET MAXGRID   SET MA

Setting: Set Maximum Grid Points
   A series of crosses and dots make up the display grid. As you zoom out in a
   view, the display grid begins to fill the screen. MicroStation would eventually fill
   the entire screen with grid marks if not for the SET MAXGRID setting. The SET
   MAXGRID setting controls the point at which the display grid stops displaying.

The maximum number of grid crosses which can be displayed are 150 and the largest number of dots is 400. The default setting for MAXGRID is 90 dots and 40 crosses.

Item Selection
On sidebar menu pick: Params Set Grid
For window, point to: User Preferences... Max. Grid Pts/View
On paper menu see: n/a
To show, key in: SET MAXGRID
  Max. Grid Pnts = 90,40
To set, key in: SET MAXGRID nnn,mmm
  Where the first field is the number of dots, and the second field is the num of crosses.

✍ *Note:* This setting cannot be saved by the FILE DESIGN tool.
See Also: ACTIVE GRIDMODE, ACTIVE GRIDRATIO, ACTIVE GRIDREF TIVE GRIDUNIT

SET MIRTEXT

Setting: Toggle Mirror Text On/Off
  The SET MIRTEXT setting controls whether or not text is mirrored any of the fence or element mirroring operations.
Item Selection
On sidebar menu pick: n/a
For window, point to: Element Text Mirror Text
On paper menu see: MODIFY TEXT
To show, key in: SET MIRTEXT
  Text Mirroring disabled
To set, key in: SET MIRTEXT on/off
  Where on/off is ON or OFF.
See Also: FENCE MIRROR tools, MIRROR tools

SET NODES

Setting: Toggle Text Node Display On/Off
  The SET NODES setting controls the display of text nodes. Each view's text node display can be independently set. SET NODES on makes the text nodes appear in the view(s). SET NODES off hides the text node numbers. Keying in SET NODES toggles the display between on and off. This setting requires one input point.

Item Selection
On sidebar menu pick: Params Disp Node
For window, point to: View Attributes Text Nodes
On paper menu see: DISPLAY
Short key-in: SET N
To show, key in: SET NODES
  Text nodes : 1,2,3,4,5,6,7,8
To set, key in: SET NODES on/off
  Where on/off is ON or OFF.
Command Window Prompts:
  1. Select the item
    Select view
  2. Pick the view
    Text nodes : 1,2,3,4,5,6,7,8
    ⇨ If the view number does not appear, that view's text node display is off.
To exit item: Click a Reset

✍ *Note:* If you don't want to see the text node numbers plot, SET NODES OFF before plotting the drawing file.
See Also: PLACE NODE tools

SET OVERVIEW

SET O

Setting: Toggle Overview On/Off

Allows toggling between a primary and secondary display. Most of the drawing is done on the primary screen. The secondary screen serves as a reference screen to view the entire drawing. To keep track of where you are working on the primary screen, use the SET OVERVIEW tool. A dashed box appears on the secondary screen showing the display of the primary screen.

Item Selection

On sidebar menu pick: | Params | View | OvrVw |

For window, point to: n/a

On paper menu see: n/a

To show, key in: `SET OVERVIEW`

Overview : ON

To set, key in: `SET OVERVIEW on/off`

Where on/off is either ON or OFF.

✍ *Note:* The overview box may be set to appear on either the left or right screen by keying in SET OVERVIEW LEFT or SET OVERVIEW RIGHT.

See Also: SWAP SCREEN, VIEW view controls, WINDOW view controls

SET PARSEALL

SET PAR

Setting: Toggle Parse All Key-ins On/Off

Enables or disables the passing of tools to MicroStation through the TEXT: prompt. For instance, while placing text, you may want to boost the size of the text. With SET PARSEALL on, the TX= key-in resets the text size. If SET PARSEALL is toggled off, the TX= key-in reads as text, not a tool.

Item Selection

On sidebar menu pick: n/a

For window, point to: n/a

On paper menu see: n/a

See Also: PLACE TEXT tools

SET PATTERN

SET PA

Setting: Toggle Pattern Display On/Off

Controls the display of patterns on the screen. Each view's pattern display can be set independent of the other views. Keying in SET PATTERN toggles the display between on and off. This setting requires one input point for each view.

Item Selection

On sidebar menu pick: | Params | Disp | Pattrn |

For window, point to: View Attributes Patterns

On paper menu see: DISPLAY

To show, key in: `SET PATTERN`

Patterns : 1,2,3,4,5,6,7,8

To set, key in: `SET PATTERN on/off`

Where on/off is ON or OFF.

Command Window Prompts:

1. Select the item

Select view

2. Pick the view

Patterns : 1,2,3,4,5,6,7,8

⇨ If the view number does not appear, that view's pattern display is off.

To exit item: `Click a Reset`

☞ *Tip:* Turning the pattern display off speeds up screen updates if there are dense patterns in the drawing file.

See Also: PATTERN AREA and PATTERN LINEAR tools, SHOW PATTERN

SET PERSPECTIVE

Obsolete

A 3D model can be viewed from different perspectives. To have the objects look more life-like, use SET PERSPECTIVE [distance]. The value of the [distance] parameter is the distance from the eye to the front of the display cube. This way objects farther from the eye will appear smaller, while objects closer to the eye appear much larger.

SET PLOTTER   SET PL

Setting: Set Plotter Type

Opens the Plotter Configuration File dialog box. This dialog box gives you the ability to select plotting devices without exiting a design session. Pick the plotter you want and click on OK. Click Cancel to close the window.

Item Selection

On sidebar menu pick: n/a
For window, point to: File   Plot...   Plotters...   Name:
On paper menu see: n/a
To show, key in: SET PLOTTER
   Plotter=C:\USTATION|PLTCFG\JDL850.PLT
To set, key in: SET PLOTTER filename
   Where filename is a valid plotter configuration file.

✍ *Note:* The plotter configuration files reside in \USTATION\PLTCFG.

✱ *Error:* No plotter config file - If this message appears, it means the plotter configuration does not exist. Run USCONFIG to configure a plotter.
See Also: PLOT @, SHOW PLOTTER

SET PROMPT   SET PR

DOS and Unix Only
Setting: Set Prompt

Changes the screen prompt "(1) USTN>" in the MicroStation Command Window. Key in SET PROMPT with no qualifier after it, and no prompt appears.

Item Selection

On sidebar menu pick: | Params | | Prompt |
For window, point to: n/a
On paper menu see: n/a
To set, key in: SET PROMPT uuu
   Where uuu is the new prompt.

Example(s):
   SET PROMPT YOU RANG>   changes the uSTN prompt to YOU RANG

✍ *Note:* This setting cannot be saved with the FILE DESIGN operation.

SET RANGE   SET RA

Setting: Toggle Range Block Display On/Off

Displays range blocks around each element. SET RANGE is a diagnostic tool.

Item Selection

On sidebar menu pick: n/a
For window, point to: n/a
On paper menu see: UTILITIES

SET RASTERTEXT   SET RA

Mac Only

Setting: Toggle Rastertext Display On/Off

Mac only – toggles the display of raster text. This item requires an input point in the view(s) where the raster text display needs changing.

Item Selection

On sidebar menu pick: n/a

For window, point to: n/a
On paper menu see: n/a
To show, key in: **SET RASTERTEXT**
    Raster Text :1
To set, key in: **SET RASTERTEXT on/off**
    Where on/off is ON or OFF.

✳ *Error:* Raster Text Disabled: - **This messages appears if this setting is attempted on MicroStation 32 or MicroStation PC.**
See Also: ACTIVE TEXTSTYLE

SET REFBOUND   SET REFB

Setting: Toggle Reference Boundary Display On/Off
    Toggles the display of the dashed polygons used to outline the reference file clipping boundaries. This item can be set or viewed through the View Attributes settings box or with a key-in.
Item Selection
On sidebar menu pick: n/a
For window, point to: <u>V</u>iew   Attributes   Ref Boundaries
On paper menu see: VIEW
To show, open: **View Attributes settings box**
To set, key in: **SET REFBOUND on/off**
    Where on/off is ON or OFF.
See Also: REFERENCE CLIP BOUNDARY

SET REFCLIP   SET REFC

Setting: Toggle Reference Clipping Display Fast/Slow
    MicroStation supports fast (rectangular) and slow (non-rectangular) reference file clipping. Use SET REFCLIP FAST when your reference file boundaries are rectangular. Use SET REFCLIP SLOW for non-rectangular reference file boundaries.
    SET REFCLIP FAST is the default setting.
Item Selection
On sidebar menu pick: n/a
For window, point to: <u>V</u>iew   Attributes   Fast Ref Clipping
On paper menu see: VIEW
To show, key in: **SET REFCLIP**
    Fast Reference Clipping : FAST
To set, key in: **SET REFCLIP opt**
    Where opt is either fast or slow.

✎ *Note:* Non-rectangular reference file boundaries take more time to refresh on the display screen.
See Also: REFERENCE CLIP settings

SET SHARECELL   SET SH

Setting: Toggle Sharecell On/Off
    Cells can share elements. To place a shared cell, SET SHARECELL on, then place a cell. The elements of this cell are placed only once in the drawing file. Each subsequent placement of the cell shares the first cell's components. Modify any cell in the shared group and all the cells change.
Item Selection
On sidebar menu pick: n/a
For window, point to: <u>V</u>iew   C<u>e</u>lls   Use Shared Cells
On paper menu see: CELLS
To show, key in: **SET SHARECELL**
    Shared Cells : ON
To set, key in: **SET SHARECELL On/Off**
    Where On/Off is either on or off.

🖉 *Note:* You can't edit the text in a shared cell unless you convert it to an unshared cell.

🖉 *Note:* Freeze shared cells before shipping them to IGDS or MicroStation version 3.3 or lower.

☞ *Tip:* Shared cells can be manipulated faster than non-shared cells.

☞ *Tip:* You can place additional shared cells in the drawing file without attaching a cell library.

☞ *Tip:* Shared cells take up less space in the drawing file.

See Also: PLACE CELL tools

SET SMALLTEXT   SET SM

Setting: Set Smallest Text

Determines the size of the smallest text MicroStation displays on the screen. The default pixel value is set to 4. When the text size is below this value, MicroStation displays the text as "Greeked"; instead of text, a range block and an X inside is shown on the screen.

Item Selection

On sidebar menu pick: n/a

For window, point to: User   Preferences...   Drawing   Smallest Text:

On paper menu see: n/a

To show, open: **Preferences dialog box**

To set, key in: **SET SMALLTEXT nnn**

Where nnn is a number between 0 and 100.

SET STACKFRACTIONS   SET STA

Setting: Toggle Stacked Fractions On/Off

Stacks a fraction as a single text character. MicroStation places fractions two ways. For example, with SET STACKFRACTIONS off, the characters three slash four (3/4) places as three separate characters. With SET STACKFRACTIONS on, the same three characters (3/4) place as one character. Allow an extra space before and after the fraction if you place a stacked fraction between other text characters.

Item Selection

On sidebar menu pick: Params   Set   Stackd

For window, point to: Element   Text   Fractions

On paper menu see: MODIFY TEXT

To show, key in: **SET STACKFRACTIONS**

Stacked Fractions : OFF

To set, key in: **SET STACKFRACTIONS on/off**

Where on/off is ON or OFF.

See Also: PLACE TEXT tools

SET STREAM   SET STR

Setting: Toggle Stream Acceptance Display On/Off

Helps evaluate the values assigned to ACTIVE STREAM TOLERANCE and ACTIVE STREAM DELTA settings during placement of a stream curve. As the setting samples the input points placed in the drawing file, a counter shown in the status field continually updates the number of points of the curve.

Item Selection

On sidebar menu pick: n/a

For window, point to: Settings   Digitizing   Acceptance Display

On paper menu see: n/a

To show, key in: **SET STREAM**

Stream Acceptance Display : OFF

To set, key in: **SET STREAM on/off**

Where on/off is ON or OFF.

See Also: PLACE CURVE STREAM

SET TASKSIZE
SET TA

Setting: Set Task size
>Displays the maximum amount of physical memory allotted to a program. The maximum task size is 32767.

Item Selection
On sidebar menu pick: | Params | Set | Task |

For window, point to: User Preferences... External Programs

On paper menu see: n/a

To show, key in: **SET TASKSIZE**
>Task Size (conventional) = 100k

See Also: SET TASKSIZE settings, SHOW MEM

SET TASKSIZE CONVENTIONAL
SET TA C

DOS Only

Setting: Set the Tasksize for Conventional Memory
>Programs can be executed from the MicroStation Command Window. This setting determines the program's physical memory size in kilo-bytes of conventional memory. The first mega-byte of memory storage is considered the conventional memory space. This setting can be changed or viewed with a key-in or through the Preferences dialog box.

Item Selection
On sidebar menu pick: | Utils | User | Pref |

For window, point to: User Preferences... External Progs. (Conv.)

On paper menu see: n/a

To show, open: **Preferences dialog box**

To set, key in: **SET TA C nnn**
>Where nnn is the amount of conventional memory (in Kb) allocated for external programs.

☞ *Tip:* SET TASKSIZE -1 for the maximum amount of memory.

See Also: SET TASKSIZE settings

SET TASKSIZE EXTENDED
SET TA E

DOS Only

Setting: Set the Tasksize for Extended Memory
>Programs can be executed from the MicroStation Command Window. This item determines the program's physical memory size in kilo-bytes of extended memory. The first mega-byte of PC memory storage is considered the conventional memory space. Extended memory is memory above the first mega-byte. Memory managers, EMS simulators and protected mode applications such as MicroStation use this extended memory location. This setting can be changed or viewed with a key-in or through the Preferences dialog box.

Item Selection
On sidebar menu pick: | Utils | User | Pref |

For window, point to: User Preferences... External Progs. (Conv.)

On paper menu see: n/a

To show, open: **Preferences dialog box**

To set, key in: **SET TA E nnn**
>Where nnn is the amount of extended memory (in Kb) allocated to external programs.

See Also: SET TASKSIZE settings

SET TEXT

Obsolete
>Toggles text display. Each view's text display can be independently set by placing an input point in each view.

See Also: PLACE TEXT tools

SET TPMODE ACSDELTA

SET TP

Setting: Set TPMode ACSDelta
Displays the distance between a tentative point and the last input point or tentative point of the current Auxiliary Coordinate System display.

Item Selection
On sidebar menu pick: n/a
For window, point to: n/a
On paper menu see: n/a
To show, key in: SET TPMODE
Tentative Point Mode - LOCATE
To set, key in: SET TPMODE ACSDELTA
See Also: SET TPMODE settings, SET FONT

SET TPMODE ACSLOCATE

SET TP

Setting: Set TPMode ACSLocate
Displays the distance between the tentative points of the current Auxiliary Coordinate System.

Item Selection
On sidebar menu pick: n/a
For window, point to: n/a
On paper menu see: n/a
To show, key in: SET TPMODE
Tentative Point Mode - LOCATE
To set, key in: SET TPMODE ACSLOCATE
See Also: SET TPMODE settings

SET TPMODE ANGLE2

Same as SET TPMODE DISTANCE

SET TPMODE ANGLE3

SET TP

Setting: Set TPMode Angle3
Displays the distance between each tentative point and the last input point. The angle displayed is based on an imaginary line through the last two input points and the tentative point. The vertex of the angle uses the last input point placed.

Item Selection
On sidebar menu pick: Params Set TPMode 3 Pt
For window, point to: n/a
On paper menu see: n/a
To show, key in: SET TPMODE
Tentative Point Mode - LOCATE
To set, key in: SET TPMODE ANGLE3
See Also: SET TPMODE settings

SET TPMODE DELTA

SET TP

Setting: Set TPMode Delta
Displays the delta-X and delta-Y distance between the tentative point and the last input point display. Rotating the view does not affect how the SET TPMODE DELTA setting measures the distance.

Item Selection
On sidebar menu pick: Params Set TPMode DELTA
For window, point to: n/a
On paper menu see: n/a
To show, key in: SET TPMODE
Tentative Point Mode - LOCATE
To set, key in: SET TPMODE DE
See Also: SET TPMODE settings

SET TPMODE DISTANCE   SET TP

Setting: Set TPMode Distance
   Displays the distance between each tentative point and the last input or
tentative point. The angle displayed is based on the tentative point and the X-axis.
The vertex of the angle uses the last input point placed.
Item Selection
   On sidebar menu pick: | Params | | Set | | TPMode | | 2 Pt |
   For window, point to: n/a
   On paper menu see: n/a
   To show, key in: **SET TPMODE**
      Tentative Point Mode - LOCATE
   To set, key in: **SET TPMODE DISTANCE**
See Also: SET TPMODE settings

SET TPMODE LOCATE   SET TP

Setting: Set TPMode Locate
   Displays a tentative point's X, Y and Z locations as it is placed.
Item Selection
   On sidebar menu pick: | Params | | Set | | TPMode | | XYZ |
   For window, point to: n/a
   On paper menu see: n/a
   To show, key in: **SET TPMODE**
      Tentative Point Mode - LOCATE
   To set, key in: **SET TPMODE L**
See Also: SET TPMODE settings

SET TPMODE VDELTA   SET TP

Setting: Set TPMode Vdelta
   Measures the delta-X and delta-Y distances between the tentative point and
the last input point. However, rotating the view changes the way the distance is
measured.
Item Selection
   On sidebar menu pick: | Params | | Set | | TPMode | | VDELTA |
   For window, point to: n/a
   On paper menu see: n/a
   To show, key in: **SET TPMODE**
      Tentative Point Mode - LOCATE
   To set, key in: **SET TPMODE V**
See Also: SET TPMODE settings

SET TUTVIEW

Obsolete
   Defines the display view for all screen tutorials. View five is the default
tutorial view. To change the tutorial view, choose this setting and place an input
point in some other view. Only one view can be set as the TUTVIEW. MicroStation
stores its tutorials in \USTATION\DATA\MSTUTLIB.CEL.
See Also: TUTORIAL

SET UNDO   SET U

Setting: Toggle Undo Buffer On/Off
   MicroStation's digital eraser. With SET UNDO on, operations can be UN-
DOne; i.e., if you delete elements by accident, you can recover them by UNDOing
the delete operation. With the setting off, operations cannot be UNDOne.
Item Selection
   On sidebar menu pick: n/a
   For window, point to: n/a
   On paper menu see: n/a

To show, key in: **SET UNDO**
```
Undo : ON (65535)
   ⇨    Undo area in bytes.
```
To set, key in: **SET UNDO on/off**
Where on/off is ON or OFF.

See Also: MARK, REDO, UNDO

SET VIEW   SET V

Setting: Set Shading View Mode
Changes the shading of a rendered view. For example, after you render a view for hidden line removal, you could key in SET VIEW SMOOTH and update the view. The objects in the view would appear smooth without rerunning the rendering process.

Choices:
   CONSTANT - displays constant shading
   FILLED - displays filled polygons
   HIDDEN - displays polygons
   PHONG - displays Phong shading
   SECTION - displays cross-section
   SMOOTH - displays Gouraud shading
   WIREFRAME - displays wireframes
   WIREMESH - displays wiremesh

Item Selection
On sidebar menu pick: | Params | | View | | Mode |
For window, point to: Yiew   Attributes   Display Mode:
On paper menu see: VIEW
Command Window Prompts:
   1. Select the item
      Select view
   2. Pick a view
To exit item: Click a Reset

✍ *Note:* A screen update is required after picking the SET VIEW setting.
See Also: RENDER commands

SET WEIGHT   SET W

Setting: Toggle Line Weight Display On/Off
Controls the weight or thickness of elements. Each view's weight display can be set independently. This setting requires one input point. Keying in SET WEIGHT toggles the display between on and off.

Item Selection
On sidebar menu pick: | Params | | Disp | | Weight |
For window, point to: Yiew   Attributes   Line Weights
On paper menu see: DISPLAY
To show, key in: SET WEIGHT
```
Line weights : 1,2,3,4,5,6,7,8
```
To set, key in: SET WEIGHT on/off
Where on/off is ON or OFF.
Command Window Prompts:
   1. Select the item
      Select view
   2. Pick the view
      ⇨   If the view number does not appear, that view's weight display is off.
To exit item: Click a Reset
See Also: ACTIVE WEIGHT

SET XOR

Setting: Set X Pointer Color

Controls the color of temporary graphics on the display screen. Examples of some of the these temporary graphics are the screen cursor, the locate tolerance circle around the cursor, and the box drawn by the WINDOW AREA control. This setting may be saved with the FILEDESIGN tool.

Item Selection

For window, point to: n/a

On sidebar menu pick: n/a

On paper menu see: n/a

To show, key in: SET XOR

   Xor color = 0

To set, key in: SET XOR nnn

   Where nnn is a color number.

Example(s):

SET XOR 3   turns the cursor red – provided your color table has color 3 set to red.

See Also: SET HILITE

SF=

Alternate Key-in

Alternate key-in for FENCE SEPARATE. Use it to move the contents of a fence into an output file, and delete the contents of the fence from the input file. For example, SF=OUTPUT.DGN.

See Also: FENCE SEPARATE

Shadelem

Obsolete

The word SHADELEM cannot be typed in to execute the shading process. This 3D shading user command shades a selected element.

See Also: Render tools

Shades

Obsolete

The word SHADES cannot be typed in to execute the shading process. This 3D shading user command shades a selected view.

See Also: Render tools

SHOW ACS

Obsolete

Opens the Auxiliary Coordinate Systems setting box. In this box, you can define, attach, delete or review the saved coordinate systems.

See Also: ATTACH ACS, DEFINE ACS Settings, DELETE ACS, SAVE ACS

SHOW AE

Setting: Show Active Entity

Displays the current active entity. Use ACTIVE ENTITY, DEFINE AE or FIND to define active entities.

Item Selection

On sidebar menu pick: dBase   ActEnt   Show

For window, point to: Palettes Database

On paper menu see: DATABASE
To show, key in: SHOW AE
⇨ Database window appears.

✳ Error: No active entity defined - If this message appears, define an active entity first.
See Also: ACTIVE ENTITY, DEFINE AE, FIND

SHOW CAMERA LENS
SH C L

3D Setting: Show Camera Lens
Displays the camera lens information in the MicroStation Command Window.
Item Selection
On sidebar menu pick: Params Camera ShoCam Lens
For window, point to: n/a
On paper menu see: CAMERA
To show, key in: SHOW CAMERA LENS
Camera Lens: 50 MM, Field of Vision: 1 degree
✍ Note: This menu item is dimmed if the file is not 3D.
See Also: SET CAMERA LENS, SHOW CAMERA settings

SHOW CAMERA POSITION
SH C P

3D Setting: Show Camera Position
Displays the position of a view camera. This setting requires an input point to select the view(s).
Item Selection
On sidebar menu pick: Params Camera ShoCam Pos
For window, point to: n/a
On paper menu see: CAMERA
To show, key in: SHOW CAMERA POSITION
Camera Position: 2:1,2:0,0:0
✍ Note: This menu item is dimmed if the file is not 3D.
See Also: SET CAMERA POSITION

SHOW CAMERA TARGET
SH C T

3D Setting: Show Camera Target
Displays the target of a view camera. This setting requires an input point in each view to display the target information.
Item Selection
On sidebar menu pick: Params Camera ShoCam Target
For window, point to: n/a
On paper menu see: CAMERA
To show, key in: SHOW CAMERA TARGET
Camera Target: 4:8/12, 4:6 1/4, -5:4 7/8
✍ Note: This menu item is dimmed if the file is not 3D.
See Also: SET CAMERA TARGET, SHOW CAMERA settings

SHOW CLIPBOARD
SH CL

Mac Only
Setting: Show Clipboard
  Brings up the clipboard view on MicroStation MAC. The clipboard is an area where different applications can share data between themselves.

Item Selection
  On sidebar menu pick: n/a
  For window, point to: Edit
  On paper menu see: n/a
See Also: CLIPBOARD commands

SHOW COLORS

Obsolete
  Shows the relationship between the color shades and the display color numbers. Any color shade can be assigned to a display color number. MicroStation has the ability of displaying 256 colors. Most graphics boards display fewer than 256 colors.
See Also: SET COLOR

SHOW DEPTH ACTIVE
SH D A

3D Setting: Show Active Depth
  Displays the Active Depth of a view on selection of an input point.
  When you place input points in a 3D drawing, you place them along a plane perpendicular to the Z-axis of any view – the active depth. Each view can have a different active depth.

Item Selection
  On sidebar menu pick: Place   3-D
  For window, point to: Palettes 3D

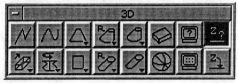

  On paper menu see: 3D
  To show, key in: SHOW DEPTH ACTIVE
    View 4: Active Depth=-1:1 1/8
See Also: DEPTH ACTIVE

SHOW DEPTH DISPLAY
SH D D

3D Setting: Show Display Depth
  Displays a view's defined depth on entry of an input point. Each view may have a different display depth. The maximum display depth of a 3D drawing is from the front to the back of the design cube.

Item Selection
  On sidebar menu pick: Place   3-D
  For window, point to: Palettes 3D

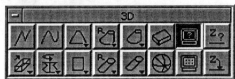

  On paper menu see: 3D
  To show, key in: SHOW DEPTH DISPLAY

```
View 1: Display Depth=-100:1, 100:0
```
✍ *Note:* See the SET DDEPTH settings to change the display depth of a view.
See Also: SET DDEPTH settings, SHOW ACTIVE DEPTH

SHOW EOF   SH E

Setting: Show End of File
Displays the End Of File mark of the drawing file.
Item Selection
On sidebar menu pick: n/a
For window, point to: n/a
On paper menu see: n/a
To show, key in: SHOW EOF
EOF = 54 68(006c44)

SHOW FONT

Obsolete
Displays the Fonts settings box. Fonts are the lettering styles you use to place text in the drawing file.
See Also: ACTIVE FONT, PLACE TEXT tools

SHOW HEADER

Obsolete
Opens the View Levels settings box. The box displays the on/off status of the 63 drawing levels. Pick the view number to show the level status of a particular view.
See Also: SET LEVELS

SHOW LIBRARY

Obsolete
Opens the Cells settings box. One window lists the names and descriptions of the cells. Another, smaller, window allows you to edit, delete, create or set a cell to Shared Cell status. You can also set the names of the active cell, line terminator, pattern and point cell.
See Also: PLACE CELL tools

SHOW MEM

Obsolete
This setting shows the amount of physical memory available for use by other programs.
See Also: SET TASKSIZE

SHOW PATTERN   SH PA

Setting: Show Pattern Attributes
Identifies a pattern element and displays its characteristics. This item requires two input points. Once you place the first point, the pattern's level and element type display. The second point confirms your element choice; the pattern's angle and scale also appear.
Item Selection
On sidebar menu pick: n/a

For window, point to: Palettes   Patterning

On paper menu see: PATTERNING
See Also: ACTIVE PATTERN settings, SET PATTERN

SHOW PLOTTER

Obsolete
    Opens Show Plotter dialog box. The current plotter configuration file displays the plotter type, number of pens and paper size.
See Also: PLOT @, SET PLOTTER

SHOW REFERENCE   SH R

Setting: Show Reference
    Opens the Reference Files settings box. This box lists the logical name and description of each reference file attached to the drawing file.
Item Selection
  On sidebar menu pick: Refrnc   Show
  For window, point to: File   Reference
  On paper menu see: n/a
Command Window Prompts:
  1. Select the item
    Reference Files settings box appears
  To exit item: Close the window
See Also: REFERENCE ATTACH, REFERENCE DETACH

SHOW STACK   SH S

Mac Only

Command: Show Stack
    Mac only – opens a standard Open File box where a hypercard stack can be selected from the help window.
Item Selection
  On sidebar menu pick: n/a
  For window, point to: n/a
  On paper menu see: n/a
  To exit item: Make next selection
See Also: HELP commands

SHOW UORS   SH U

Setting: Show UORs
    Shows the coordinates, in units of resolution (UORs), of an input point.
Item Selection
  On sidebar menu pick: n/a
  For window, point to: n/a
  On paper menu see: n/a
  To show, key in: SHOW UORS
Command Window Prompts:
  1. Select the item
    Enter point or RESET to complete
  2. Place an input point on the screen
    UORs = 1440000,3120000
  To exit item: Click a Reset
See Also: SET COORDINATES

SHOW VIEWS

Obsolete
> Lists the name and description of the Saved Views attached to the current drawing file. You can also attach or detach views through this setting box.

See Also: DELETE VIEW, SAVE VIEW

SPIN COPY   SP C

Drawing Tool: Spin Element (Copy)
> Rotates and copies an element by using an input point to define the swing angle. This tool requires three input points. The first point identifies the element. The second point sets the pivot point. The third point defines the angle of spin.
> Cells and graphic groups can be rotated with this tool.

Item Selection

| On sidebar menu pick: | Manip | Spin | Copy |
| For window, point to: | Palettes | Main | Copy Element |

On paper menu see: COPY ELEMENT

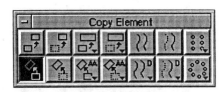

☞ *Tip:* Place the third input point with a precision key-in.
   See Also: FENCE ROTATE COPY, ROTATE COPY

SPIN ORIGINAL   SP

Drawing Tool: Spin Element (Original)
> Rotates an element by using an input point to define the swing angle. This tool requires three input points. The first point identifies the element. The second point sets the pivot point. The third point defines the angle of spin.
> Cells and graphic groups can be rotated with this tool.

Item Selection

| On sidebar menu pick: | Manip | Spin | Orig |
| For window, point to: | Palettes | Main | Copy Element |

On paper menu see: COPY ELEMENT

☞ *Tip:* Place the third input point with a precision key-in.
   See Also: FENCE ROTATE ORIGINAL, ROTATE ORIGINAL

ST=

Alternate Key-in
> Alternate key-in for ACTIVE STREAM TOLERANCE. Use it to set the distance between input points placed in stream mode.

See Also: ACTIVE STREAM TOLERANCE

START ST

Unix Only

Command: Start MicroCSL application

Unix only – Activates a MicroCSL application from the command window. Key in START *application* to start up the application program.

Item Selection

On sidebar menu pick: n/a
For window, point to: n/a
On paper menu see: n/a

STOP DRAWING

Obsolete

The display of a drawing file can be stopped before it finishes by using the STOP DRAWING tool.

SURFACE PROJECTION SUR

3D Drawing Tool: Construct Surface/Solid of Projection

Creates a surface by projecting 3D elements and complex shapes. This tool requires two input points. The first point identifies the element to project. The second point defines the distance to project the element or shape.

Item Selection

On sidebar menu pick: | Place | 3-D |
For window, point to: Palettes 3D

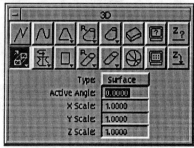

On paper menu see: 3D

Command Window Prompts:

1. Select the item
   Identify element
2. Pick element to project
   Define projection distance
3. Place input point
   ⇨ This may be a precision key-in.
   ⇨ The element is projected.

To exit item: **Make next selection**

☞ *Tip:* If the elements are skewed when projected, try snapping to the element when identifying it. Then use a precision key-in for the projection distance.

See Also: FENCE SURFACE tools, PLACE SLAB, SURFACE REVOLUTION

SURFACE REVOLUTION SUR R

3D Drawing Tool: Construct Surface/Solid of Revolution

Creates a surface of revolution by rotating a 3D element or complex shape around a pivot point. This tool requires a rotation angle key-in and two input points. The first point identifies the element or complex shape to rotate. The second point defines the axis of rotation.

Item Selection

On sidebar menu pick: | Place | 3-D |

For window, point to: Palettes 3D

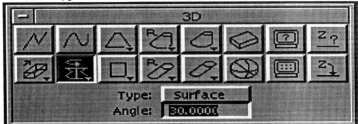

On paper menu see: 3D

Command Window Prompts:
1. Select the item
   Enter the revolution angle.
2. Key in the rotation
   Identify element
   ⇨ Snap to the element.
   Define axis of revolution
3. Place another input point
   ⇨ This may be a precision key-in.
   ⇨ The element rotates.

To exit item: **Make next selection**

☞ *Tip:* If the elements are skewed when rotated, try snapping to the element when identifying it. Then use a precision key-in for the projection distance.

☞ *Tip:* If you want to revolve the element again and the angle and axis remain the same, use the DX=0,0,0 key-in.

See Also: FENCE SURFACE tools, PLACE SPHERE, SURFACE PROJECTION

SV=

Alternate Key-in
Alternate key-in for SAVE VIEW. Use it to save a view for recall.
See Also: SAVE VIEW

SX=

Alternate Key-in
Alternate key-in for SAVE ACS. Use it to save the current Auxiliary Coordinate System for referral.
See Also: SAVE ACS

SWAP SCREEN   SW

DOS Only
View Control: Swap Screen
Toggles between two display screens on a single-screen system. Use the primary screen for close work, such as detailing a part. The secondary screen provides an overview screen to see the entire drawing sheet.

Item Selection
On sidebar menu pick: Window Swap
For window, point to: View Swap
On paper menu see: VIEW

☞ *Tip:* Watch the "(1) uSTN" prompt to tell whether you're on the primary screen or the secondary screen. The 1 clicks to a 2 when the secondary screen displays.

See Also: COPY VIEW, SET OVERVIEW, SWAP VIEWS

SWAP VIEWS

Obsolete
>   SWAP VIEWS swaps the view(s) on the primary screen with the view(s) on
> the secondary screen. Display characteristics such as SET GRID on, SET CURVES
> and SET FONT transfer between the views when selecting SWAP VIEWS.
> *See Also:* SWAP SCREEN

TB=

Alternate Key-in
>   Alternate key-in for ACTIVE TAB. Use it when merging text with the
> INCLUDE tool to include text tabs in the drawing file.
> *See Also:* ACTIVE TAB

TH=

Alternate Key-in
>   Alternate key-in for ACTIVE TXHEIGHT. Use it to set the height of text
> placed in the drawing.
> *See Also:* ACTIVE TXHEIGHT

THAW   TH

Drawing Tool: Thaw Element
>   Thaws previously frozen multi-line elements, shared cells and dimension
> data that were frozen. Once thawed, these elements can be manipulated. This item
> requires an input point to identify the frozen element and a second input point to
> complete the thawing process.

Item Selection
On sidebar menu pick: n/a
For window, point to: n/a
On paper menu see: LINEAR DIMENSION
Command Window Prompts:
  1. Select the item
     `Identify element`
  2. Pick an element
     `Accept/Reject (select next input)`
  3. Give an accept point
To exit item: Make next selection
See Also: FENCE FREEZE, FENCE THAW, FREEZE

TI=

Alternate Key-in
>   Alternate key-in for ACTIVE TAG. Use it while placing text to increment
> numerics such as lot numbers or tag numbers.
> *See Also:* ACTIVE TAG

TILE ALL

Obsolete
>   Mac only – tiles the display screen with windows and palettes. Normally,
> windows and tool palettes might overlay each other. With the TILE view control,
> MicroStation rectifies the size of each window and palette until they all fit on the
> screen so that no window overlaps the other.
> *See Also:* WINDOW TILE

TILE VIEWS

Obsolete
>   Mac only – tiles the display screen with windows and palettes. Normally,
> windows and tool palettes might overlay each other. With the TILE VIEWS
> control, MicroStation rectifies the size of each window and palette until they all

fit on the screen so that no window overlaps the other. The TILE VIEWS option
tiles just the views and not the palettes.
See Also: WINDOW TILE

Tmatrx

Tmatrx stands for transformation matrix. Element are stored in the drawing
file by XYZ coordinates. In order to display or transform the elements so they may
be seen in different view perspectives (top, right, front, etc), they need to run
through a transformation matrix to orient the element correctly. MicroStation,
and application programs typically take advantage of orientating and scaling
elements through tmatrx operations. A transformation matrix is built with a user
command.

TRANSFORM   TRA

Command: Transform Element by Transformation Matrix
Transforms an element by the values stored in the transformation matrix.
This command requires an input point to identify the target element for change
and an accept point to complete the transformation process.

Item Selection
On sidebar menu pick: n/a
For window, point to: n/a
On paper menu see: n/a
Command Window Prompts:
1. Select the item
   Identify element
2. Pick an element
   Accept/Reject (select next input)
3. Give an accept point
To exit item: Make next selection
See Also: FENCE TRANSFORM, Tmatrix

TRESET   TRE

DOS Only
Command: Treset
Resets or initializes a digitizing tablet to the tablet's default settings. Digi-
tizing tablets allow you to attach paper tool menus or to trace existing drawings.

Item Selection
On sidebar menu pick: n/a
For window, point to: n/a
On paper menu see: n/a
See Also: DIGITIZER tools and settings

TS=

Alternate Key-in
Alternate key-in for ACTIVE TSCALE. Use it to set the scale factor for
placing an ACTIVE TERMINATOR.
See Also: ACTIVE TSCALE

TUTORIAL   AT=

Command: Tutorial
Tutorials are old-style menus that prompted for input. Design Options was
the most common group of tutorials. Call a tutorial into view by using the
TUTORIAL tool followed by a tutorial name.

Item Selection
On sidebar menu pick: n/a
For window, point to: n/a
On paper menu see: n/a
To show, key in: AT=uuu

Where uuu is a tutorial name.

To exit item: **Follow prompts in tutorial**

Example(s):

AT=DESOPT	brings up the Bentley Systems version of the Design Options tutorial.
AT=GRAOPT	brings up the Intergraph version of the Design Options tutorial.

✍ *Note:* Keying in AT= erases the tutorial from the display screen.

See Also: DESIGN, DIMENSION TUTORIAL, SET TUTVIEW

TV=

Alternate Key-in

Alternate key-in for DIMENSION TOLERANCE SCALE. Use it to set tolerance values that appear with dimension text.

See Also: DIMENSION TOLERANCE SCALE.

TW=

Alternate Key-in

Alternate key-in for ACTIVE TXWIDTH. Use it to set the width of text placed in the drawing.

See Also: ACTIVE TXWIDTH

TX=

Alternate Key-in

Alternate key-in for ACTIVE TXSIZE. Use it to set the size (height and width) of text placed in the drawing.

See Also: ACTIVE TXSIZE

TYPE   DR=

Command: Type

Opens an ASCII text file window while remaining in the graphics session.

Item Selection

On sidebar menu pick: n/a

For window, point to: File   Display   Text

On paper menu see: n/a

Command Window Prompts:

1. TYPE filename

Where filename is a text character file.

The Type window appears

To exit item: **Close the window**

See Also: %, !, DOS

UC=

Alternate Key-in

Alternate key-in for USERCOMMAND. Use it to starts up a user command.

See Also: USERCOMMAND

UCC   UCC filename

Command: User Command Compiler

Compiles user commands into a format that runs quicker in the computer.

Item Selection

On sidebar menu pick: n/a

For window, point to: n/a

On paper menu see: n/a

Where filename is a user command input file.

Command Window Prompts:
   To exit item: **Exits by itself**

Example(s):
   UCC CGSI.UCM
      Creating <C:\USTATION\UCM\CGSI.UCC>
      Compiling <C:\USTATION\UCM\CGSI.UCM>

✍ *Note:* The default directory for storing user commands is \USTATION\UCM.
See Also: UCI, USERCOMMAND

UCI   UC=filename

Command: UCI
      Opens the Run User Command dialog box. After the dialog box appears,
      either pick the UCM name followed by OK or pick Cancel to abort running a user
      command.

Item Selection
   On sidebar menu pick: n/a
   For window, point to: n/a
   On paper menu see: n/a
      Where filename is a user command input file.

See Also: USERCOMMAND

Uncut Component Lines

Drawing Tool: Uncut Component Lines
      Restores multi-line elements which were previously broken apart. This MDL
      application tool requires one input point to identify the multi-line element which
      needs restoration.

Item Selection
   On sidebar menu pick: n/a
   For window, point to: Palettes   Multi-line Joints

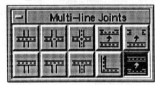

   On paper menu see: n/a

Command Window Prompts:
   1. Select the item
      Identify element
   2. Pick a multi-line element
         ⇨   the prompts will not change
         ⇨   continue to pick other multi-line elements

   To exit item: **Make next selection**

✍ *Note:* This MDL application is loaded automatically when you start the design
session.
See Also: Cut All Component Lines, Cut Single Component Line, PLACE MLINE

UNDO   Alt-Backspace opt

Drawing Tool: Undo
      MicroStation's digital eraser. UNDO could be renamed OOPS. Any element
      placement or manipulation done since the design session started can be UNDOne.
      You cannot UNDO an operation from a past design session.
      The UNDO tool can UNDO the last operation, several operations or every
      operation since the design session started. You can also put a MARK in the
      drawing file, continue drawing and then UNDO every operation back to that
      MARK.

Item Selection
    On sidebar menu pick: `UnDo`
    For window, point to: Edit   Undo
    On paper menu see: TOP BORDER
        Where opt is either last, steps:, marked, or all. The word steps: is in lower case letters because you don't type in steps:, you type in the number of steps you want undone. See the steps: example shown.
Command Window Prompts:
    1. Select the item
      . <Last operation> undone
    To exit item: **Make next selection**
Example(s):

UNDO	undoes the last command operation
UNDO steps:	undoes the last so many operations. (Example: UNDO 5, will undo the last five operations)
UNDO MARKED	undo the operations up to the file MARK
UNDO ALL	undo all operations since staring this design session
UNDO ALL NOCONFIRM	undo all operations since staring this design session without warning message

See Also: MARK, REDO, SET UNDO

UNGROUP   Ctrl+U

Command: Ungroup
        Discontinues elements grouping within selection sets. Once you drop this relationship, the elements can be manipulated as single elements.
Item Selection
    On sidebar menu pick: n/a
    For window, point to: Edit   Ungroup
    On paper menu see: TOP BORDER
See Also: GROUP SELECTION

UNIX   !

Unix Only
Command: Push to Unix
        Suspends the design session and places you in the UNIX Operating System Environment. This item is handy if you need to jump out of the design session to run a subprocess. For example, to check on a filename, key in UNIX and do an ls for the file in question. Then key in EXIT to terminate the UNIX session and return to MicroStation. You can also jump to UNIX by keying in the exclamation character (!).
Item Selection
    On sidebar menu pick: `Utils`   `Unix`
    For window, point to: n/a
    On paper menu see: UTILITIES
Command Window Prompts:
    1. Select the item
      [uSTN] /usr/leavy - >
        ⇨   You may now execute your operating system command.
    To Return to graphics: **Key in EXIT**
See Also: !, %

UPDATE   UP opt

View Control: Update View
        Refreshes a section of the display screen or the entire screen. The UPDATE view control has a variety of UPDATE options.
Item Selection
    On sidebar menu pick: `Update`
    For window, point to: View   Update

On paper menu see: VIEW
  Where opt is one of the valid UPDATE options.
Command Window Prompts:
1. Select the item
   Select view
2. Pick a view
   Update in progress
   Display complete
   ⇨ Others view can be selected by placing additional points in those views.
To exit item: Make next selection
Example(s):
  UPDATE
    ⇨ Asks you to pick the view to update.
  UPDATE [viewnumber]
    ⇨ Updates a specified view number (1,2,3,etc).
  UPDATE [screen] (DOS and UNIX only)
    ⇨ Updates all views on the RIGHT or LEFT screen.
  UPDATE BOTH (DOS and UNIX only)
    ⇨ Updates both the primary and secondary screens.
  UPDATE VIEW
    ⇨ Updates a view based on a "Select view" data point.
  UPDATE GRID
    ⇨ Updates the display grid for a view.
  UPDATE FENCE [choice]
    ⇨ Updates a fence area (INSIDE or OUTSIDE).
  UPDATE FILE
    ⇨ Updates the active drawing file.
  UPDATE TUTORIAL
    ⇨ Updates the active tutorial.
  UPDATE REFERENCE
    ⇨ Updates attached reference files.
  UPDATE ALL
    ⇨ Updates the whole shootin' match.
See Also: FIT view controls, IUPDATE, VIEW view controls, WINDOW view controls

|UPDATE AE |UPDATE AE

Command: Edit Active Entity
  Edits an existing active entity in a database. Key in |UPDATE AE followed
  by the SET information.
Item Selection
  On sidebar menu pick: n/a
  For window, point to: n/a
  On paper menu see: n/a
See Also: DEFINE AE, SHOW AE

UR=

Alternate Key-in
  Alternate key-in for ACTIVE UNITROUND. Each input point rounds off to
  the nearest multiple of the unit round-off when turning on the LOCK UNIT
  setting.
See Also: ACTIVE UNITROUND

USERCOMMAND   UC=filename

Command: User Command
  Opens the Run User Command dialog box. After the dialog box appears,
  either pick the UCM name followed by OK or pick Cancel to abort running a user
  command.
Item Selection
  On sidebar menu pick: Utils   User   UCM
  For window, point to: User   User Command...

On paper menu see: RIGHT BORDER
Where filename is a user defined file.
Command Window Prompts:
1. Select the item
The Run User Command dialog box appears
To exit item: Close the window
Example(s):
UC=LEAVY.UCM

✍ *Note:* MicroStation stores it's user commands in \USTATION\UCM.

✳ *Error:* File :\USTATION\UCM\PIPE.UCM not found - If this message appears, MicroStation cannot find the user command file.
See Also: ACTIVE INDEX, FENCE LOCATE

VERSION   VE

Command: MicroStation Software Version
Displays the current release number and date of the MicroStation software.
Item Selection
On sidebar menu pick: n/a
For window, point to: Help On Version
On paper menu see: n/a
To show, key in: VERSION
Version 4.0.n PRELIM November, 1990
To exit item: Exits by itself

VI=

Alternate Key-in
Alternate key-in for VIEW. Use it to recall a named view or to declare a view perspective in a 3D file.
See Also: VIEW

VIEW   VI

View Control: View
Performs two chores.
It recalls named views previously saved. For instance, if you had saved a view named TITLE with SAVE VIEW, you could key in VIEW TITLE to recall the saved view.
In 3D, the VIEW <orientation> view control also declares the common 3D view orientations such as: TOP, BOTTOM, RIGHT, LEFT, FRONT, BACK and ISO.
Item Selection
On sidebar menu pick: Window View VI:
For window, point to: View Saved
On paper menu see: n/a
To set, key in: VIEW uuu
Where uuu is a user defined view name.
Command Window Prompts:
1. Select the item
Saved Views setting box appears
To exit item: Close the window
See Also: SAVE VIEW

VIEW CLEAR   VI C

View Control: Clear View
Clears a view. VIEW CLEAR is intended for rolling demos. This item expects an input in the view(s) you want wiped clean.
Item Selection
On sidebar menu pick: n/a

For window, point to: n/a
On paper menu see: VIEW
Command Window Prompts:
1. Select the item
   Select view
2. Pick a view
   ⇨   The view clears, but the prompt does not change.
3. Give an accept point
To exit item: Exits by itself
See Also: SAVE IMAGE, VIEW IMAGE

VIEW IMAGE   VI I

Command: Recall Saved Image
Recalls a saved image of a shaded view. This image can be stored in either Intergraph RGB, PICT, Targa or TIFF. The image can be saved as mono on stereo (3D appearance). Keying in VIEW IMAGE followed by a filename displays the saved image.

Item Selection
On sidebar menu pick: n/a
For window, point to: File   Display   Images...
On paper menu see: n/a
Command Window Prompts:
1. Select the item
   Display Image File dialog box appears
To exit item: Click Save or Cancel
✍ *Note:* To save an image click on File, Save Image As... to open the Save Image dialog box.
See Also: SAVE IMAGE, VIEW CLEAR

VIEW OFF   VIEW OF

View Control: Set View(s) Off
Turns off individual screen views. Select the control and click each view number you want turned off. Each display screen divides into four views or windows. Each window acts as an independent camera, viewing the action in the drawing file. Certain views may have close up perspectives of an object while others stand back, observing the entire drawing sheet.
One view must remain active on each display screen. Keying in VIEW TOGGLE (number) toggles the view on or off. For example, if View 5 were on, VIEW TOGGLE 5 would toggle view five off.

Item Selection
On sidebar menu pick: |Window| |View| |Off|
For window, point to: View   Open/Close
On paper menu see: VIEW
See Also: VIEW ON

VIEW ON   VIEW ON

View Control: Set View(s) On
Turns on individual screen views. Select the control and click each view number you want turned on. Each display screen divides into four views or windows. Each window acts as an independent camera, viewing the action in the drawing file. Certain views may have close up perspectives of an object while others stand back, observing the entire drawing sheet.
One view must remain active on each display screen. Keying in VIEW TOGGLE (number) toggles the view on or off. For example, if View 5 were off, VIEW TOGGLE 5 would toggle view five on.

Item Selection
On sidebar menu pick: |Window| |View| |On|
For window, point to: View   Open/Close
On paper menu see: VIEW
To set, key in: VIEW ON opt

Where opt is 1 thru 8.
See Also: VIEW OFF

WINDOW AREA   WI

View Control: Window Area
"Windows in" on a specific area of the drawing file. This control requires three input points. The first two points place a rectangular shape around the area you want to see. Place the third point in the view where you want to display the window area.

Item Selection
On sidebar menu pick: `Window` `Area`
For window, point to: Ⴟiew   Ⴟindow   Ⴟrea
On paper menu see: VIEW
Command Window Prompts:
1. Select the item
   Select window origin
2. Pick the lower left corner
   Select window corner
3. Select the upper right corner
   Select view
4. Select view to display new window
To exit item: Click a Reset

☞ *Tip:* Use the secondary screen to pick the window area points and the primary screen to choose a view to display the new area.
See Also: FIT view controls, WINDOW view controls, ZOOM view controls

WINDOW BACK

View Control: Lower Window
Lower a view, tool palette, settings box or the MicroStation Command Window one level. Lower a window by striking Alt+F3 (or key in WINDOW BACK) and clicking in the target window. Performs the same action as clicking on the window menu button in the upper left corner of the window and selecting Lower.

✍ *Note:* You cannot lower a tool palette below a view.
See Also: WINDOW view controls

WINDOW BOTTOMTOTOP

View Control: Bottom View to Top
Raise a view from the bottom of the pile. Key in WINDOW BOTTOMTOTOP and click in the target view. Performs the same action as clicking on the window title bar at the top of the window.
See Also: WINDOW view controls

WINDOW CASCADE   WI CA

View Control: Window Cascade
Adjusts each view to the full size of the display screen. The control then shuffles the view in numerical order behind each other with View 1 being first in line.

Item Selection
On sidebar menu pick: `Window` `Cascad`
For window, point to: Ⴟiew   Cascade
On paper menu see: n/a
Command Window Prompts:
1. Select the item
   ⇨ the views cascade
To exit item: Exits by itself
See Also: WINDOW view controls

WINDOW CENTER

View Control: Window Center

Selects a different viewing area or recenters an existing viewing area. Selecting the new viewing area requires two input points. The first point declares the center of the window. The second point tells MicroStation in which view to display the new window center.

Item Selection

On sidebar menu pick: `Window` `Center`

For window, point to: <u>V</u>iew <u>W</u>indow <u>C</u>enter

On paper menu see: VIEW

Command Window Prompts:

1. Select the item
   Select new window center
2. Pick the new center of the view
   Select view
3. Select view to display new window

To exit item: `Click a Reset`

☞ *Tip:* Use the secondary screen to pick the window center point and the primary screen to choose a view to display the new area.

See Also: FIT view controls, WINDOW view controls, ZOOM controls

WINDOW CLOSE

View Control: Close Window

Closes a dialog box, settings box or view. Close a window by striking Alt+F4 (or key in WINDOW CLOSE) and clicking in target window. Performs the same action as clicking on the window menu button in the upper left corner of the window and selecting Close.

✍ *Note:* The Close selection is dimmed on the MicroStation Command Window.

See Also: WINDOW view controls

WINDOW FRONT

View Control: Raise Window

Raises a dialog box, settings box or the MicroStation Command Window to the top. Raise a window by keying in WINDOW FRONT and clicking in the target window. Performs the same action as clicking on the window title bar in the top of the window.

See Also: WINDOW view controls

WINDOW MAXIMIZE

View Control: Maximize Window

Expand a view to its maximum size. Strike Alt+F10 (or key in WINDOW MAXIMIZE) and place an input point in the view you want expanded. The control performs the same action as clicking on the large window control button in the upper right corner of the window and selecting Maximize.

✍ *Note:* The Maximize selection is dimmed on the MicroStation Command Window and all settings boxes and tool palettes.

See Also: WINDOW view controls

WINDOW MINIMIZE

View Control: Minimize Window

Shrink a view to its smallest size. Strike Alt+F10 (or key in WINDOW MINIMIZE) and place an input point in the view you want collapsed. This control performs the same action as clicking on the small window control button in the upper right corner of the window and selecting Minimize.

✍ *Note:* The Restore selection is dimmed on the MicroStation Command Window and all settings boxes and tool palettes.

See Also: WINDOW view controls

WINDOW MOVE

View Control: Move Window Edge
Use this control to move the edge of any window. Key in WINDOW MOVE [choice] and click in the target view. The window is resized, but the graphics in the window remains the same. This selection has no use today and is only displayed to abide by motif standards.

✍ *Note:* The Move selection is dimmed on all tool palettes, settings boxes, views and the MicroStation Command Window.
See Also: WINDOW view controls

WINDOW ORIGIN   WO=

Setting: Window Origin
The lower left hand corner of every view. Specified coordinates in the design plane may be set to the Window Origin.

Item Selection
On sidebar menu pick: Window   Orig

For window, point to: n/a
On paper menu see: n/a
To show, do this: Click a tentative point   102:3:004, 101:4.479
   ⇨   Place the point near the lower left corner of a view.
   ⇨   The origin location displays in X and Y coordinates.
To set, key in: WI O xxx,yyy
Where xxx is the X-axis location and yyy is the Y-axis location. Enter xxx and yyy in working units.
Command Window Prompts:
1. Select the item
   Select view
2. Pick view for new window origin
To exit item: Click a Reset

Example(s):
WO=1000,1000
See Also: ACTIVE ORIGIN, WINDOW view controls

WINDOW RESTORE

Use this control to restore a view to its custom size after it has been minimized or maximized. Strike Alt+F5 (or key in WINDOW RESTORE) and click in the target view. Performs the result as clicking on the Window Menu Button in the upper left corner of the view and select Restore.

✍ *Note:* The Restore selection is dimmed on the MicroStation Command Window and all settings boxes and tool palettes.
See Also: WINDOW view controls

WINDOW SINK

View Control: Sink Window
Places the MicroStation Command Window or a settings box on the base plane. Key in WINDOW SINK and click in the target window. Performs the same function as clicking on the Window Menu Button in the upper left corner of the settings box or command window and selecting Sink.

✍ *Note:* The Sink selection is dimmed on all views and tool palettes.
See Also: WINDOW view controls

WINDOW SIZE

View Control: Size Window
Use this control to resize a view. Strike Alt+F8 (or key in WINDOW SIZE) and click in the target view. Performs the same result as clicking on the Window Menu Button in the upper left corner of the view and select Size.

Because MicroStation gives you the ability to resize a view graphically this selection is only displayed to abide by motif standards.

📝 *Note:* The Size selection is dimmed on all tool palettes, settings boxes and the MicroStation Command Window.
See Also: WINDOW view controls

WINDOW TILE   WI T

View Control: Window Tile
Adjusts all views on the display screen until each view is the same size.
Item Selection
On sidebar menu pick: | Window | Tile |
For window, point to: View   Tile
On paper menu see: VIEW
Command Window Prompts:
1. Select the item
   ⇨   the view are tiled on the display screen.
To exit item: Exits by itself
See Also: TILE commands, WINDOW view controls

WINDOW VOLUME   WI V

3D View Control: Window Volume
Adjusts the display depth and volume of a view. This view control asks for two input points to define the window volume, and a third input point placed in the view where the volume is being changed.
Item Selection
On sidebar menu pick: | Window | Vol |
For window, point to: View   Window   Volume
On paper menu see: VIEW
To set key in: WINDOW VOLUME
Command Window Prompts:
1. Select the item
   Select window origin
2. Pick first window point
   Select window corner
3. Select opposite corner
   Select view
4. Pick view for new window volume
To exit item: Click a Reset
See Also: DEPTH DISPLAY, SET DDEPTH settings

WO=

Alternate Key-in
Alternate key-in for WINDOW ORIGIN. Use it to set the origin of the lower left corner in an active view.
See Also: WINDOW ORIGIN

WSET ADD   WS

Drawing Tool: Add to Working Set
Temporarily groups elements into a Working SET. Working Sets can be manipulated using the FENCE tools. This tool expects an input point to identify the target element, and an accept point to add it to the working set.
Item Selection
On sidebar menu pick: | Manip | Groups | WS Elm | Add |
For window, point to: n/a
On paper menu see: W-SET
Command Window Prompts:
1. Select the item
   Identify element

2. Accept/Reject
   To exit item: **Make next selection**

✍ *Note:* Any elements placed while a working set is active becomes part of that working set.
See Also: FENCE tools, Graphic Group

WSET COPY   WS C

Drawing Tool: Add Copy to Working Set
   Copies an existing element and adds it to a working set. The original element
is not part of the working set. Elements in a working set can be manipulated using
a FENCE tool. This tool requires an input point to identify the element and a
second point to accept it for copying.
Item Selection
   On sidebar menu pick: | Manip | | Groups | | WS Elm | | Copy |
   For window, point to: n/a
   On paper menu see: W-SET
Command Window Prompts:
   1. Select the item
      Identify element
   2. Accept/Reject
   To exit item: **Make next selection**

✍ *Note:* Any elements placed while a working set is active becomes part of that working
set.
   See Also: WSET tools

WSET DROP   WS D

Drawing Tool: Drop Working Set
   Discontinues the active working set.

Item Selection
   On sidebar menu pick: | Manip | | Groups | | WS Elm | | Drop |
   For window, point to: n/a
   On paper menu see: GROUP
Command Window Prompts:
   1. Select the item
      Working set dropped
   To exit item: Exits by itself
   See Also: WSET tools

WT=

Alternate Key-in
   Alternate key-in for ACTIVE WEIGHT. Use it to set the thickness, or weight,
of elements placed in the drawing.
See Also: ACTIVE WEIGHT

XD=

Alternate Key-in
   The XD= key-in, which is an alternate key-in for EXCHANGEFILE, opens
another drawing file. The view configuration remains the same.
See Also: EXCHANGEFILE

XS=

Alternate Key-in
   Alternate key-in for ACTIVE XSCALE. Use it to control the X-axis scale
factor of elements placed in the drawing.
See Also: ACTIVE XSCALE

XY=

Alternate Key-in
>Alternate key-in for the POINT ABSOLUTE. Use it to place an input point at X,Y coordinates.

See Also: POINT ABSOLUTE

YS=

Alternate Key-in
>Alternate key-in for ACTIVE YSCALE. Use it to control the Y-axis scale factor of elements placed in the drawing.

See Also: ACTIVE YSCALE

ZOOM IN   ZO I

View Control: Zoom In
>Enlarges the size of objects on the screen, causing them to appear closer. Objects remain their true size, but the scale factor at which you view them changes. The zoom factor controls the amount of change.

>The default zoom-in factor is 2. This means the objects appear twice, or 2X, their size with each ZOOM IN picked. This zoom factor can be set as high as 100. The factor may be set once for the design session by keying in ZOOM IN [factor]. For example, ZOOM IN 5.

Item Selection
On sidebar menu pick: | Window | Zoom I |
For window, point to: View   Zoom   In
On paper menu see: n/a
Command Window Prompts:
1. Select the item
        Select point to zoom about
2. Select view to zoom in
To exit item: Click a Reset

✻   *Error:* Minimum window - If this messages appears, it means you cannot move any closer to the object.

See Also: FIT, WINDOW and ZOOM view controls

ZOOM IN 2   ZO I 2

View Control: Zoom In Twice
>Enlarges the size of objects on the screen, causing them to appear twice their size. Objects remain their true size, but the scale factor at which you view them changes.

Item Selection
For window, point to: n/a
On sidebar menu pick: n/a
On paper menu see: VIEW
Command Window Prompts:
1. Select the item
        Select view
2. Pick a view
To exit item: Click a Reset

✻   *Error:* Minimum window - If this messages appears, it means you cannot move any closer to the object.

See Also: FIT, WINDOW and ZOOM view controls

ZOOM IN CENTER   ZO I C

View Control: Zoom In About View Center
>Enlarges the size of objects on the screen, causing them to appear closer. Objects remain their true size, and centered in the view, but the scale factor at which you view them changes. The zoom factor controls the amount of change.

The default zoom-in factor is 2. This means the objects appear twice, or 2X, their size with each ZOOM IN CENTER picked. This zoom factor can be set as high as 100. The factor may be set once for the design session by keying in ZOOM IN CENTER [factor]. For example, ZOOM IN CENTER 5.

Item Selection
On sidebar menu pick: n/a
For window, point to: n/a
On paper menu see: VIEW
Command Window Prompts:
1. Select the item
   Select view
2. Pick a view
To exit item: Click a Reset

✳ **Error:** Minimum window - If this messages appears, it means you cannot move any closer to the object.
See Also: FIT, WINDOW and ZOOM view controls

ZOOM OUT   ZO

View Control: Zoom Out
Shrinks the size of objects, causing them to appear farther away. Objects remain their true size, but the scale factor at which you view them changes. The zoom factor controls the amount of change.

The zoom factor controls the amount an object changes. The default zoom-out factor is 2. This means the objects appear half their size with each ZOOM OUT picked. This zoom factor can be set as high as 100. The factor may be set once for the design session by keying in ZOOM OUT [factor]. For instance, ZOOM OUT 5.

Item Selection
On sidebar menu pick: Window Zoom O
For window, point to: View Zoom Out
On paper menu see: n/a
Command Window Prompts:
1. Select the item
   Select point to zoom about
2. Select view to zoom out
To exit item: Click a Reset

✳ **Error:** Maximum window - If this message appears, it means you cannot move any farther away from the object.
See Also: FIT, WINDOW and ZOOM view controls

ZOOM OUT 2   ZO O 2

View Control: Zoom Out Twice
Shrinks the size of objects, causing them to appear twice as far. Objects remain their true size, but the scale factor at which you view them changes.

Item Selection
On sidebar menu pick: n/a
For window, point to: n/a
On paper menu see: VIEW
Command Window Prompts:
1. Select the item
   Select point to zoom about
2. Select view to zoom out
To exit item: Click a Reset

✳ **Error:** Maximum window - If this message appears, it means you cannot move any farther away from the object.
See Also: FIT, WINDOW and ZOOM view controls

ZOOM OUT CENTER

View Control: Zoom Out About View Center

Shrinks the size of objects, causing them to appear farther away. Objects remain their true size, but the scale factor at which you view them changes. The objects remain centered in the view. The zoom factor controls the amount of change.

The zoom factor controls the amount an object changes. The default zoom-out factor is 2. This means the objects appear half their size with each ZOOM OUT CENTER picked. This zoom factor can be set as high as 100. The factor may be set once for the design session by keying in ZOOM OUT CENTER [factor]. For instance, ZOOM OUT CENTER 5.

Item Selection

For window, point to: n/a

On sidebar menu pick: n/a

On paper menu see: VIEW

Command Window Prompts:
1. Select the item
   Select view
2. Pick a view
To exit item: Click a Reset

* Error: Maximum window - If this message appears, it means you cannot move any farther away from the object.

See Also: FIT, WINDOW and ZOOM view controls

ZS=

Alternate Key-in

Alternate key-in for ACTIVE ZSCALE. Use it to control the Z-axis scale factor of elements placed in the drawing.

See Also: ACTIVE ZSCALE

Appendix A: MicroStation Directory Highlights

Following is a list of directories that come with the 4.0 version of MicroStation. The directories are listed here in a hierarchy found in a default installation of MicroStation. If you are a DOS user, remember to change the name to ustation and reverse the slash (/ for UNIX) to a backslash (\ for DOS) in front of each directory name. Directories may differ between the MicroStation platforms.

```
/mstation        .......... Parent directory and system files.
   /acct         .......... Accounting files.
      /sum       .......... Summary files produced by accounting
                                 program.
   /atp          .......... Automated testing procedures.
   /cell         .......... Cell and pattern cell libraries.
   /cfg          .......... Contains files for configuring MicroStation.
   /data         .......... Contains default color tables.
   /database     .......... Parent directory for database software
                                 packages.
      /informix  ......... Parent directory for Informix.
         /doors  ........ Door example support files.
            /doors.dbs .. Actual door example database files.
         /parts  ........ Parts example support files.
            /parts.dbs .. Actual parts example database files.
         /usa    .......... USA example support files.
            /usa.dbs .... Actual USA example database files.
      /ris        .......... RIS database interface to MicroStation.
   /docs          .......... MicroStation documentation text files.
   /dxf           .......... Parent directory for Drawing Exchange
                                 Format (DXF) utility.
      /data       .......... Data files necessary for converting drawing
                                 files.
      /out        .......... Destination directory of DXF output files.
   /edg           .......... Edit Graphics (EDG) utility.
   /examples      .......... Parent directory for example files.
      /cell       .......... Cell library example files.
      /dgn        .......... Drawing file examples.
   /font          .......... Parent directory for text fonts.
      /cell       .......... Base cell libraries for developing fonts.
      /libs       .......... The Font Librarian and sample font
                                 libraries.
   /gui           .......... Parent directory for the graphical user
                                 interface.
```

/mce	....	MicroStation Command Environment.
/mgr	....	MicroStation Manager.
/mstlg	....	MicroStation interface to Looking Glass.
/bin	....	Program directory.
/forms	....	Display screen forms.
/scripts	....	Script file which drive the forms.
/help	....	Parent directory for help files.
/english	....	Parent directory for English help files. (Language type may differ outside the USA.)
/language	....	Parent directory for the different languages.
/english	....	English language resource file. (Language type may differ outside the USA.)
/mdl	....	Parent directory for MDL applications.
/bin	....	MDL program files.
/examples	....	Parent directory for MDL examples.
/calculat	......	Calculator example files.
/english	....	Done in English.
/chngtxt	....	Change text example files.
/colortst	....	Color test example files.
/doc	....	Documentation.
/misc	....	Miscellaneous.
/objects	....	Object example files.
/plashape	.....	Placing shape example files.
/rasticon	......	Raster icon example files.
/robot	....	Robot example files.
/trumpet	......	Trumpet example files.
/include	....	Header files for compiled MDL applications.
/library	....	Compiled libraries for MDL programs.
/objects	....	Object files.
/mdlapps	....	MDL application files.
/menus	....	Hierarchical, control strip, function key and sidebar menus
/menudgns	....	Drawing files used to create menu graphics.
/menulibs	....	Cell libraries for screen and paper menus.
/plotting	....	Parent directory for MicroStation plotting.
/nqsfiles	....	MicroStation's NQS plotting directory.
/nqsreq	....	Default path for NQS plot daemon.
/pltcfg	....	Files used to configure plotters.
/pltfiles	....	Plotting output files.
/seeds	....	Sample seed files.
/tmp	....	Scratch directory for temporary files.
/ucm	....	Main user command directory.
/tutucms	....	User commands that drive screen tutorials.
/umenu	....	Micro menu interface.
/bin	....	Executable program files.
/cfg	....	Configuration files for micro menu interface.
/forms	....	Screen menu forms for micro menu interface.
/lib	....	MicroCSL program interface.
/mstn	....	Actual pocket and panel micro menus.

/cfg . . . . Configuration files for pocket and panel menus.
/menus . . . . Menu forms for pocket and panel menus.
/sym . . . . Symbol files for pocket and panel menus.
/sym   . . . . Symbol files for micro menu interface.
/utils   . . . . Various support utilities.

Appendix B: Speed Keys

^A	Unassigned
^B	Opens the View Attributes settings box
^C	Unassigned
^D	Duplicates a selection set (DUPLICATE)
^E	Opens the View Levels settings box
^F	Saves the current drawing settings (FILEDESIGN)
^G	Groups selected elements together (GROUP)
^H	Unassigned
^I	Asks you to identify an element and then opens an Element Information window (ANALYZE)
^J	Unassigned
^K	Unassigned
^L	Locks chosen elements (CHANGE LOCK)
^M	Unlocks chosen elements (CHANGE UNLOCK)
^N	Opens the Create Design File dialog box
^O	Opens the Open Design File dialog box
^P	Unassigned
^Q	Ends the design session (QUIT)
^R	Unassigned
^S	Unassigned
^T	Opens the Preview Plot dialog box (PLOT)
^U	Discontinues selected element group (UNGROUP)
^V	Unassigned
^W	Closes the design session and activates an MDL application (CLOSE DESIGN)
^X	Unassigned
^Y	Unassigned
^Z	Unassigned
Alt+F3	Lowers a view window (WINDOW LOWER)
Alt+F4	Closes a view window (WINDOW CLOSE)
Alt+F5	Restores a custom size view (WINDOW RESTORE)
Alt+F7	Moves a view edge (WINDOW MOVE)
Alt+F8	Sizes a view (WINDOW SIZE)
Alt+F9	Shrinks a view window (WINDOW MINIMIZE)
Alt+F10	Expands a view window (WINDOW MAXIMIZE)
Alt+Backspace	Undo last Redo (UNDO LAST)
Ctrl+Ins	Copy
Shift+Del	Cut selected text from a setting or dialog box
Shift+Ins	Paste select text from a buffer

Appendix C: Alternate Key-ins

AA=   Sets the angle at which elements are placed in the drawing file.

AC=   Sets the ACTIVE CELL name and reactivates the last PLACE
        CELL tool.

AD=   Places an input point at coordinates relative to the previous tenta-
        tive or data point placed in the current Auxiliary Coordi-
        nate System.

AE=   Define an active database entity without indicating an existing row.

AM=   Attaches screen menus or paper menus to the PC or workstation.

AP=   Controls the name of the cell employed in the patterning process.

AR=   Sets the replacement cell name and calls the PLACE CELL
        RELATIVE tool.

AS=   Sets the scale factor when placing elements or cells in a drawing
        file.

AT=   Displays a screen tutorial menu.

AX=   Places absolute coordinate points for an Auxiliary Coordinate Sys-
        tem based on key-ins.

AZ=   Sets the active depth in a three dimensional drawing file.

CC=   Takes the elements contained within a fence and files them in a li-
        brary as a cell.

CD=   Deletes a cell from a cell library.

CM=   Takes an existing cell in the drawing file and creates an array, or
        matrix, of several more cells.

CO=   Determines the color of the elements placed in a drawing file.

CR=   Changes the name of an existing cell in a cell library.

CT=   Attaches a color table to a display screen.

DA=   Determines the displayable attribute type.

DB=   Attaches a database file to a drawing file.

DD=   Alters the display depth in a three dimensional file.

DF=   Displays the Fonts settings box.

DI=   Places an input point at some known distance and direction in a
        2D or 3D file.

DL=   Places an input point at some known distance and direction in a
        2D or 3D file.

DP=   Allows you to control how much depth you see on the screen.

DR=   Displays a text file while remaining in the drawing file.

DS=   Restricts or establishes the database search criteria for elements
        being considered during a fence operation.

DV=   Deletes a named view.

DX=   Places an input point along the view axis relative to the last data
        or tentative point placement.

DZ=   Sets a new active depth in 3D drawing file relative to the current
        active depth.

EL=   Creates an element list file.

FF=   Copies the contents of a fence to an output file.

FI=   Defines an active entity in a database.

FT=   Sets the active font at which the text is placed in the drawing.

GO=   Determines the relative origin of all input points placed.

GR=   Controls the number of dots between each reference cross on the display grid.

GU=   Defines the number of working units between each grid unit on the display grid.

KY=   Divides an element into segments for snapping purposes.

LC=   Determines the current style, or line code, at which an element will be created.

LD=   Sets the level on which the dimensioning data are placed.

LL=   Controls the number of characters in a line of text of a text node.

LS=   Sets the spacing between lines of text.

LT=   Defines the name of the cell used by the PLACE TERMINATOR tool.

LV=   MicroStation creates the elements on this level.

NN=   Sets the active text node counter.

OF=   Turns the display of a drawing level off.

ON=   Turns the display of a drawing level on.

OX=   Attaches a user command index to a drawing file.

PA=   Sets the angle of pattern cells.

PD=   Controls the spacing between patterns.

PS=   Determines the scale factor of patterns.

PT=   Defines a line without length, text, or a cell as the active point.

PX=   Deletes a previously stored Auxiliary Coordinate System.

RA=   Controls the name of an SQL SELECT statement used during the review process.

RC=   Attaches a cell library to a drawing file.

RD=   Jumps you to a new drawing file without ending the design session.

RF=   Attaches a reference file to the drawing file.

RS=   Controls the name of the report table created during a fence operation.

RV=   Rotates an active view.

RX=   Attaches a previously saved Auxiliary Coordinate System.

SD=   Controls the distance between sampled input points.

SF=   Moves the contents of a fence to an output file.

ST=   Controls the distance, or tolerance, saved between input points while in stream mode.

SV=   Saves a named view perspective for later recall.

SX=   Saves the current Auxiliary Coordinate System for referral later.

TB=   Allows text tabs to be included in a design file when merging text in with the INCLUDE command.

TH=   Determines the height of text placed in a drawing.

TI=   Determines the text increment number.

TS=   Set the scale factor of the active terminator.

TV=   Sets the dimensioning tolerance limits.

TW=   Determines the width of text placed in a drawing.

TX=   Sets the height and width of text placed to the same value.

UC=   Starts up a user command.

UCC=   Compiles a user command so it will run faster.

UCI=   Runs a user command.

UR=   When the LOCK UNIT is on, every point placed is forced to the nearest multiple of the unit round-off.

VI=   Recalls a named view, or declare an orientation view in 3D files.

WO=   Sets the window origin of a view.

WT=   Determines the current weight, or thickness, at which an element will be created.

XD=   Exchanges the active file with a reference file. The reference file becomes the active drawing and vice versa.

XS=   Controls the X scale factor at which elements are placed.

XY=   Places an input point at some known X, Y, Z location.

YS=   Controls the Y scale factor at which elements are placed.

ZS=   Controls the Z scale factor at which elements are placed.

Appendix D: MicroStation Environment Variables

Check the current MicroStation environment variable settings by opening the Environment Variables settings box. Click on User, Environment Variables.

DBPATH	Default path for INFORMIX database.
INFORMIXDIR	Default path for INFORMIX programs.
MDL_COMP	Generally spells out (to MDL compiler) the default path to include files.
MS	Base directory path used by MDL make file.
MS_ACCOUNT	Default path to accounting program files.
MS_ACCT	Turns accounting on or off.
MS_APP	Default path for TSK statement of a user command.
MS_APPMEN	Default path for control strip and hierarchical menus.
MS_ARGS	For internal use by MCE.
MS_BACKUP	Default path for backup files.
MS_CACHE	Sets the maximum memory size for a drawing file edit session.
MS_CELL	Default path for example cell libraries.
MS_CELLSEED	Name of default cell library seed file.
MS_CMDWINDRSC	Alternate filename for resource file. (used for command window)
MS_CODESET	MDL application to handle multi-byte character sets.
MS_CS	Command window pull down menu switch.
MS_DATA	Default path for MicroStation's data files.
MS_DBASE	Default path for database files.
MS_DBGOUT	Name of output file used by MDL debugger.
MS_DBMODE	Allows multiple network users to access the same database.
MS_DEBUG	Used for debugging application programs.
MS_DEBUGSOURCE	Used to debug MDL application programs.
MS_DEF	See MS_DGN.
MS_DEFF	For internal use.
MS_DEFCTBL	Default colortable name.
MS_DESIGNSEED	Name of default drawing seed file.
MS_DGN	Default path for users drawing files.
MS_DGNAPPS	MDL application called by MicroStation when entering a new drawing.
MS_DIR	Default path for Microstation program files.
MS_DOSWINDOW	Enables MicroStation PC DOS window.
MS_DRIVERS	Default path for display card drivers.

MS_DXF	Default path for existing and newly created DXF files.
MS_DXFCFG	Name of default DXF configure file.
MS_DXFDATA	Default path containing table files and standard headers used during translation.
MS_EDG	Default path to edit graphics (EDG) executables.
MS_EMMRSV	Memory required by application programs.
MS_EXE	Default path for Microstation executables.
MS_EXITUC	File executed by user commands upon exiting.
MS_FKMENU	See MS_FKYMNU.
MS_FKEYMNU	Name of default function key menu file.
MS_FNTLB	Font library used during drawing session.
MS_FONT	*Default path to font librarian files.
MS_FONTCEL	Default path for font cell libraries.
MS_FOSI	Obsolete
MS_GUI	*Default path to Graphic User Interface (GUI) files.
MS_HELP	Default path for MCE help files.
MS_HELPFILE	Name of help file.
MS_HELPPATH	Default path for help files.
MS_HLINE	Obsolete
MS_HOST	Default host node for receiving and sending files.
MS_INIT	Filename executed by user command upon entering a design file.
MS_INITAPPS	MDL application called during MicroStation startup.
MS_LANGUAGE	Default path to language file.
MS_LEFTFONTS	Names of raster fonts for left screen
MS_LINKTYPE	Type of database interface used to link elements to a database.
MS_MDL	Default path for MDL applications.
MS_MENU	Name of default cell library for menus.
MS_MENUS	Default menu path.
MS_MTBL	Default path for material files.
MS_NEWFILE	Filename for user command which is executed whenever user changes drawing files during a design session.
MS_NO66	Tells MicroStation not to create Type 66 elements.
MS_NODE	Network node name.
MS_NODOSWIN	If present, MicroStation will not create DOS windows.
MS_NORMBUFFER	Used for debugging MicroCSL applications.
MS_ORACLE	Default path to Oracle interface.
MS_PLOTTING	Default plotting path.
MS_PLTFILES	Default path for plot files.
MS_PLTPORT	Number of serial ports to be used by plotter.
MS_PLTR	Plotter configuration filename.
MS_PMGRAPH	File of protected mode graphics drives loaded by MicroStation.
MS_QUEOPT	Options for Network Queueing System (NQS).

MS_READO	See MS_READONLY
MS_READONLY	Causes graphic files to be read only.
MS_REFSYM	Obsolete
MS_RENDER	Default path for shading and hidden line executables.
MS_RFDIR	Default path for reference files.
MS_RIS	Default path to RIS interface.
MS_RVHL	Obsolete
MS_RFDIR	Default path for reference files.
MS_RFONT	Name of raster font library.
MS_RSRC	Microstation resource file.
MS_RSRCPATH	Path for MDL applications to load resource files.
MS_SCR	Scratch directory.
MS_SEED	See MS_SEEDFILES.
MS_SEEDFILES	Default path for seed files.
MS_SERVER	Application to load a database server.
MS_STATIONERY	Default path for Microstation MAC seed files.
MS_SUPL	Default path for supplemental programs.
MS_SYSFONTS	Names of raster fonts used by MicroStation.
MS_SYSTEM	Controls the ability to push to the UNIX operating system.
MS_TMP	Default path for temporary files.
MS_TRAP	Program exception handling flag.
MS_TUTLIB	Cell library containing graphic tutorial menus.
MS_TUTUCMS	Default path for tutorial drivers.
MS_UCM	Default path for user commands.
MS_UCMSND	Default path for gr_ucmsnd file. (MicroCSL only)
MS_UCMRCV	Default path for gr_ucmrcv file. (MicroCSL only)
MS_UMENU	Default path to micro menu interface.
MS_UNDO	Sets the size of the UNDO buffer.
MS_UNIX	See MS_SYSTEM
MS_USER	Username at log in.
MS_USERDIR	Home directory.
MS_USERID	User ID number (Unix only).
MS_USERPREF	Name of user preference resource file.
MS_UTILS	Default path for utility programs.
MS_VT200	VT200 window catalog file.
PRO_DD_CSL	Points to MS_DIR (for internal use)
PRO_DD_VPLOT	Points to MS_DIR (For internal use)
RSC_COMP	Generally spells out (to resource compiler) the default path to the include files.
RVHL_SEED	Points to MS_DIR (For internal use).
UM_AUTO	Switch for umenus (For internal use).
UM_CMDS	Commands for umenus.
UM_DIR	Points to MS_DIR (For internal use)
WRK_DD_IGDS	Points to MS_DIR (For internal use)
WRK_DD_VPLOT	Points to MS_DIR (For internal use)

Tool names / Key-in equivalents

Activate an MDL Application Command	MDL COMMAND
Active Angle	ACTIVE ANGLE
Active Angle by Three Points	ACTIVE ANGLE PT3
Active Angle by Two Points	ACTIVE ANGLE PT2
Active Area Hole	ACTIVE AREA HOLE
Active Area Solid	ACTIVE AREA SOLID
Active Axis	ACTIVE AXIS
Active AXOrigin	ACTIVE AXORIGIN
Active B-spline Closed	ACTIVE BSPLINE CLOSED
Active B-spline Open	ACTIVE BSPLINE OPEN
Active B-spline Order	ACTIVE BSPLINE ORDER
Active B-spline Pole	ACTIVE BSPLINE POLE
Active B-spline Polygon	ACTIVE BSPLINE POLYGON
Active B-spline Tolerance	ACTIVE BSPLINE TOLERANCE
Active B-spline U Closed	ACTIVE BSPLINE UCLOSED
Active B-spline U Open	ACTIVE BSPLINE UOPEN
Active B-spline U Order	ACTIVE BSPLINE UORDER
Active B-spline U Pole	ACTIVE BSPLINE UPOLE
Active B-spline U Rules	ACTIVE BSPLINE URULES
Active B-spline V Closed	ACTIVE BSPLINE VCLOSED
Active B-spline V Open	ACTIVE BSPLINE VOPEN
Active B-spline V Order	ACTIVE BSPLINE VORDER
Active B-spline V Pole	ACTIVE BSPLINE VPOLE
Active B-spline V Rules	ACTIVE BSPLINE VRULES
Active Background	ACTIVE BACKGROUND
Active Capmode	ACTIVE CAPMODE
Active Cell	ACTIVE CELL
Active Class Construction	ACTIVE CLASS CONSTRUCTION
Active Class Primary	ACTIVE CLASS PRIMARY
Active Color	ACTIVE COLOR
Active Database	ACTIVE DATABASE
Active DAType	ACTIVE DATYPE
Active Entity	ACTIVE ENTITY
Active Fill	ACTIVE FILL
Active Font	ACTIVE FONT
Active Gridmode Isometric	ACTIVE GRIDMODE ISOMETRIC
Active Gridmode Offset	ACTIVE GRIDMODE OFFSET
Active Gridmode Orthogonal	ACTIVE GRIDMODE ORTHOGONAL
Active Gridratio	ACTIVE GRIDRATIO
Active Gridref	ACTIVE GRIDREF
Active Gridunit	ACTIVE GRIDUNIT
Active Index	ACTIVE INDEX
Active Keypoint	ACTIVE KEYPNT
Active Level	ACTIVE LEVEL
Active Line Length	ACTIVE LINE LENGTH
Active Line Space	ACTIVE LINE SPACE
Active Linewidth	ACTIVE LINEWIDTH
Active Linkage	ACTIVE LINKAGE
Active Linkage Duplicate	ACTIVE LINKAGE DUPLICATE
Active Linkage Information	ACTIVE LINKAGE INFORMATION
Active Linkage New	ACTIVE LINKAGE NEW
Active Linkage None	ACTIVE LINKAGE NONE
Active Node	ACTIVE NODE
Active Origin	ACTIVE ORIGIN
Active Pattern Angle	ACTIVE PATTERN ANGLE
Active Pattern Cell	ACTIVE PATTERN CELL
Active Pattern Delta	ACTIVE PATTERN DELTA
Active Pattern Scale	ACTIVE PATTERN SCALE
Active Point	ACTIVE POINT

Active Rcell	ACTIVE RCELL
Active Report	ACTIVE REPORT
Active Review	ACTIVE REVIEW
Active Scale	ACTIVE SCALE
Active Scale by Three Points	ACTIVE SCALE DISTANCE
Active Stream Angle	ACTIVE STREAM ANGLE
Active Stream Area	ACTIVE STREAM AREA
Active Stream Delta	ACTIVE STREAM DELTA
Active Stream Tolerance	ACTIVE STREAM TOLERANCE
Active Style	ACTIVE STYLE
Active Tab	ACTIVE TAB
Active Tag	ACTIVE TAG
Active Terminator	ACTIVE TERMINATOR
Active Textstyle	ACTIVE TEXTSTYLE
Active TNJ	ACTIVE TNJ
Active Tolerance Lower	DIMENSION TOLERANCE LOWER
Active TScale	ACTIVE TSCALE
Active TXHeight	ACTIVE TXHEIGHT
Active TXJ	ACTIVE TXJ
Active TXSize	ACTIVE TXSIZE
Active TXWidth	ACTIVE TXWIDTH
Active Unitround	ACTIVE UNITROUND
Active Weight	ACTIVE WEIGHT
Active XScale	ACTIVE XSCALE
Active YScale	ACTIVE YSCALE
Active ZScale	ACTIVE ZSCALE
Add Copy of Fence Contents to Working Set	FENCE WSET COPY
Add Copy to Working Set	WSET COPY
Add Fence Contents to Working Set	FENCE WSET ADD
Add to Graphic Group	GROUP ADD
Add to Working Set	WSET ADD
Align View	ALIGN
Alternate Key-in	OF=
Alternate Key-in	ON=
Alternate Key-in	OX=
Analyze Element	ANALYZE
Attach Active Entity	ATTACH AE
Attach Active Entity to Fence Contents	FENCE ATTACH
Attach Cell Library	ATTACH LIBRARY
Attach Colortable	ATTACH COLORTABLE
Attach Colortable (Create)	ATTACH COLORTABLE CREATE
Attach Colortable (Write)	ATTACH COLORTABLE WRITE
Attach Default Colortable	COLORTABLE DEFAULT
Attach Displayable Attributes	ATTACH DA
Attach Menu	ATTACH MENU
Attach Reference File	ATTACH REFERENCE
Attach Reference File	REFERENCE ATTACH
Automatic Create Complex Chain	CREATE CHAIN AUTOMATIC
Automatic Create Complex Shape	CREATE SHAPE AUTOMATIC
Automatic Fill in Enter Data Fields	EDIT AUTO
Automatic Fill in Enter Data Fields	EDIT AUTO DIALOG
Backup	BACKUP
Beep	BEEP
Bottom View to Top	WINDOW BOTTOMTOTOP
Center Justify Enter Data Field	JUSTIFY CENTER
Chamfer	CHAMFER
Change B-spline Polygon Display On/Off	CHANGE BSPLINE POLYGON
Change B-spline Surface to Active U-Order	CHANGE BSPLINE UORDER
Change B-spline Surface to Active U-Rules	CHANGE BSPLINE URULES
Change B-spline Surface to Active V-Order	CHANGE BSPLINE VORDER
Change B-spline Surface to Active V-Rules	CHANGE BSPLINE VRULES
Change B-spline to Active Order	CHANGE BSPLINE ORDER
Change Dimension to Active Settings	CHANGE DIMENSION SYMBOLOGY ALTERNATE
Change Element to Active Area (Solid/Hole)	CHANGE AREA HOLE
Change Element to Active Area (Solid/Hole)	CHANGE AREA SOLID

Change Element to Active Class	CHANGE CLASS CONSTRUCTION
Change Element to Active Class	CHANGE CLASS PRIMARY
Change Element to Active Color	CHANGE COLOR
Change Element to Active Level	CHANGE LEVEL
Change Element to Active Line Style	CHANGE STYLE
Change Element to Active Line Weight	CHANGE WEIGHT
Change Element to Active Symbology	CHANGE SYMBOLOGY
Change Fence Contents to Active Class	FENCE CHANGE CLASS CONSTRUCTION
Change Fence Contents to Active Class	FENCE CHANGE CLASS PRIMARY
Change Fence Contents to Active Color	FENCE CHANGE COLOR
Change Fence Contents to Active Level	FENCE CHANGE LEVEL
Change Fence Contents to Active Style	FENCE CHANGE STYLE
Change Fence Contents to Active Symbology	FENCE CHANGE SYMBOLOGY
Change Fence Contents to Active Weight	FENCE CHANGE WEIGHT
Change Fill	CHANGE FILL
Change Text to Active Attributes	MODIFY TEXT
Change to Alternate Dimension Symbology	CHANGE DIMENSION SYMBOLOGY STANDARD
Change to Standard Dimension Symbology	CHANGE DIMENSION
Choose All	CHOOSE ALL
Circular Fillet (No Truncation)	FILLET NOMODIFY
Circular Fillet and Truncate Both	FILLET MODIFY
Circular Fillet and Truncate Single	FILLET SINGLE
Clear View	VIEW CLEAR
Clipboard	CLIPBOARD
Close Design	CLOSE DESIGN
Close Element	CLOSE ELEMENT
Close Window	WINDOW CLOSE
Closed Cross Joint	Closed Cross Joint
Closed Tee Joint	Closed Tee Joint
Colortable Invert	COLORTABLE INVERT
Complete Cycle Linear Pattern	PATTERN LINEAR UNTRUNCATED
Compress Design	COMPRESS DESIGN
Compress Library	COMPRESS LIBRARY
Connect	|CONNECT
Constant Shading (All Views)	RENDER ALL CONSTANT
Constant Shading (Fence)	RENDER FENCE CONSTANT
Constant Shading (View)	RENDER VIEW CONSTANT
Construct Active Point at Distance Along Element .	CONSTRUCT POINT DISTANCE
Construct Active Point at Intersection	CONSTRUCT POINT INTERSECTION
Construct Active Points Along Element	CONSTRUCT POINT ALONG
Construct Active Points Between Data Points	CONSTRUCT POINT BETWEEN
Construct Angle Bisector	CONSTRUCT BISECTOR ANGLE
Construct Arc Tangent to Three Elements	CONSTRUCT TANGENT ARC 3
Construct B-spline Curve by Least Squares	CONSTRUCT BSPLINE CURVE LEASTSQUARE
Construct B-spline Curve by Points	CONSTRUCT BSPLINE CURVE POINTS
Construct B-spline Curve by Poles	CONSTRUCT BSPLINE CURVE POLES
Construct B-spline Surface by Cross-section	CONSTRUCT BSPLINE SURFACE CROSS
Construct B-spline Surface by Edges	CONSTRUCT BSPLINE SURFACE EDGE
Construct B-spline Surface by Least Squares	CONSTRUCT BSPLINE SURFACE LEASTSQUARE
Construct B-spline Surface by Points	CONSTRUCT BSPLINE SURFACE POINTS
Construct B-spline Surface by Poles	CONSTRUCT BSPLINE SURFACE POLES
Construct B-spline Surface by Skin	CONSTRUCT BSPLINE SURFACE SKIN

Construct B-spline Surface by Tube	CONSTRUCT SURFACE TUBE
Construct B-spline Surface of Projection	CONSTRUCT BSPLINE SURFACE PROJECTION
Construct B-spline Surface of Revolution	CONSTRUCT BSPLINE SURFACE REVOLUTION
Construct Circle Tangent to Element	CONSTRUCT TANGENT CIRCLE 1
Construct Circle Tangent to Three Elements	CONSTRUCT TANGENT CIRCLE 3
Construct Line at Active Angle from Point	CONSTRUCT LINE AA3
Construct Line at Active Angle from Point (Key-in)	CONSTRUCT LINE AA4
Construct Line at Active Angle to Point	CONSTRUCT LINE AA1
Construct Line at Active Angle to Point (Key-in)	CONSTRUCT LINE AA2
Construct Line Bisector	CONSTRUCT BISECTOR LINE
Construct Minimum Distance Line	CONSTRUCT LINE MINIMUM
Construct Perpendicular from Element	CONSTRUCT PERPENDICULAR FROM
Construct Perpendicular to Element	CONSTRUCT PERPENDICULAR TO
Construct Surface/Solid of Projection	SURFACE PROJECTION
Construct Surface/Solid of Revolution	SURFACE REVOLUTION
Construct Tangent Arc by Keyed-in Radius	CONSTRUCT TANGENT ARC 1
Construct Tangent from Element	CONSTRUCT TANGENT FROM
Construct Tangent to Circular Element and Perpendicular to Linear Element	
	CONSTRUCT TANGENT PERPENDICULAR
Construct Tangent to Element	CONSTRUCT TANGENT TO
Construct Tangent to Two Elements	CONSTRUCT TANGENT BETWEEN
Convert Element to B-spline (Copy)	CONSTRUCT BSPLINE CONVERT COPY
Convert Element to B-spline (Original)	CONSTRUCT BSPLINE CONVERT ORIGINAL
Copy and Increment Enter Data Field	INCREMENT ED
Copy and Increment Text	INCREMENT TEXT
Copy Element	COPY ELEMENT
Copy Enter Data Field	COPY ED
Copy Fence Contents	FENCE COPY
Copy Parallel by Distance	COPY PARALLEL DISTANCE
Copy Parallel by Key-in	COPY PARALLEL KEYIN
Copy View	COPY VIEW
Corner Joint	Corner Joint
Create Cell	CREATE CELL
Create Chain EOF	CREATE CHAIN EOF
Create Complex Chain	CREATE CHAIN MANUAL
Create Complex Shape	CREATE SHAPE MANUAL
Create Drawing	CREATE DRAWING
Create Entity	CREATE ENTITY
Create Library	CREATE LIBRARY
Create Shape EOF	CREATE SHAPE EOF
Cross Section Display (All Views)	RENDER ALL SECTION
Cross Section Display (Fence)	RENDER FENCE SECTION
Cross Section Display (View)	RENDER VIEW SECTION
Crosshatch Element Area	CROSSHATCH
Cut All Component Lines	Cut All Component Lines
Cut Single Component Line	Cut Single Component Line
Cut to Clipboard	CLIPBOARD CUT
Debug an MDL Application	MDL DEBUG
Define ACS (Aligned with Element)	DEFINE ACS ELEMENT
Define ACS (Aligned with View)	DEFINE ACS VIEW
Define ACS (By Points)	DEFINE ACS POINTS
Define Active Entity Graphically	DEFINE AE
Define Cell Origin	DEFINE CELL ORIGIN
Define Cylindrical ACS (Aligned with Element)	DEFINE ACS ELEMENT CYLINDRICAL
Define Cylindrical ACS (Aligned with View)	DEFINE ACS VIEW CYLINDRICAL

Define Cylindrical ACS (By Points) .... DEFINE ACS POINTS
CYLINDRICAL
Define Levels .... DEFINE LEVELS
Define Lights .... DEFINE LIGHTS
Define Materials .... DEFINE MATERIALS
Define Monument Points .... DIGITIZER SETUP
Define Rectangular ACS (Aligned with Element) .... DEFINE ACS ELEMENT
RECTANGULAR
Define Rectangular ACS (Aligned with View) .... DEFINE ACS VIEW
RECTANGULAR
Define Rectangular ACS (By Points) .... DEFINE ACS POINTS
RECTANGULAR
Define Reference File Back Clipping Plane .... REFERENCE CLIP BACK
Define Reference File Clipping Boundary .... REFERENCE CLIP BOUNDARY
Define Reference File Clipping Mask .... REFERENCE CLIP MASK
Define Reference File Front Clipping Plane .... REFERENCE CLIP FRONT
Define Search .... DEFINE SEARCH
Define Spherical ACS (Aligned with Element) .... DEFINE ACS ELEMENT
SPHERICAL
Define Spherical ACS (Aligned with View) .... DEFINE ACS VIEW SPHERICAL
Define Spherical ACS (By Points) .... DEFINE ACS POINTS
SPHERICAL
Define True North .... DEFINE NORTH
Delete 66Elements TCB .... DELETE 66ELEMENTS TCB
Delete 66Elements View .... DELETE 66ELEMENTS VIEW
Delete ACS .... DELETE ACS
Delete Cell .... DELETE CELL
Delete Element .... DELETE ELEMENT
Delete Fence Contents .... FENCE DELETE
Delete Part of Element .... DELETE PARTIAL
Delete Vertex .... DELETE VERTEX
Delete View .... DELETE VIEW
Detach Database Linkage .... DETACH
Detach Database Linkage from Fence Contents ...... FENCE DETACH
Detach Reference File .... REFERENCE DETACH
Dialog .... DIALOG
Dimension Angle Between Lines .... DIMENSION ANGLE LINES
Dimension Angle from X-Axis .... DIMENSION ANGLE X
Dimension Angle from Y-Axis .... DIMENSION ANGLE Y
Dimension Angle Location .... DIMENSION ANGLE LOCATION
Dimension Angle Size .... DIMENSION ANGLE SIZE
Dimension Arc Location .... DIMENSION ARC LOCATION
Dimension Arc Size .... DIMENSION ARC SIZE
Dimension Arclength .... DIMENSION ARCLENGTH
Dimension Axis Drawing .... DIMENSION AXIS ARBITRARY
Dimension Axis Drawing .... DIMENSION AXIS DRAWING
Dimension Axis True .... DIMENSION AXIS TRUE
Dimension Axis View .... DIMENSION AXIS VIEW
Dimension Center Mark .... DIMENSION CENTER
Dimension Center Size .... DIMENSION CENTER SIZE
Dimension Color .... DIMENSION COLOR
Dimension Diameter .... DIMENSION DIAMETER POINT
Dimension Diameter (Extended Leader) .... DIMENSION DIAMETER
EXTENDED
Dimension Diameter Parallel .... DIMENSION DIAMETER
PARALLEL
Dimension Diameter Perpendicular .... DIMENSION DIAMETER
PERPENDICULAR
Dimension Element .... DIMENSION ELEMENT
Dimension File Active .... DIMENSION FILE ACTIVE
Dimension File Reference .... DIMENSION FILE REFERENCE
Dimension Font .... DIMENSION FONT
Dimension Font Active .... DIMENSION FONT ACTIVE
Dimension Justification Center .... DIMENSION JUSTIFICATION
CENTER
Dimension Justification Left .... DIMENSION JUSTIFICATION
LEFT

Dimension Justification Right	DIMENSION JUSTIFICATION RIGHT
Dimension Level	DIMENSION LEVEL
Dimension Level Active	DIMENSION LEVEL ACTIVE
Dimension Location	DIMENSION LOCATION SINGLE
Dimension Location (Stacked)	DIMENSION LOCATION STACKED
Dimension Ordinates	DIMENSION ORDINATE
Dimension Placement Automatic	DIMENSION PLACEMENT AUTO
Dimension Placement Manual	DIMENSION PLACEMENT MANUAL
Dimension Placement Semi-automatic	DIMENSION PLACEMENT SEMIAUTO
Dimension Post Diameter	DIMENSION POST DIAMETER
Dimension Post Off	DIMENSION POST OFF
Dimension Post Radius	DIMENSION POST RADIUS
Dimension Post Square	DIMENSION POST SQUARE
Dimension Pre Diameter	DIMENSION PRE DIAMETER
Dimension Pre Off	DIMENSION PRE OFF
Dimension Pre Radius	DIMENSION PRE RADIUS
Dimension Pre Square	DIMENSION PRE SQUARE
Dimension Radius	DIMENSION RADIUS POINT
Dimension Radius (Extended Leader)	DIMENSION RADIUS EXTENDED
Dimension Scale	DIMENSION SCALE
Dimension Size (Custom)	DIMENSION LINEAR
Dimension Size with Arrows	DIMENSION SIZE ARROW
Dimension Size with Strokes	DIMENSION SIZE STROKE
Dimension Stacked	DIMENSION STACKED
Dimension Terminator First Arrow	DIMENSION TERMINATOR FIRST ARROW
Dimension Terminator First Origin	DIMENSION TERMINATOR FIRST ORIGIN
Dimension Terminator First Stroke	DIMENSION TERMINATOR FIRST STROKE
Dimension Terminator Left Arrow	DIMENSION TERMINATOR LEFT ARROW
Dimension Terminator Left Origin	DIMENSION TERMINATOR LEFT ORIGIN
Dimension Terminator Left Stroke	DIMENSION TERMINATOR LEFT STROKE
Dimension Terminator Right Arrow	DIMENSION TERMINATOR RIGHT ARROW
Dimension Terminator Right Origin	DIMENSION TERMINATOR RIGHT ORIGIN
Dimension Terminator Right Stroke	DIMENSION RIGHT TERMINATOR STROKE
Dimension Text Box	DIMENSION TEXT BOX
Dimension Text Capsule	DIMENSION TEXT CAPSULE
Dimension Text Color	DIMENSION TEXT COLOR
Dimension Text Color Active	DIMENSION TEXT COLOR ACTIVE
Dimension Text Weight	DIMENSION TEXT WEIGHT
Dimension Text Weight Active	DIMENSION TEXT WEIGHT ACTIVE
Dimension Tolerance	DIMENSION TOLERANCE
Dimension Tolerance Scale	DIMENSION TOLERANCE SCALE
Dimension Tolerance Upper	DIMENSION TOLERANCE UPPER
Dimension Tutorial	DIMENSION TUTORIAL
Dimension Units Degrees	DIMENSION UNITS DEGREES
Dimension Units Length	DIMENSION UNITS LENGTH
Dimension Vertical Mixed	DIMENSION VERTICAL MIXED
Dimension Vertical Off	DIMENSION VERTICAL OFF
Dimension Vertical On	DIMENSION VERTICAL ON
Dimension Weight	DIMENSION WEIGHT

Dimension Weight Active	DIMENSION WEIGHT ACTIVE
Dimension Witness Bottom	DIMENSION WITNESS BOTTOM
Dimension Witness Left	DIMENSION WITNESS LEFT
Dimension Witness Off	DIMENSION WITNESS OFF
Dimension Witness Right	DIMENSION WITNESS RIGHT
Dimension Witness Top	DIMENSION WITNESS TOP
Disconnect	|DISCONNECT
Display Attributes of Text Element	IDENTIFY TEXT
Display Erase	DISPLAY ERASE
Display Hilite	DISPLAY HILITE
Display Set	DISPLAY SET
Display SQL*Forms Screen Form	FORMS DISPLAY
DOS	DOS
Drop Association	DROP ASSOCIATION
Drop Associations in Fence	FENCE DROP ASSOCIATION
Drop Complex Status	DROP COMPLEX
Drop Complex Status of Fence Contents	FENCE DROP COMPLEX
Drop Dimension Element	DROP DIMENSION
Drop Dimension Elements in Fence	FENCE DROP DIMENSION
Drop from Graphic Group	GROUP DROP
Drop Line String/Shape Status	DROP STRING
Drop Multi-line	DROP MLINE
Drop Multi-lines in Fence	FENCE DROP MLINE
Drop Shared Cell	DROP SHARECELL
Drop Text	DROP TEXT
Drop Working Set	WSET DROP
Duplicate	DUPLICATE
Echo	ECHO
Edit Active Entity	|UPDATE AE
Edit AE	EDIT AE
Edit Text	EDIT TEXT
Element List	ELEMENT LIST
Element Selection	CHOOSE ELEMENT
Enable SQL*Forms	FORMS
Enter input point - absolute auxiliary coordinates	POINT ACSABSOLUTE
Enter input point - absolute coordinates	POINT ABSOLUTE
Enter input point - delta auxiliary coordinates	POINT ACSDELTA
Enter input point - delta coordinates	POINT DELTA
Enter input point - delta view coordinates	POINT VDELTA
Enter input point - distance, direction	POINT DISTANCE
Exchange Drawing File	EXCHANGEFILE
Exclamation point	!
Exit	EXIT
Exit NOClear	EXIT NOCLEAR
Exit NOUnClear	EXIT NOUC
Extend Element to Intersection	EXTEND ELEMENT INTERSECTION
Extend Line	EXTEND LINE
Extend Line by Key-in	EXTEND LINE KEYIN
Extend Two Elements to Intersection	EXTEND ELEMENT 2
Extract B-spline Surface Boundary	EXTRACT BSPLINE SURFACE BOUNDARY
Fence File	FENCE FILE
Fence Locate	FENCE LOCATE
Fence Separate	FENCE SEPARATE
Fence Stretch	FENCE STRETCH
File Design	FILEDESIGN
Fill in Single Enter Data Field	EDIT SINGLE
Fill in Single Enter Data Field	EDIT SINGLE DIALOG
Filled Hidden Line Removal (All Views)	RENDER ALL FILLED
Filled Hidden Line Removal (Fence)	RENDER FENCE FILLED
Filled Hidden Line Removal (View)	RENDER VIEW FILLED
Find	FIND
Fit Active Design	FIT ACTIVE
Fit Design and Reference Files	FIT ALL
Fit Reference	FIT REFERENCE
Fit Reference	REFERENCE FIT

Flush	FLUSH
Free	FREE
Freeze Element	FREEZE
Freeze Elements in Fence	FENCE FREEZE
Generate Report Table	FENCE REPORT
Group Holes	GROUP HOLES
Group Selection	GROUP SELECTION
Hatch Element Area	HATCH
Help	HELP
Help on Context	HELP CONTEXT
Help on Help	HELP HELP
Help on Keys	HELP KEYS
Help on Topics	HELP TOPICS
Hidden Line Removal (All Views)	RENDER ALL HIDDEN
Hidden Line Removal (Fence)	RENDER FENCE HIDDEN
Hidden Line Removal (View)	RENDER VIEW HIDDEN
Horizontal Parabola (No Truncation)	PLACE
Horizontal Parabola and Truncate Both	PLACE
Icons	ICONS
Identify Cell	IDENTIFY CELL
Import Text File	INCLUDE
Impose B-spline Surface Boundary	IMPOSE BSPLINE SURFACE BOUNDARY
Index Cell Library	INDEX
Insert Vertex	INSERT VERTEX
Iupdate	IUPDATE
Label Line	LABEL LINE
Left Justify Enter Data Field	JUSTIFY LEFT
Listen	LISTEN
Load an MDL Application	MDL LOAD
Load Displayable Attributes	LOAD DA
Load Displayable Attributes to Fence Contents	FENCE LOAD
Load MDL Applications	MDL DLOGLOAD
Locate Element	LOCELE
Lock	LOCK
Lock Angle	LOCK ANGLE
Lock Association	LOCK ASSOCIATION
Lock Auxiliary Coordinate System Plane	LOCK ACS
Lock Axis	LOCK AXIS
Lock Boresite	LOCK BORESITE
Lock Cellstretch	LOCK CELLSTRETCH
Lock Construction Plane	LOCK CONSPLANE
Lock Element	CHANGE LOCK
Lock Fence Clip	LOCK FENCE CLIP
Lock Fence Contents	FENCE CHANGE LOCK
Lock Fence Inside	LOCK FENCE INSIDE
Lock Fence Overlap	LOCK FENCE OVERLAP
Lock Fence Void Clip	LOCK FENCE VOID CLIP
Lock Fence Void Outside	LOCK FENCE VOID OUTSIDE
Lock Fence Void Overlap	LOCK FENCE VOID OVERLAP
Lock Graphic Group	LOCK GGROUP
Lock Grid	LOCK GRID
Lock Isometric	LOCK ISOMETRIC
Lock Level	LOCK LEVEL
Lock Scale	LOCK SCALE
Lock Selection	LOCK SELECTION
Lock Snap Auxiliary Coordinate System	LOCK SNAP ACS
Lock Snap Keypoint	LOCK SNAP KEYPOINT
Lock Snap Project	LOCK SNAP PROJECT
Lock Text Node	LOCK TEXTNODE
Lock Unit	LOCK UNIT
Lower Window	WINDOW BACK
Match Pattern Attributes	ACTIVE PATTERN MATCH
Match Text Attributes	ACTIVE TEXT
Maximize Window	WINDOW MAXIMIZE
Measure Angle Between Lines	MEASURE ANGLE
Measure Area by Points	MEASURE AREA POINTS

Measure Area of Element	MEASURE AREA ELEMENT
Measure Distance Along Element	MEASURE DISTANCE ALONG
Measure Distance Between Points	MEASURE DISTANCE POINTS
Measure Minimum Distance Between Elements	MEASURE DISTANCE MINIMUM
Measure Perpendicular Distance From Element	MEASURE DISTANCE PERPENDICULAR
Measure Radius	MEASURE RADIUS
Menu Check	MC
Merged Cross Joint	Merged Cross Joint
Merged Tee Joint	Merged Tee Joint
MicroStation Software Version	VERSION
Minimize Window	WINDOW MINIMIZE
Mirror Element About Horizontal	MIRROR COPY HORIZONTAL
Mirror Element About Horizontal (Copy)	MIRROR COPY HORIZONTAL
Mirror Element About Horizontal (Original)	MIRROR ORIGINAL HORIZONTAL
Mirror Element About Line (Copy)	MIRROR COPY LINE
Mirror Element About Line (Original)	MIRROR ORIGINAL LINE
Mirror Element About Vertical (Copy)	MIRROR COPY VERTICAL
Mirror Element About Vertical (Original)	MIRROR ORIGINAL VERTICAL
Mirror Fence Content About Vertical (Copy)	FENCE MIRROR COPY VERTICAL
Mirror Fence Contents About Horizontal (Copy)	FENCE MIRROR COPY HORIZONTAL
Mirror Fence Contents About Horizontal (Original)	FENCE MIRROR ORIGINAL HORIZONTAL
Mirror Fence Contents About Line (Copy)	FENCE MIRROR COPY LINE
Mirror Fence Contents About Line (Original)	FENCE MIRROR ORIGINAL LINE
Mirror Fence Contents About Vertical (Original)	FENCE MIRROR ORIGINAL VERTICAL
Mirror Reference File About Horizontal	REFERENCE MIRROR HORIZONTAL
Mirror Reference File About Vertical	REFERENCE MIRROR VERTICAL
Modify Arc Angle	MODIFY ARC ANGLE
Modify Arc Axis	MODIFY ARC AXIS
Modify Arc Radius	MODIFY ARC RADIUS
Modify Element	MODIFY ELEMENT
Modify Fence	MODIFY FENCE
Move ACS	MOVE ACS
Move Down (Scroll)	MOVE DOWN
Move Element	MOVE ELEMENT
Move Fence Block/Shape	MOVE FENCE
Move Fence Contents	FENCE MOVE
Move Left (Scroll)	MOVE LEFT
Move Parallel by Distance	MOVE PARALLEL DISTANCE
Move Parallel by Key-in	MOVE PARALLEL KEYIN
Move Reference File	REFERENCE MOVE
Move Right (Scroll)	MOVE RIGHT
Move Up (Scroll)	MOVE UP
Move Window Edge	WINDOW MOVE
Multi-Cycle Segment Linear Pattern	PATTERN LINEAR MULTIPLE
New File	NEWFILE
No Echo	NOECHO
Null	NULL
Open Cross Joint	Open Cross Joint
Open Tee Joint	Open Tee Joint
Page Setup	PAGE SETUP
Partition Tablet Surface	DIGITIZER PARTITION
Paste From Clipboard	CLIPBOARD PASTE
Paste PICT file data	PICTFILE PASTE
Pattern Element Area	PATTERN AREA ELEMENT
Pattern Fence Area	PATTERN AREA FENCE
Pause	PAUSE
Percent sign	%
Phong Shading (All Views)	RENDER ALL PHONG
Phong Shading (Fence)	RENDER FENCE PHONG
Phong Shading (View)	RENDER VIEW PHONG

Place Active Cell	PLACE CELL ABSOLUTE
Place Active Cell (Interactive)	PLACE CELL INTERACTIVE ABSOLUTE
Place Active Cell by Transformation Matrix	PLACE CELL ABSOLUTE TMATRX
Place Active Cell Matrix	MATRIX CELL
Place Active Cell Relative	PLACE CELL RELATIVE
Place Active Cell Relative (Interactive)	PLACE CELL INTERACTIVE RELATIVE
Place Active Cell Relative by Transformation Matrix	PLACE CELL RELATIVE TMATRX
Place Active Line Terminator	PLACE TERMINATOR
Place Active Point	PLACE POINT
Place Arc by Center	PLACE ARC CENTER
Place Arc by Edge	PLACE ARC EDGE
Place Arc by Keyed-in Radius	PLACE ARC RADIUS
Place B-spline Curve by Least Squares	PLACE BSPLINE CURVE LEASTSQUARE
Place B-spline Curve by Points	PLACE BSPLINE CURVE POINTS
Place B-spline Curve by Poles	PLACE BSPLINE CURVE POLES
Place B-spline Surface by Least Squares	PLACE BSPLINE SURFACE LEASTSQUARE
Place B-spline Surface by Points	PLACE BSPLINE SURFACE POINTS
Place B-spline Surface by Poles	PLACE BSPLINE SURFACE POLES
Place Block	PLACE BLOCK ORTHOGONAL
Place Center Mark	DIMENSION CENTER MARK
Place Circle by Center	PLACE CIRCLE CENTER
Place Circle by Diameter	PLACE CIRCLE DIAMETER
Place Circle by Edge	PLACE CIRCLE EDGE
Place Circle by Keyed-in Radius	PLACE CIRCLE RADIUS
Place Circumscribed Polygon	PLACE POLYGON CIRCUMSCRIBED
Place Continuous Point String	PLACE POINT STRING CONTINUOUS
Place Dialog Note	PLACE NOTE DIALOG
Place Disjoint Point String	PLACE POINT STRING DISJOINT
Place Ellipse by Center and Edge	PLACE ELLIPSE CENTER
Place Ellipse by Edge Points	PLACE ELLIPSE EDGE
Place Ellipse Quarter	PLACE ELLIPSE QUARTER
Place Fence Block	PLACE FENCE BLOCK
Place Fence Shape	PLACE FENCE SHAPE
Place Fitted Text	PLACE DIALOGTEXT FITTED
Place Fitted Text	PLACE TEXT FITTED
Place Fitted View Independent Text	PLACE DIALOGTEXT FVI
Place Fitted View Independent Text	PLACE TEXT FVI
Place Half Ellipse	PLACE ELLIPSE HALF
Place Helix	PLACE HELIX
Place Inscribed Polygon	PLACE POLYGON INSCRIBED
Place Isometric Block	PLACE BLOCK ISOMETRIC
Place Isometric Circle	PLACE CIRCLE ISOMETRIC
Place Line	PLACE LINE
Place Line at Active Angle	PLACE LINE ANGLE
Place Line String	PLACE LSTRING POINT
Place Multi-line	PLACE MLINE
Place Note	PLACE NOTE
Place Orthogonal Shape	PLACE SHAPE ORTHOGONAL
Place Parabola by Endpoints	PLACE PARABOLA ENDPOINTS
Place Point Curve	PLACE CURVE POINT
Place Polygon by Edge	PLACE POLYGON EDGE
Place Right Cone	PLACE CONE RIGHT
Place Right Cone by Keyed-in Radii	PLACE CONE RADIUS
Place Right Cylinder	PLACE CYLINDER RIGHT
Place Right Cylinder by Keyed-in Radius	PLACE CYLINDER RADIUS
Place Rotated Block	PLACE BLOCK ROTATED
Place Shape	PLACE SHAPE

Place Skewed Cone	PLACE CONE SKEWED
Place Skewed Cylinder	PLACE CYLINDER SKEWED
Place Slab	PLACE SLAB
Place Space Curve	PLACE CURVE SPACE
Place Space Line String	PLACE LSTRING SPACE
Place Sphere	PLACE SPHERE
Place Spiral by Endpoints	PLACE SPIRAL ENDPOINTS
Place Spiral by Length	PLACE SPIRAL LENGTH
Place Spiral by Sweep Angle	PLACE SPIRAL ANGLE
Place Stream Curve	PLACE CURVE STREAM
Place Stream Line String	PLACE LSTRING STREAM
Place Text	PLACE DIALOGTEXT
Place Text	PLACE TEXT
Place Text Above Element	PLACE DIALOGTEXT ABOVE
Place Text Above Element	PLACE TEXT ABOVE
Place Text Along Element	PLACE DIALOGTEXT ALONG
Place Text Along Element	PLACE TEXT ALONG
Place Text Below Element	PLACE DIALOGTEXT BELOW
Place Text Below Element	PLACE TEXT BELOW
Place Text by Transformation Matrix	PLACE DIALOGTEXT TMATRIX
Place Text by Transformation Matrix	PLACE TEXT TMATRIX
Place Text Node	PLACE NODE
Place Text Node by Transformation Matrix	PLACE NODE TMATRX
Place Text On Element	PLACE DIALOGTEXT ON
Place Text On Element	PLACE TEXT ON
Place View Independent Text	PLACE DIALOGTEXT VI
Place View Independent Text	PLACE TEXT VI
Place View Independent Text Node	PLACE NODE VIEW
Plot	PLOT @
Polar Array	ARRAY POLAR
Polar Array Fence Contents	FENCE ARRAY POLAR
Print	PRINT
Project Active Point Onto Element	CONSTRUCT POINT PROJECT
Project Fence Contents	FENCE SURFACE PROJECTION
Push to Unix	UNIX
Quit	QUIT
Quit NoClear	QUIT NOCLEAR
Quit NoUnclear	QUIT NOUC
Raise Window	WINDOW FRONT
Recall Saved Image	VIEW IMAGE
Record Input on/off	RECORD
Rectangular Array	ARRAY RECTANGULAR
Rectangular Array Fence Contents	FENCE ARRAY RECTANGULAR
Redo	REDO
Reference File Levels On	REFERENCE LEVELS
Reload Reference File	REFERENCE RELOAD
Remap Digitizer Buttons	DIGITIZER BUTTONS
Remap Digitizer Buttons	SET BUTTON
Remove All 66Elements	DELETE 66ELEMENTS ALL
Remove Levelname 66Elements	DELETE 66ELEMENTS LEVELNAME
Remove MicroStation 66Elements	DELETE 66ELEMENTS MS
Remove Reference 66Elements	DELETE 66ELEMENTS REFERENCE
Remove Start 66Elements	DELETE 66ELEMENTS START
Rename Cell name	RENAME CELL
Replace Cell	REPLACE CELL
Reset	RESET
Reset Dimension Scale Factor	DIMENSION SCALE RESET
Review Database Attributes of Element	REVIEW
Revolve Fence Contents	FENCE SURFACE REVOLUTION
Right Justify Enter Data Field	JUSTIFY RIGHT
Rotate ACS Absolute	ROTATE ACS ABSOLUTE
Rotate ACS Relative	ROTATE ACS RELATIVE
Rotate Element by Active Angle (Copy)	ROTATE COPY
Rotate Element by Active Angle (Original)	ROTATE ORIGINAL
Rotate Fence Contents by Active Angle (Copy)	FENCE ROTATE COPY

Rotate Fence Contents by Active Angle (Original) ...	FENCE ROTATE ORIGINAL
Rotate Reference File	REFERENCE ROTATE
Rotate View (Absolute)	ROTATE VIEW ABSOLUTE
Rotate View (Aligned with Element)	ROTATE VIEW ELEMENT
Rotate View (Relative)	ROTATE VIEW RELATIVE
Rotate View by Points	ROTATE VIEW POINTS
Rotate View by Transformation Matrix	ROTATE VMATRX
Save	SAVE
Save Auxiliary Coordinate System	SAVE ACS
Save Function Keys	SAVE FUNCTION_KEY
Save Image	SAVE IMAGE
Save Named View	SAVE VIEW
Save Raster Data to a PICT file	PICTFILE SAVE RASTER
Save Vector Data to a PICT file	PICTFILE SAVE VECTOR
Scale Element (Copy)	SCALE COPY
Scale Element (Original)	SCALE ORIGINAL
Scale Fence Contents (Copy)	FENCE SCALE COPY
Scale Fence Contents (Original)	FENCE SCALE ORIGINAL
Scale Reference File	REFERENCE SCALE
Select ACS	ATTACH ACS
Select and Place Cell	SELECT CELL ABSOLUTE
Select and Place Cell (Relative)	SELECT CELL RELATIVE
Select Cell and Place Cell by Transformation Matrix	SELECT CELL ABSOLUTE TMATRX
Select Cell and Place Cell Relative by Transformation Matrix ..	SELECT CELL RELATIVE TMATRX
Select View	SELVIEW
Set Active Depth	ACTIVE ZDEPTH ABSOLUTE
Set Active Depth	DEPTH ACTIVE
Set Active Depth (Relative)	ACTIVE ZDEPTH RELATIVE
Set Active Text Height by Two Points	ACTIVE TXHEIGHT PT2
Set Active Text Width by Two Points	ACTIVE TXWIDTH PT2
Set Auxiliary Input	SET AUXINPUT
Set Camera Definition	SET CAMERA DEFINITION
Set Camera Focal Length	SET CAMERA LENGTH
Set Camera Lens Angle	SET CAMERA LENS ANGLE
Set Camera on/off	SET CAMERA
Set Compatibility On/Off	SET COMPATIBLE
Set Cursor Isometric	SET CURSOR ISOMETRIC
Set Cursor Orthogonal	SET CURSOR ORTHOGONAL
Set Database	SET DATABASE
Set Debug	SET DEBUG
Set Display Depth	DEPTH DISPLAY
Set Display Depth (Absolute)	SET DDEPTH ABSOLUTE
Set Display Depth (Relative)	SET DDEPTH RELATIVE
Set Enter Data Field Character	SET EDCHAR
Set Full View Pointer	SET CURSOR FULL
Set Hilite Color	SET HILITE
Set Isometric Drawing Plane (Left)	SET ISOPLANE LEFT
Set Isometric Drawing Plane (Right)	SET ISOPLANE RIGHT
Set Isometric Drawing Plane (Top)	SET ISOPLANE TOP
Set Locate Tolerance	SET LOCATE
Set Mark	MARK
Set Maximum Grid Points	SET MAXGRID
Set Normal Pointer	SET CURSOR SMALL
Set Plotter Type	SET PLOTTER
Set Prompt	SET PROMPT
Set Shading View Mode	SET VIEW
Set Smallest Text	SET SMALLTEXT
Set Task size	SET TASKSIZE
Set the Tasksize for Conventional Memory	SET TASKSIZE CONVENTIONAL
Set the Tasksize for Extended Memory	SET TASKSIZE EXTENDED
Set TPMode ACSDelta	SET TPMODE ACSDELTA
Set TPMode ACSLocate	SET TPMODE ACSLOCATE
Set TPMode Angle3	SET TPMODE ANGLE3
Set TPMode Delta	SET TPMODE DELTA
Set TPMode Distance	SET TPMODE DISTANCE

Set TPMode Locate	SET TPMODE LOCATE
Set TPMode Vdelta	SET TPMODE VDELTA
Set View Camera Lens	SET CAMERA LENS
Set View Camera Position	SET CAMERA POSITION
Set View Camera Target	SET CAMERA TARGET
Set View(s) Off	VIEW OFF
Set View(s) On	VIEW ON
Set X Pointer Color	SET XOR
Show Active Depth	SHOW DEPTH ACTIVE
Show Active Entity	SHOW AE
Show Camera Distance	SET CAMERA DISTANCE
Show Camera Lens	SHOW CAMERA LENS
Show Camera Position	SHOW CAMERA POSITION
Show Camera Target	SHOW CAMERA TARGET
Show Clipboard	SHOW CLIPBOARD
Show Display Depth	SHOW DEPTH DISPLAY
Show End of File	SHOW EOF
Show Pattern Attributes	SHOW PATTERN
Show Reference	SHOW REFERENCE
Show Stack	SHOW STACK
Show UORs	SHOW UORS
Single Cycle Segment Linear Pattern	PATTERN LINEAR SINGLE
Sink Window	WINDOW SINK
Size Window	WINDOW SIZE
Smooth Shading (All Views)	RENDER ALL SMOOTH
Smooth Shading (Fence)	RENDER FENCE SMOOTH
Smooth Shading (View)	RENDER VIEW SMOOTH
Snap Lock Intersection	LOCK SNAP INTERSECTION
Spin Element (Copy)	SPIN COPY
Spin Element (Original)	SPIN ORIGINAL
Spin Fence Contents (Copy)	FENCE SPIN COPY
Spin Fence Contents (Original)	FENCE SPIN ORIGINAL
Start MicroCSL application	START
Stereo Image (All Views)	RENDER ALL STEREO
Stereo Image (Fence)	RENDER FENCE STEREO
Stereo Image (View)	RENDER VIEW STEREO
Swap Screen	SWAP SCREEN
Symmetric Parabola (No Truncation)	PLACE PARABOLA NOMODIFY
Symmetric Parabola and Truncate Both	PLACE PARABOLA MODIFY
Thaw Element	THAW
Thaw Elements in Fence	FENCE THAW
Toggle ACS Display On/Off	SET ACSDISPLAY
Toggle Background Display On/Off	SET BACKGROUND
Toggle Compatibility Dimension On/Off	SET COMPATIBLE DIMENSION
Toggle Compatibility Multi-line On/Off	SET COMPATIBLE MLINE
Toggle Construction Display On/Off	SET CONSTRUCT
Toggle Curve Display Fast/Slow	SET CURVES
Toggle Delete On/Off	SET DELETE
Toggle Depth Cueing On/Off	SET DEPTHCUE
Toggle Dimension Display On/Off	SET DIMENSION
Toggle Dynamic Update On/Off	SET DYNAMIC
Toggle Enter Data Field Display On/Off	SET ED
Toggle Fill Display On/Off	SET FILL
Toggle Fill Display On/Off	SET LINEFILL
Toggle Font Fast/Slow	SET FONT
Toggle Grid Display On/Off	SET GRID
Toggle Help On/Off	SET HELP
Toggle Level Display On/Off	SET LEVELS
Toggle Level Symbology On/Off	SET LVLSYMB
Toggle Line Weight Display On/Off	SET WEIGHT
Toggle Line Width On/Off	SET LINEWIDTH
Toggle Mirror Text On/Off	SET MIRTEXT
Toggle Overview On/Off	SET OVERVIEW
Toggle Parse All Key-ins On/Off	SET PARSEALL
Toggle Pattern Display On/Off	SET PATTERN
Toggle Range Block Display On/Off	SET RANGE
Toggle Rastertext Display On/Off	SET RASTERTEXT

Toggle Reference Boundary Display On/Off	SET REFBOUND
Toggle Reference Clipping Display Fast/Slow	SET REFCLIP
Toggle Reference Display On/Off	REFERENCE DISPLAY
Toggle Reference Locate On/Off	REFERENCE LOCATE
Toggle Reference Snap On/Off	REFERENCE SNAP
Toggle Sharecell On/Off	SET SHARECELL
Toggle Stacked Fractions On/Off	SET STACKFRACTIONS
Toggle Stream Acceptance Display On/Off	SET STREAM
Toggle Text Node Display On/Off	SET NODES
Toggle Undo Buffer On/Off	SET UNDO
Transform Element by Transformation Matrix	TRANSFORM
Transform Fence Contents	FENCE TRANSFORM
Treset	TRESET
Truncated Cycle Linear Pattern	PATTERN LINEAR TRUNCATED
Tutorial	TUTORIAL
Type	TYPE
UCI	UCI
Uncut Component Lines	Uncut Component Lines
Undo	UNDO
Ungroup	UNGROUP
Unload an MDL Application	MDL UNLOAD
Unlock Element	CHANGE UNLOCK
Unlock Fence Contents	FENCE CHANGE UNLOCK
Update View	UPDATE
User Command	USERCOMMAND
User Command Compiler	UCC
View	VIEW
Window Area	WINDOW AREA
Window Cascade	WINDOW CASCADE
Window Center	WINDOW CENTER
Window Origin	WINDOW ORIGIN
Window Tile	WINDOW TILE
Window Volume	WINDOW VOLUME
Wiremesh Display (All Views)	RENDER ALL WIREMESH
Wiremesh Display (Fence)	RENDER FENCE WIREMESH
Wiremesh Display (View)	RENDER VIEW WIREMESH
Zoom In	ZOOM IN
Zoom In About View Center	ZOOM IN CENTER
Zoom In Twice	ZOOM IN 2
Zoom Out	ZOOM OUT
Zoom Out About View Center	ZOOM OUT CENTER
Zoom Out Twice	ZOOM OUT 2

Commands Grouped by Function

Entries marked with an asterisk are new to this edition of the MicroStation Reference Guide.

REFERENCE FILES

RENDERING

SPLINES

THREE DIMENSIONAL

USER COMMAND

VIEW

Alphabetical Command Listing

Entries marked with an asterisk are new to this edition of the MicroStation Reference Guide.

Check Out These Other MicroStation Titles Available from OnWord Press

OnWord Press Products are Available Directly From

1. Your Local MicroStation Dealer or Intergraph Education Center
2. Your Local Bookseller
3. In Australia, New Zealand, and Southeast Asia from:
   Pen & Brush Publishers
   2nd Floor, 94 Flinders Street
   Melbourne Victoria 3000
   Australia
   Phone 61 (0)3 818 6226
   Fax   61(0)3 818 3704

4. .... Or Directly From OnWord Press
   (see ordering information on the last page.)

Upgrading to MicroStation 4.X from an earlier version? OnWord Press has Upgrade Tools!

The MicroStation 4.X Delta Book

Frank Conforti
A Quick Guide To Upgrading to 4.X on all Platforms from Earlier Versions of MicroStation

This short and sweet book takes you from 3.X versions of MicroStation into the new world of MicroStation 4.X.

Did you know that there are now over 950 commands in MicroStation? Have you seen the new graphical user interface (GUI)? Are you ready for dialog boxes? Associative dimensioning? Palettes? The new MicroStation Manager?

Intergraph's release notes don't tell the whole story. Sure they let you know about a few new features.

But the MicroStation 4.X Delta Book gets you up and running with these new productivity tools fast.

Highly graphic, the MicroStation 4.X Delta Book doesn't just tell you, it shows you!

Written by Frank Conforti, author of best-selling INSIDE MicroStation, the MicroStation 4.X Delta Book will have you turning out MicroStation 4.X productivity in no time.

Price: **US$19.95** (Australia A$29.95)
Pages: 110
Illustrations: 75+
ISBN: 0-934605-34-3

Cut Your Learning Curve... With

MicroStation 4.X Upgrade Training Video Series

MicroStation version 4.0 is the most comprehensive release to date, enabling users to shorten their production cycles and gain a competitive edge. Now you can cut your learning curve on all the newest features in version 4.0 by using the MicroStation Version 4.0 Upgrade Training Video Series.

The training video contains 12 lessons designed for the experienced MicroStation user. Written by qualified Microstation trainers at Intergraph Corporation, this complete training video series provides hands-on training on the latest MicroStation has to offer. It is a convenient way for MicroStation users to come up to speed on many of the capabilities in version 4.0.

MicroStation users will notice a completely new look in version 4.0 with the new GUI. You will find everything you need on the screen at your fingertips from icon-based tool palettes and dialog boxes to context sensitive help. The training video will walk you through the user interface, showing you a new way of accessing old and new commands and features, plus how to make them work for you.

The lessons in the video will show you how to use Microstation's B-spline surfacing tools, new 2D element placement commands, associated dimensioning, expanded rendering capabilities, multi-line placement, shared cells, named levels, plus view and file manipulation and control, and much more.

Price: **US$149.00**
Three 1/2 Hour VHS Tapes
Video Tapes, Index

Tap the power of MicroStation 4.X
with these new books from
On Word Press.

INSIDE MicroStation

Frank Conforti
The Complete MicroStation Guide
Second Edition, Release 4.X, Supports all 4.X and 3.X Platforms including DOS, Unix, Mac, and VMS

Completely updated for release 4.0!

This easy to use book serves as both a tutorial and a lasting reference guide. Learn to use every MicroStation command as well as time saving drawing techniques and tips. Includes coverage of 3-D, modelling and shading. This is the book that lets you keep up and stay in control with MicroStation.

With a complete update for MicroStation 4.X, this book covers all the new GUI features, additional 2D element placement, associative dimensioning, multi-line placement, shared cells, and more.

The first half of the book concentrates on personal productivity with lessons on basic MicroStation screen, menu, drawing and file control. Here you'll find tutorial information on creating, editing, detailing, and plotting your design work.

The second half of the book gives you the concepts and practical tools for workgroup productivity including advanced editing, cell library work, reference files, and 3D design. You'll also find examples of menu control, DXF transfers, networking, and applications development.

INSIDE MicroStation takes you beyond stand-alone commands. Using real world examples it gives you practical and effective methods for building good working habits with MicroStation.

Frank Conforti is an independent MicroStation consultant based in Delray Beach Florida. Formerly CAD Manager at Keith and Schnarrs Engineers, Frank managed a network of VAX-based IGDS and MicroStation systems running workstations, PCs and MACs. Conforti has been using CAD, IGDS and MicroStation for 16 years. He is an avid writer, trainer, and user group sponsor. He is co-author of the Macintosh CAD/CAM book.

◑ Optional INSIDE MicroStation Disk: **US$14.95** (Australia A$29.95) Includes the tutorial example design files, menus, listings, examples and more.

Price: **US$29.95**   (Australia A$54.95)
Pages: 550
Illustrations: 220+
ISBN: 0-934605-49-1

The INSIDE MicroStation
Companion Workbook

Michael Ward and Support From Frank Conforti
32 Steps to MicroStation -- A self-paced tutorial workbook
for individual or classroom use.
First Edition, Release 4.X, Supports all 4.X and 3.X Platforms including DOS, Unix, Mac, and VMS

This highly readable hands-on text is your guide to learning MicroStation. As a companion to the INSIDE MicroStation book, the workbook gives you practical exercises for developing MicroStation skills.

The INSIDE MicroStation Companion Workbook is set up for self-paced work alone, or in classroom or group training. The workbook comes setup for three types of training scenarios. You can work through the exercises and test yourself. Or you can organize the course into a three or four day professional or 10 - 14 week semester course. Or you can make up your own grouping from the 32 basic training units.

The workbook course takes you through at least two paths to develop your proficiency with MicroStation. First, you are taught basic functions and given exercises to develop your skills. Second, you can select from an architectural, civil, or electrical engineering "real life" professional drawing that you work on during the length of the course.

◑ The INSIDE MicroStation Companion Workbook training Disk comes with the book (5.25" DOS standard, other formats available on request). On the disk you will find exercise tools, exercise drawings, plot files, even a diploma!

✍ Also included in the package are several professional blueprints used as part of the training course.

Price: **US$34.95** (Australia A$54.95)
Pages: 400 (Includes DOS 5.25" Disk and Blueprints)
Illustrations: 75+
ISBN: 0-934605-42-4

Optional INSIDE MicroStation Companion Workbook Teacher/Trainer's Guide.

This 75 page instructor's guide to the INSIDE MicroStation Companion Workbook is geared for the professional trainer, college professor, or corporate training department. Filled with teaching tricks, organization tools and more, the instructor's guide is a must for anyone in an instructor's position.

Price: **US$9.95** (Australia A$18.95)
NOTE: The Teacher/Trainer's Guide is **FREE** to any group ordering 10 copies or more of the workbook. Contact OnWord Press for ordering information.
Pages: 75
ISBN: 0-934605-39-4

The MicroStation Productivity Book

Kincaid, Steinbock, Malm
Tapping the Hidden Power of MicroStation
Second Edition, Release 4.X, Supports all 4.X and 3.X Platforms including DOS, Unix, Mac, and VMS

With this book beginning and advanced users alike can take big leaps in productivity, job security and personal satisfaction. Thirty-six step-by-step chapters show you how to take charge of MicroStation.

Completely updated for MicroStation 4.X, the MicroStation Productivity book now includes complete information on MDL, using MicroStation with Oracle, 100 pages of new 3D tools, and more.

This book is really two books in one. The first half is "The Power Users Guide to MicroStation". Use the tools and tutorials in these chapters to go beyond the basic MicroStation menus and commands. Power drawing, power editing, automation of repetitive tasks, working in 3D, this section teaches you how to get the most out of MicroStation.

Turn to the second half "The Unofficial MicroStation Installation Guide" to take charge of your MicroStation installation. Here you'll find tools to supercharge your software installation, customize command environment, build menus, manage your files and make DXF transfers. There is even a chapter on the undocumented EDG editor that tells you how to fix corrupted design files.

Learn how to write user commands and immediately put to work the "Ten User Commands for Everybody" and "Our Ten Favorite User Commands". Take advantage of attribute data with links to database programs including dBASE and ORACLE. This book shows you how.

Appendices include listing of all TCB variables, command names, and the syntax you need to know to be a power user.

John Kincaid and Bill Steinbock are MicroStation installation managers for the U.S. Army Corps of Engineers, Rock Island and Louisville districts respectively. They are avid writers and MicroStation bulletin board aficionados. Rich Malm managed the Corp's Intergraph/MicroStation procurement efforts until he recently retired to devote his time to consulting and MicroStation database applications. Between them, the authors have over 40 years of CAD, IGDS and MicroStation experience.

☯ Optional Productivity Disk: **US$49.95** (Australia A$79.95) Includes all of the user commands in the book, design file and tutorial examples, menus for user commands, custom batch files, and the exclusive MicroStation 3D menu.

Price: **US$39.95** (Australia A$69.95)
Pages: 600
Illustrations: 220+
ISBN: 0-934605-53-X

MicroStation Reference Guide, Pocket Edition

John Leavy
Everything you want to know about MicroStation -- Fast!
Second Edition, Release 4.X, Supports all 4.X and 3.X Platforms including DOS, Unix, Mac, and VMS

Finally, all of MicroStation's commands are in one easy to use reference guide. Important information on every MicroStation command is at your fingertips (including some that even Intergraph/Bentley don't tell you about!).

This book gives you everything you need to know about a command including:

How to find it
How to use it
What it does
What happens in an error situation

The book includes all commands, key-ins, ACTIVES, and environmental settings. Also you'll find background on such important concepts as coordinate entry, PLACE commands, and drawing construction techniques. Actual screens and examples help you get up and drawing now.

Now completely updated for 4.X, the MicroStation Reference Guide covers over 950 commands (The last edition of the book had only 400 commands!). Every user and every workstation should have one of these books handy.

The new edition features hundreds of command illustrations including the new 4.X GUI palettes for command operations.

John Leavy is president and chief Intergraph/MicroStation consultant with Computer Graphic Solutions, Inc. His specialty is MicroStation training and applications development. Leavy spent 12 years with Intergraph before moving into the private consulting world.

☯ Optional MicroStation Reference Guide, Pocket Edition Disk: **US$14.95** (Australia A$19.95) The MicroStation Online Reference Guide disk puts this book online with MicroStation.

Price: **US$18.95** (Australia A$26.95)
Pages: 320
Illustrations: 200+
ISBN: 0-934605-55-6

The Complete Guide To MicroStation 3D

David Wilkinson
First Edition, Release 4.X, Supports all 4.X and 3.X Platforms including DOS,
Unix, Mac, and VMS
(Published by Pen & Brush Publishers, Distributed by OnWord Press)

If you are a MicroStation 2D user wanting to advance into the "real-world" of 3D,
then this is the book for you. It is both a tutorial and a reference guide. Written
especially for MicroStation 2D users, it teaches you how to use the 3D capabilities
of MicroStation. Diagrams (over 220 in total) are used extensively throughout
this book to illustrate the various points and topics.

Chapters 1 and 2 incorporate the basic tutorial section. Here you are introduced
to the 3D environment of MicroStation and shown how to place elements in 3D.
This is explained with simple exercises, graphically illustrated to avoid any confu-
sion.

From this basic introduction you can advance, step-by-step, through complex prob-
lem solving techniques. This can be accomplished at your own pace. Having
learned to create a 3D model, you are shown methods for extracting drawings and
how to create rendered (shaded) images.

MicroStation version 4 tools are covered comprehensively, along with those for version 3.3 (PC version for 286 machines).

Among topics covered are the following:

* Basic placement and manipulation of elements
* Views and view rotation
* Projection and Surface of Revolution
* B-Spline Surface Constructions (version 4)
* Rendering (for both versions 3.3 and 4)

Ⓐ   Optional Complete Guide to MicroStation 3D Disk: **US$19.95** (Australia
A$29.95) Includes examples from the book as well as tips and tricks for 3D
work, model building, shading and rendering.

Price: **US$39.95**   (Australia A$69.95)
Pages: 400
Illustrations: 200+
ISBN: 0-934605-66-1 (Australia ISBN: 0-646-01678-4)

101 MDL Commands

Bill Steinbock
First Edition, Release 4.X, Supports all 4.X and 3.X Platforms including DOS,
Unix, Mac, and VMS

This is the book you need to get started with MDL!

With MicroStation 4.0 comes the MicroStation Development Language or MDL.
MDL is a powerful programming language built right in to MicroStation.

MDL can be used to add productivity to MicroStation or to develop complete applications using MicroStation tools. Virtually all of the 3rd party applications vendors are already using MDL for their development.

Now you can too, with 101 MDL Commands.

The first part of this book is a 100 page introduction to MDL including a guide to how source code is created, compiled, linked, and run. This section includes full discussion of Resource Files, Source Codes, Include Files, Make Files, dependencies, conditionals, interference rules, command line options and more.

Learn to control MicroStation's new GUI with dialog boxes, state functions, element displays and file control.

The second part of the book is 101 actual working MDL commands ready-to-go. Here you will find about 45 applications with over 101 MDL tools. Some of these MDL commands replace user commands costing over $100 a piece in the 3.0 market!

Here's a sampling of the MDL applications in the book and on the book:

MATCH - existing element parameters
Creation - create all the new element types from MDL
Multi-line - convert existing lines and linestrings to multi-line elements
CALC - Dialog box calculator
DATSTMP - Places and updates filename and in-drawing date stamp
PREVIEW - previews a design file within a dialog box
Text - complete text control - underline, rotate, resize, upper/lower, locate text
  string, import ASCII columns, extract text
Fence - complete fence manipulations including patterning, group control,
  circular fence and more
3D surfaces - complete projection and surface of revolution control
Cell routines - place along, place view dependent cell, scale cell, extract to
  cell library
Dialog boxes - make your own using these templates!
Search Criteria - delete, fence, copy, etc based on extensive search criteria.

Use these MDL commands to get you started with the power of MicroStation MDL. You can put these tools to work immediately, or use the listings to learn about MDL and develop your own applications.

☞ Optional 101 MDL Commands Disk: **US$101.00** (Australia A$155.00) Includes all of the MDL commands from the book in executable form, ready to be loaded and used.

Price: **US$49.95** (Australia A$85.00)
Pages: 680
Illustrations: 75+
ISBN: 0-934605-61-0

Bill Steinbock's
Pocket MDL Programmers Guide

Bill Steinbock
First Edition, Release 4.X, Supports all 4.X and 3.X Platforms including DOS, Unix, Mac, and VMS

Intergraph/Bentley's MDL documentation is over 1000 pages!

Bill's Steinbock's Pocket MDL Programmers Guide gives you all the MDL tools you need to know for most applications in a brief, easy-to-read format.

All the MDL tools, all the parameters, all the definitions, all the ranges -- in a short and sweet pocket guide.

If you're serious about MDL, put the power of MicroStation MDL in your hands with this complete quick guide.

Includes all the MDL commands, tables, indexes, and a quick guide to completing MDL source for MDL compilation.

Price: **US$24.95** (Australia A$39.95)
Pages: 256
Illustrations: 75+
ISBN: 0-934605-32-7

MDL-Guides

CAD Perfect
First Edition, Release 4.X, Supports all MDL 4.X Platforms Runs under DOS only.

1000 Pages of Intergraph MDL Documentation On-Line at Your Fingertips!

The Intergraph MDL Documentation is voluminous, to say the least. MDL-GUIDES puts the MDL and MicroCSL documentation in a hypertext TSR for reference access while you are programming or debugging in MDL.

This terminate-and-stay-ready program was sanctioned by Intergraph as the practical way to find out all the MDL information you need in an easy format.

The program is environment friendly, works with high DOS memory space to leave room for your other applications, and is quick.

The package includes the hypertext software, the complete set of MDL and MicroCSL Intergraph documentation in hypertext format, and a proper set of installation instructions.

MDL-GUIDES
For DOS Formats only (Other formats available on request from CAD Perfect)
Includes Disk and User's Manual
Disk Includes all MDL Documentation formatted for use on-line
Price: **US$295.00**
ISBN 0-934605-71-8

Programming With User Commands

Mach Dinh-Vu
Second Edition
For Intergraph IGDS and MicroStation
Release 3.X, Supports all 4.X and 3.X Platforms including DOS, Unix, Mac, and VMS. Will work with all 4.X versions of MicroStation.
(Published by Pen & Brush Publishers, Distributed by OnWord Press)

Programming With User Commands is an indispensable tool for user command newcomers and programmers alike. This book serves as both a tutorial guide and handbook to the ins-and-outs of UCMs.

Step-by-step explanations and examples help you create menus and tutorials to speed the design and drafting process. Learn how to attach "intelligence" to the drawing with or without database links. Take control of your menu and command environment and customize it for your own application.

Learn how to add your own functions to MicroStation's built in commands. You can do things like: save the current cell library, attach another, and then restore the previous one; or locate, add, move and modify elements -- all from User Commands.

The User Command language is already built into your copy of MicroStation or IGDS. Put it to work for you today with Programming With User Commands. Most users think UCMs are too complicated, too much like "programming". This book shatters that myth and makes User Commands accessible to every user.

Mach Dinh-Vu is an Intergraph and MicroStation CAD specialist in Engineering, Architectural, and Public Utility applications. His background includes six years of Intergraph experience, first on the VAX, but now on MicroStation and MicroStation DOS, VAX, and UNIX networks.

⊛ Programming With User Commands Disk: **$40.00** (Australia A$55.00) Includes all of the User Command examples in the book in a ready-to-use form. Use them as they are, or modify them with your own editor to get a jump-start on User Command programming.

Price: **US$65.00** (Australia A$80.00)
Pages: 320
Illustrations, Tables, Examples: 120+
ISBN: 0-934605-45-9 (Australia ISBN: 0-7316-5883-3)

101 User Commands

Brockway, Dinh-Vu, Steinbock
For Intergraph MicroStation & IGDS Users

Putting user commands to work on all Intergraph MicroStation and IGDS Platforms.

Supports all versions of MicroStation and IGDS under DOS, Mac OS, UNIX and VMS

With this book user and programmers alike can jump ahead with MicroStation or IGDS productivity. 101 User Commands gives you 101 programs to automate your CAD environment.

Never programmed with user commands before?
This book shows you how. With the program listings, input and output variables, prompting sequences and more, you will be using user commands in no time. Or copy the program listings from the optional disk into your word processor or line editor and you will be programming in no time.

Already an experienced user command programmer?
Here you'll find some of the finest in the business. Each user command is built from basic building blocks to help you organize your programs.
Mix and Match programs or subroutines to put together your own set of user commands.

This book has seven types of user commands:
Element Placement, Element Manipulations, Symbology and Attributes, Sub-Routines, Utilities, Feature Codes, and Civil Engineering Applications.

⊛ The Optional 101 User Command Disk US$**101.00** (Australia A$155.00) Includes all the user commands in the book in ready-to-go format. Edit them with your own word processor or compile them for immediate run-time programs.

Price: **US$49.95**   (Australia A$85.00)
Pages: 400
Illustrations, Tables, Examples: 75+
ISBN: 0-934605-47-5

Also Available from CAD News Bookstore:
Teaching Assistant for MicroStation

An Online Tutorial
This computer-aided-instruction package will have you learning Micro~Station by
using MicroStation! The Teaching Assistant is a series of five lessons that runs
within MicroStation (PC Version 3.0 and later).

All aspects of the program are covered from screen layout and menus to drawing
layout and concepts, to advanced editing and dimensioning. Self-paced and reusable, this courseware is a great tool for beginning users and a good refresher for ca-
sual users.

This is the perfect training material for IGDS users who need to know more about
Micro~Station. The course takes average users six to ten hours to complete.
Published: May 1990 Through Version DOS 3.3 (Call for availability of 4.X Ver-
sion)
Two 5 1/4 or 3 1/2" disks plus workbook.
Price: **US$449.95**

CAD Managers: Know Before You Hire!
The MicroStation Evaluator

MicroStation Operator Proficiency Training
The MicroStation Evaluator is an on-screen, reusable MicroStation test, designed
to help employers select quality computer personnel. The evaluator asks 100 multiple choice questions, covering basic to advanced knowledge.

The Evaluator automatically grades the test and creates a report for the employer,
including test scores, time taken, and work history. Because each operator's training is unique and every company's needs are different, the questions are placed in
specific categories. The report gives scores in each category indicating the
operator's strengths and weaknesses.

Even employers who are not familiar with Micro~Station can easily interpret cate-
gory scores and compare candidates. The Evaluator produces graphic and written
reports.
Through Version DOS 3.3 (Call for availability of 4.X Version)
Published: June 1990
Price: **US$149.00**

Order MicroStation 4.X Tools From OnWord Press Now!

Ordering Information:

On Word Press Products are Available From

1. Your Local MicroStation Dealer
2. Your Local Bookseller
3. In Australia, New Zealand, and Southeast Asia from:
    Pen & Brush Publishers
    2nd Floor, 94 Flinders Street
    Melbourne Victoria 3000
    Australia
    Phone 61(0)3 818 6226
    Fax   61(0)3 818 3704

4. .... Or Directly From OnWord Press:

To Order From On Word Press:

Three Ways To Order from OnWord Press

1. Order by **FAX** 505/587-1015
2. Order by **PHONE:** 1-800-CAD NEWS™ Outside the U.S. and Canada
    call 505/587-1010.
3. Order by **MAIL:** OnWord Press/CAD NEWS Bookstore, P.O. Box 500,
    Chamisal NM 87521-0500 USA.

Shipping and Handling Charges apply to all orders: 48 States: $4.50 for the first item, $2.25 each additional item. Canada, Hawaii, Alaska, Puerto Rico: $8.00 for the first item, $4.00 for each additional item. International: $46.00 for the first item, $15.00 each additional item. **Diskettes are counted as additional items.** New Mexico delivery address, please add 5.625% state sales tax.

Rush orders or special handling can be arranged, please phone or write for details. Government and Educational Institution POs accepted. Corporate accounts available.

MicroStation Books and Tools

Use This Form If Ordering Directly From OnWord Press

Quantity	Title	Price	Extension
	MicroStation 4.X Delta Book	$19.95	
	MicroStation 4.X Upgrade Video Series	$149.00	
	INSIDE MicroStation	$29.95	
	INSIDE MicroStation Disk	$14.95	
	INSIDE MicroStation Companion Workbook	$34.95	
	Instructor's Guide: INSIDE MicroStation Companion Workbook	$9.95	
	MicroStation Productivity Book	$39.95	
	MicroStation Productivity Disk	$49.95	
	MicroStation Reference Guide	$18.95	
	MicroStation Reference Disk	$14.95	
	The Complete Guide to MicroStation 3D	$39.95	
	The Complete Guide to MicroStation 3D Disk	$19.95	
	101 MDL Commands	$49.95	
	101 MDL Commands Disk	$101.00	
	Bill Steinbock's MDL Pocket Programmer's Guide	$24.95	
	MDL-GUIDES	$295.00	
	Programming With User Commands	$65.00	
	Programming With User Commands Disk	$40.00	
	101 User Commands	$49.95	
	101 User Commands Disk	$101.00	
	Teaching Assistant for Micro~Station	$449.95	
	The MicroStation Evaluator	$149.00	
	Shipping & Handling *		
	5.625% Tax - State of New Mexico Delivery Only		
	Total		

AD CODE M44

Name _____

Company _____

Street _____
(No P.O. Boxes Please)

_____

City, State _____

Country, Postal Code _____

Phone _____

Fax _____

If Ordering Disks, Please Note

Disk Type _____

Payment Method

___ Cash ___ Check ___ Amex

___ MasterCARD ___VISA

_____
Card Number

Expiration Date _____

_____
Signature

FAX TO: 505/587-1015

or MAIL TO:

OnWord Press
Box 500
Chamisal NM 87521 USA